# Lecture Notes in Physics

## Volume 969

# The Lecture Notes in Physics

The series Lecture Notes in Physics (LNP), founded in 1969, reports new developments in physics research and teaching - quickly and informally, but with a high quality and the explicit aim to summarize and communicate current knowledge in an accessible way. Books published in this series are conceived as bridging material between advanced graduate textbooks and the forefront of research and to serve three purposes:

- to be a compact and modern up-to-date source of reference on a well-defined topic.
- to serve as an accessible introduction to the field to postgraduate students and nonspecialist researchers from related areas.
- to be a source of advanced teaching material for specialized seminars, courses and schools.

Both monographs and multi-author volumes will be considered for publication. Edited volumes should however consist of a very limited number of contributions only. Proceedings will not be considered for LNP.

Volumes published in LNP are disseminated both in print and in electronic formats, the electronic archive being available at springerlink.com. The series content is indexed, abstracted and referenced by many abstracting and information services, bibliographic networks, subscription agencies, library networks, and consortia.

Proposals should be sent to a member of the Editorial Board, or directly to the managing editor at Springer:

Dr Lisa Scalone
Springer Nature
Physics Editorial Department
Tiergartenstrasse 17
69121 Heidelberg, Germany
lisa.scalone@springernature.com

More information about this series at http://www.springer.com/series/5304

Sergio M. Rezende

# Fundamentals
# of Magnonics

 Springer

Sergio M. Rezende
Departamento de Física
Universidade Federal de Pernambuco
Recife, Brazil

ISSN 0075-8450          ISSN 1616-6361   (electronic)
Lecture Notes in Physics
ISBN 978-3-030-41316-3          ISBN 978-3-030-41317-0   (eBook)
https://doi.org/10.1007/978-3-030-41317-0

This Springer imprint is published by the registered company Springer Nature Switzerland AG.
The registered company address is: Gewerbestrasse 11, 6330 Cham, Switzerland

*To Adélia*

# Preface

Magnetism and magnetic materials constitute one of the oldest fields of science and technology that keeps renewing itself with new discoveries and unique technological applications. Centuries before the time of Christ, ancient civilizations studied the wondrous properties of magnetite, the famed loadstone, and used it for orientation in the Earth's magnetic field. The magnetic compass became an essential instrument of navigation for the Chinese in ancient times and for the Europeans in the Late Middle Ages. This application motivated one of the oldest books in experimental physics, *De Magnete*, written by William Gilbert and published in 1600. After the discoveries of the fundamental laws of electromagnetism in the nineteenth century by Ampére, Oersted, Faraday, and Henry, magnetic materials became essential for the fabrication of generators, motors, and transformers, building blocks of electricity that revolutionized the costumes of humanity. Then, in the twentieth century, they made possible the invention of loudspeakers, phones, relays, and other magnetic devices essential for telegraphy and telephony, as well as the development of magnetic recording for audio, video, and digital information.

With the understanding of the atomic origin of magnetism of matter, made possible by the formulation of quantum mechanics in the beginning of the twentieth century, new magnetic phenomena were proposed theoretically or discovered experimentally, and new applications were developed. In 1930, Felix Bloch showed that the low-lying excitations of the spin system consisted of nonlocalized, collective spin deviations, which he named spin waves, whose quanta are called *magnons*. Few decades later, spin waves were observed experimentally and became the subject of intense research. Phenomena involving the $k = 0$ magnon, or ferromagnetic resonance, laid the groundwork for novel applications in microwave ferrite devices used in telecommunications, in radar systems, and in a variety of industrial equipment. Despite this long history of scientific activity and technological applications, recently Professor Chia-Ling Chien of the University of Baltimore stated that in the last 30 years, activity in magnetism is facing its Golden Age.

The recent vigorous impulse to magnetism was provided by spin-based phenomena that occur in nanoscale magnetic structures, such as the giant magnetoresistance (GMR) discovered in 1989 by Albert Fert and Peter Grünberg, Physics Nobel Prize winners in 2007. The discovery of GMR and other spin-dependent phenomena led to new sensing devices that boosted the capacity of storage media and made possible new storage devices such as the magnetic random-access memory. These and other developments led to the new field of spintronics, devoted to the investigation of basic phenomena and their application in devices for transport, storage, and processing of information, in which the main physical entity is the electron spin. The subfield of spintronics in which the phenomena are based on the properties of magnons is called magnonics, or magnon spintronics, which is the subject of this book. The unique properties of magnons in magnetic materials with very low damping, such as yttrium iron garnet (YIG), make them suitable for use as information carriers and logic processing without the need of electric current, overcoming an important fundamental limitation of conventional electronics: a power consumption which scales linearly with increasing number of individual processing elements.[1,2]

This book is intended to serve as a text for beginning engineering and physics graduate students in the areas of magnetism and spintronics. The level of presentation assumes only basic knowledge of the origin of magnetism, electromagnetism, and quantum mechanics. The book utilizes relatively simple mathematical derivations, aimed mainly at explaining the physical concepts involved in the phenomena studied and for the understanding of the experiments presented. We use in the book both SI and CGS units, because they are equally employed in research articles. Key topics include the basic phenomena of ferromagnetic resonance in bulk materials and in thin films, semi-classical theory of spin waves, quantum theory of spin waves and magnons, magnons in antiferromagnets, parametric excitation of magnons, magnon nonlinear dynamics and chaotic phenomena, Bose–Einstein condensation of magnons, and the very recent field of magnon spintronics. This breath of topics is not covered in any other single book. Also, no other textbook on magnetism has the material presented in the last three chapters.

I would like to thank many collaborators with whom I worked in several magnonic phenomena presented in the book. In particular, I thank Frederick R. Morgenthaler who introduced me to spin wave phenomena over 50 years ago, during my Ph.D. program at MIT. I am also grateful to Nicim Zagury, Robert M. White, Vincent Jaccarino, Carlos A. dos Santos, Wido R. Schreiner, Stuart S. P. Parkin, and Doug L. Mills, with whom I had profitable collaborations for many years. I also thank my colleagues, students, and post-docs at UFPE, my coauthors in papers on magnonics and other fields, Cid B. de Araújo, Ivon P. Fittipaldi, Mauricio D. Coutinho-Filho, José Rios Leite, Jairo R. L. de Almeida, Sandra S. Vianna, Fernando L. A. Machado, Erivaldo Montarroyos, J. Marcílio Ferreira, Eduardo Fontana, Flávio M. de Aguiar, Osiel A. Bonfim, Antonio

[1]Serga, A. A., Chumak, A. V., Hillebrands, B.: YIG Magnonics. J. Phys. D Appl. Phys. **43**, 264002 (2010).

[2]Chumak, A. V., Vasyuchka, V. I., Serga, A. A., Hillebrands, B.: Magnon spintronics. Nat. Phys. **11**, 453 (2015).

Azevedo, José R. Fermin, Carlos Chesman, Roberto L. Rodríguez-Suárez, Eduardo Padrón-Hernández, Joaquim B. S. Mendes, Rafael O. R. Cunha, Luis H. Vilela-Leão, Javier D. C. López Ortiz, Gabriel A. Fonseca Guerra, José Holanda, Daniel S. Maior, Matheus Gamino, Obed Alves Santos, Pablo R. T. Ribeiro, and Mercedes Arana. Last but not least, I am also grateful to Alberto P. Guimarães and Anderson S. L. Gomes for helpful suggestions about the book, to Pavel Kabos for a careful review of the manuscript, and to Sérgio Mascarenhas for his constant stimulus and support.

Recife, Brazil                                                           Sergio M. Rezende
October 3, 2019

# About the Book

*Fundamentals of Magnonics* is intended to serve as a text for beginning engineering and physics graduate students in the areas of magnetism and spintronics. The level of presentation assumes only basic knowledge of the origin of magnetism, electromagnetism, and quantum mechanics. The book utilizes relatively simple mathematical derivations, aimed mainly at explaining the physical concepts involved in the phenomena studied and for the understanding of the experiments presented. Key topics include the basic phenomena of ferromagnetic resonance in bulk materials and in thin films, semiclassical theory of spin waves, quantum theory of spin waves and magnons, magnons in antiferromagnets, parametric excitation of magnons, magnon nonlinear dynamics and chaotic phenomena, Bose–Einstein condensation of magnons, and the very recent field of magnon spintronics. This breadth of topics is not covered in any other single book. Also, no other textbook on magnetism has the material presented in the last three chapters. Also, end-of-chapter problems are included.

# Contents

# About the Author

**Sergio Machado Rezende** graduated in Electrical Engineering in Rio de Janeiro (1963) and received MSc (1965) and PhD (1967) degrees from the Massachusetts Institute of Technology, both in Electrical Engineering and Materials Science. He is one of the founders and first chairman of the Physics Department of the Federal University of Pernambuco (UFPE) (1972–1976), in Recife, where he is a Full Professor. He was twice a Visiting Professor at the University of California, Santa Barbara (1975–1976 and 1982–1984). He has published over 270 scientific papers in international journals on dynamic excitation phenomena in magnetic materials, magnetic nanostructures, magnonics and spintronics and has supervised over 40 MSc and PhD students. His scientific activities have never been interrupted by science managing positions he held, Dean of the Center for Exact Sciences of UFPE (1984–1988), Scientific Director of the Pernambuco Science Foundation (1990–1993), State Secretary for Science and Technology of Pernambuco (1995–1998), President of FINEP, the main federal agency for funding of S&T in Brazil (2003–2005), and Minister for Science and Technology of Brazil (2005–2010), under President Luiz Inácio Lula da Silva. He is a member of several scientific societies, Honorary President of the Brazilian Society for the Advancement of Science (SBPC), and has received several prizes and awards in Brazil and other countries.

# The Zero Wave Number Magnon: Ferromagnetic Resonance

1

This initial chapter is devoted to a basic phenomenon in magnetism, the ferromagnetic resonance (FMR). When driven by a microwave field with frequency of the FMR, the magnetization precesses about its equilibrium direction without variation in space. This corresponds to a spin wave with infinite wavelength, or zero wave number. Since the quanta of spin waves are called magnons, the FMR mode corresponds to a zero wave number magnon. Initially, we introduce the main features of magnetic materials necessary to the remainder of the book. Details about the origins of magnetism in matter and the spin interactions can be found in several books listed at the end of the chapter. Then we treat the magnetic resonance phenomenon and go on to study in detail the ferromagnetic resonance in bulk samples and in thin films.

## 1.1  Brief Introduction to Magnetic Materials

We assume that the reader is familiar with the basic concepts of magnetism in matter as presented in textbooks on solid-state physics or on magnetism. In this section, we briefly review the basic concepts and properties of strongly magnetic materials, ferromagnets, antiferromagnets, and ferrimagnets.

The ability of matter to produce strong macroscopic magnetic effects has been known to mankind for many centuries. However, only at the end of the nineteenth century, with the work of Pierre Curie, systematic investigations of the properties of magnetic substances began to be made. In the early 1900s, it was recognized that strongly magnetic materials, all of which were called ferromagnets, exhibited a *spontaneous magnetization* at temperatures below a critical value $T_c$, called Curie temperature. It was realized that this magnetization was due to a strong coupling between microscopic magnetic moments, but the origins of the moments and their interactions were unknown. With a phenomenological molecular field approach, Pierre Weiss proposed a theory that gave a variation of the spontaneous

© Springer Nature Switzerland AG 2020
S. M. Rezende, *Fundamentals of Magnonics*, Lecture Notes in Physics 969,
https://doi.org/10.1007/978-3-030-41317-0_1

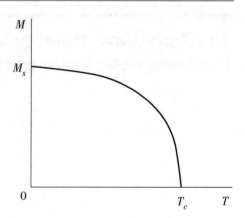

**Fig. 1.1** Variation of the spontaneous magnetization of a ferromagnet with temperature

magnetization $M$ with temperature in satisfactory agreement with the existing measurements, shown qualitatively in Fig. 1.1.

The theory assumes that each microscopic magnetic moment $\mu$ interacts with an effective molecular field $H_E$ created by the neighboring moments. With this assumption one can obtain an order of magnitude estimate for $H_E$ from the measured Curie temperature. This is the temperature at which the thermal energy $k_B T_c$ is comparable to the ordering energy of each moment. Using for $\mu$ the value of the Bohr magneton and $T_c \sim 1000$ K, the approximate value for iron, this gives $H_E \sim 10^7$ Oe, a value orders of magnitude larger than the one for the classical magnetic dipolar field. With the development of quantum mechanics in the 1920s and the detailed understanding of the atoms, the origin of the microscopic magnetic moment was explained and a possible mechanism for their strong interaction was advanced by Heisenberg. Today, it is well known that the moments originate in unpaired electronic spins of ions with partially filled inner shells of some elements of the periodic table. This is the case of elements in the 3d Fe transition group, such as Cr, Mn, Fe, Co, and Ni, and rare earth elements such as Nd, Sm, Eu, Gd, Tb, and Dy. The strong interaction between neighboring atomic moments has its origin in the quantum mechanical exchange interaction between the spins. The magnetic moment $\vec{\mu}$ of an atom, or ion, is related to its angular momentum by

$$\vec{\mu} = -g\mu_B \vec{J}, \tag{1.1}$$

where the total angular momentum $\hbar\vec{J}$ is the sum of the orbital $\hbar\vec{L}$ and spin $\hbar\vec{S}$ angular momenta, $\mu_B$ is the Bohr magneton, and $g$ is the spectroscopic splitting factor, given by the Landé equation

$$g = 1 + \frac{J(J+1) + S(S+1) - L(L+1)}{2J(J+1)}. \tag{1.2}$$

Note that for ions with half-filled shells, the orbital angular momentum vanishes, $L = 0$, so that $J = S$ and Eq. (1.2) shows that $g = 2$. This is the case of the $Mn^{2+}$ and

$Fe^{3+}$ ions of the Fe transition group. The Bohr magneton in Eq. (1.1) is given by $\mu_B = e\hbar/2m$ in the SI system, and $\mu_B = e\hbar/2mc$ in CGS units, where $e$ and $m$ are the charge and mass of the electron and $c$ the speed of light. The values of $\mu_B$ in the two systems are $9.27 \times 10^{-24}$ A m$^2$ and $9.27 \times 10^{-21}$ G cm$^3$.

The strong interaction between magnetic moments of neighboring ions is due to the overlap of their electronic orbital wave functions. According to the Pauli exclusion principle, the total wave function with orbital and spin components must be antisymmetric. Thus, if the two spins are parallel, the orbital wave function is antisymmetric in space, whereas if the two spins are antiparallel, the orbital wave function is symmetric. Since the orbital wave function represents the charge distribution, the two states have different electrostatic Coulomb energies. The difference between the two energies depends on the relative orientations of the spins and is called exchange energy. In many materials, the exchange energy of interaction of atoms $i, j$ having electron spins $\vec{S}_i$, $\vec{S}_j$ can be written in the form

$$U_{exci,j} = -2J\vec{S}_i \cdot \vec{S}_j, \tag{1.3}$$

where $J$ is the exchange constant of the interaction (not to be confused with $J$ of the angular momentum) and is related to the overlap of the charge distributions of the two atoms. Equation (1.3) is called Heisenberg energy. If $J > 0$ the exchange energy is minimum when the two spins are parallel to each other, in the so-called ferromagnetic ordering. On the other hand, if $J < 0$ the energy is minimum when the two spins are anti-parallel, in the antiferromagnetic ordering. Besides the exchange interaction, the atomic spins in a magnetic material interact with neighbors by the long-range magnetic dipolar energy, interact with external magnetic fields and also with the electric fields of the lattice structure through the spin–orbit coupling (anisotropy energy). These interactions will be considered in detail later in this book.

*Ferromagnets* are materials in which the interactions with $J > 0$ dominate so that all atomic spins tend to align parallel to each other at temperatures below $T_c$. However, in the absence of external magnetic fields, the competition among the various energies breaks the ferromagnetic order into complex patterns of the spin arrangement, with the parallel alignment within small regions called magnetic domains. When an external field is applied with increasing intensity $H$, the domains change in size and rotate so that the total magnetization $M$ eventually saturates. If the field now decreases, the magnetization decreases but the variation $M(H)$ follows a different path from the previous one due to the irreversibility of the domain structure. The full curve with increasing and decreasing $H$ is called the hysteresis loop, which has a shape that depends strongly on the material composition, structure, and physical shape. The shape of the hysteresis loop of a material is essential for determining its possible application. Throughout this book, we will consider that the material under study is magnetized to saturation, in a state of single domain.

In materials with $J < 0$, neighboring atomic spins tend to align in opposite directions. In a simple arrangement, the long-range order is attained with two opposing ferromagnetic sublattices. If the net magnetizations of the two sublattices are equal, the material is said to be *antiferromagnetic*. Antiferromagnets have no net macroscopic magnetization and thus are almost insensitive to external magnetic fields. The properties of antiferromagnets were revealed by Louis Néel in the 1930s, and for this reason the critical temperature at which the sublattice magnetizations vanish due to thermal fluctuations is called Néel temperature $T_N$. Materials with $J < 0$ that have opposite sublattices with different magnetizations are called *ferrimagnets*. They have spontaneous net magnetization below the ordering temperature $T_c$ and in many respects behave like ferromagnets.

The three types of magnetic ordering, ferro, ferri, and antiferromagnetic, can occur in metals and in insulators. In insulators, the origin of the magnetic moments lies in the localized ionic spins as we have discussed. Very few simple ferromagnets are insulators, $CrBr_3$, $EuO$, and $EuS$ are the only known ones. By far most ferromagnets are metals, such as Mn, Fe, Co, and Ni and their alloys. The origin of the magnetism in metals is more complicated than in insulators, because the magnetic moments are associated with delocalized or itinerant electrons, that are described by band theory. Most insulating magnetic materials are ferri- or antiferromagnetic. The reason is that usually the magnetic ions are not located in nearest neighbor lattice sites, so that there are ligand atoms between them. In this case, the exchange interaction is indirect, mediated by the ligand ions, favoring an antiferromagnetic alignment. Ferrimagnetic insulators, also called *ferrites*, have many important applications in high-frequency electronics.

A very important ferrite for magnonics is *yttrium iron garnet*, or simply YIG. With chemical formula $Y_3Fe_5O_{12}$, YIG has ferrimagnetic arrangement with two sublattices, with $T_c = 559$ K. Of the five $Fe^{3+}$ ions in the formula unit, three are in one sublattice and two in the other, so that the net spin per formula unit is $S = 5/2$, the spin of the half-filled 3d shell. The fact that the magnetism of YIG is due to $Fe^{3+}$ ions is of special significance, because they have no orbital angular momentum, so the spins are essentially decoupled from the lattice. This results in very low magnetic damping at microwave frequencies, making single crystal YIG the best material for studying the dynamics of spin excitations such as in spin waves.

## 1.2    Magnetization Dynamics

### 1.2.1    Free Precession in a Magnetic Field

The motion of the atomic magnetic moments in the presence of a magnetic field is governed by the classical equations of electromagnetism and mechanics. We consider initially a magnetic material with infinite size (so that the effects of the sample boundaries can be neglected) subject to a uniform static magnetic field induction $\vec{B}$. The energy of a magnetic moment $\vec{\mu}_i$ at site $i$ is

$$U_i = -\vec{\mu}_i \cdot \vec{B}. \tag{1.4}$$

Equation (1.4) shows that the energy is minimum when $\vec{\mu}_i$ is aligned with the field. If the moment is deflected from this direction by some external perturbation, it becomes subject to a torque given by

$$\vec{\tau} = \vec{\mu}_i \times \vec{B}. \tag{1.5}$$

Since the moment has an associated angular momentum $\hbar \vec{J}$, its motion is governed by Newton's law

$$\frac{d\hbar \vec{J}}{dt} = \vec{\tau}. \tag{1.6}$$

Using Eqs. (1.1, 1.5, and 1.6), we obtain

$$\frac{d\vec{\mu}_i}{dt} = -\gamma \vec{\mu}_i \times \vec{B}, \tag{1.7}$$

where

$$\gamma = \frac{g\mu_B}{\hbar} \tag{1.8}$$

is called the gyromagnetic ratio of the atom or ion. Since it is better to work with a macroscopic quantity, we shall use the magnetization vector, defined as the magnetic moment per unit volume

$$\vec{M} = \frac{1}{V} \sum_i \vec{\mu}_i, \tag{1.9}$$

where the volume $V$ is chosen so that it is large enough to have good macroscopic average, but small compared to the sample dimensions so that $\vec{M}$ represents a local quantity. The equation of motion for the magnetization is obtained from Eqs. (1.7 and 1.9). Using the relation $\vec{B} = \mu_0 (\vec{H} + \vec{M})$, where $\vec{H}$ is the magnetic field intensity and $\mu_0$ the permeability of vacuum, and considering $\vec{M} \times \vec{M} = 0$, we have

$$\frac{d\vec{M}}{dt} = -\gamma \mu_0 \vec{M} \times \vec{H}. \tag{1.10}$$

This is the Landau–Lifshitz equation [1] that governs the motion of the magnetization in the absence of magnetic damping. It shows that if $\vec{M}$ is parallel to $\vec{H}$, the time derivative is zero and the magnetization is in the equilibrium direction. However, if $\vec{M}$ is deflected from this direction by some external perturbation, the time derivative becomes nonzero and points in a direction perpendicular to the plane of $\vec{M}$ and $\vec{H}$. Thus, instead of deflecting back toward the equilibrium direction, $\vec{M}$ follows

**Fig. 1.2** Magnetization precession about a static magnetic field

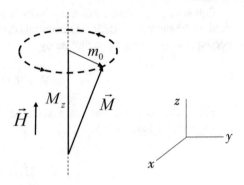

a motion of precession about $\vec{H}$. In order to describe this motion we use a Cartesian system of coordinates, with the $z$-axis pointing in the direction of the magnetic field $\vec{H}$, and consider the magnetization in the form

$$\vec{M} = \hat{x}\,m_x(t) + \hat{y}\,m_y(t) + \hat{z}\,M_z, \tag{1.11}$$

where we use small letters for the components $x$ and $y$ because they vary in time and are expected to be small compared to the component $z$, $m_x$, $m_y \ll M_z$. Using Eq. (1.11) in Eq. (1.10), we readily obtain the equations of motion for the transverse components of $\vec{M}$

$$\frac{dm_x}{dt} = -\gamma\mu_0\,m_y H, \qquad \frac{dm_y}{dt} = \gamma\mu_0\,m_x H. \tag{1.12}$$

A solution for Eq. (1.12) is

$$m_x(t) = m_0 \cos(\omega_0 t), \qquad m_y(t) = m_0 \sin(\omega_0 t). \tag{1.13}$$

The motion of the magnetization described by Eq. (1.13) is illustrated in Fig. 1.2. The magnetization vector $\vec{M}$ precesses about the field $\vec{H}$, with its tip in a circular motion with amplitude $m_0$ defined by the initial condition. This motion is similar to that of a gyroscope precessing about the gravitational field. The angular frequency of precession, which is readily obtained by replacing one of the two equations in Eq. (1.13) into (1.12), is given by

$$\omega_0 = \gamma\mu_0 H. \tag{1.14}$$

In the CGS, since $\mu_0 = 1$, the precession frequency is given by $\omega_0 = \gamma H$. This is also the *magnetic resonance frequency* at which the magnetization response is maximum when driven by a *rf* field, as will be shown later. The resonance frequency is proportional to the magnetic field, with a factor that is the gyromagnetic ratio given by Eq. (1.8). For $g = 2$ we obtain from Eq. (1.8)

$$\gamma = 2\pi \times 28 \text{ GHz/T (SI)}, \qquad \gamma = 2\pi \times 2.8 \text{ GHz/kOe (CGS)}. \qquad (1.15)$$

Thus, for a typical laboratory field of 0.1 T, or 1 kOe, the precession frequency is 2.8 GHz, which lies in the microwave region. Of course, the free precession illustrated in Fig. 1.2 corresponds to an idealized situation in which there are no losses. In the real world, there is always magnetic damping, or relaxation. Thus, if the magnetization is deflected from the direction of the field and released, the precession amplitude decays in time so that the magnetization spirals about the field and reaches equilibrium in a characteristic time, called relaxation time, that depends on the material properties.

## 1.2.2 *rf*-Driven Magnetic Resonance

We now consider that a magnetic sample is subjected to a static magnetic field $\vec{H}$ and an alternating field with frequency $\omega$ applied perpendicularly to the static field. Using the coordinate system in Fig. 1.2, we write the total field intensity as

$$\vec{H} = (\hat{x}h_x + \hat{y}h_y)e^{-i\omega t} + \hat{z}H, \qquad (1.16)$$

where $h_x$, $h_y \ll H$, since $H$ is of the order of 1 kOe (in CGS units) while the *rf* microwave field amplitude is of the order of a fraction or a few Oe. We anticipate that the magnetization response can be written as in Eq. (1.11), with a large component in the $z$-direction and small time-varying components in the $x$- and $y$-directions. Thus, using Eqs. (1.11 and 1.16) in Eq. (1.10), we obtain the equations of motion for the transverse magnetization components

$$\frac{dm_x}{dt} = -\gamma\mu_0 m_y H + \gamma\mu_0 M_z h_y e^{-i\omega t}, \qquad (1.17)$$

$$\frac{dm_y}{dt} = \gamma\mu_0 m_x H - \gamma\mu_0 M_z h_x e^{-i\omega t}. \qquad (1.18)$$

Since we are interested in the steady-state response to the driving field we take

$$m_x(t) = m_x e^{-i\omega t}, \quad m_y(t) = m_y e^{-i\omega t}, \qquad (1.19)$$

where $m_x$, $m_y$ are complex quantities. Using (1.19) in Eqs. (1.17 and 1.18) and considering that $\omega_0 = \gamma\mu_0 H$ we obtain

$$-i\omega m_x = -\omega_0 m_y + \gamma\mu_0 M_z h_y, \qquad (1.20)$$

$$-i\omega m_y = \omega_0 m_x - \gamma\mu_0 M_z h_x. \tag{1.21}$$

The solutions of Eqs. (1.20 and 1.21) can be written in a tensorial form

$$\vec{m} = \overline{\overline{\chi}} \cdot \vec{h}, \tag{1.22}$$

where the vectors $\vec{m}$ and $\vec{h}$ are represented by the column matrices

$$\vec{m} = \begin{pmatrix} m_x \\ m_y \end{pmatrix}, \qquad \vec{h} = \begin{pmatrix} h_x \\ h_y \end{pmatrix}, \tag{1.23}$$

and $\overline{\overline{\chi}}$ is the *rf* magnetic susceptibility tensor, represented by the square matrix

$$\overline{\overline{\chi}} = \begin{pmatrix} \chi_{xx} & \chi_{xy} \\ \chi_{yx} & \chi_{yy} \end{pmatrix}, \tag{1.24}$$

where

$$\chi_{xx}(\omega) = \chi_{yy}(\omega) = \frac{\omega_M \omega_0}{\omega_0^2 - \omega^2}, \tag{1.25}$$

$$\chi_{yx}(\omega) = -\chi_{xy}(\omega) = i\frac{\omega_M \omega}{\omega_0^2 - \omega^2}, \tag{1.26}$$

where we have introduced the material parameter $\omega_M = \gamma\mu_0 M_z \approx \gamma\mu_0 M$. Note that in the CGS this has to be written as $\omega_M = \gamma 4\pi M$, since $\mu_0 = 1$ and the factor $4\pi$ comes from the definition of the susceptibility in that system, $\vec{m} = 4\pi\overline{\overline{\chi}} \cdot \vec{h}$. Equations (1.23–1.26) reveal that the application of a *rf* field either in the *x*- or *y*-direction generates components of the magnetization in both *x*- and *y*-directions. This is due to the fact that the natural motion of $\vec{M}$ is the precession about the *z*-axis. Thus, the application of an *rf* field either in the *x*- or *y*-direction produces the precession motion with frequency $\omega$ with associated $m_x$ and $m_y$ components. For this reason, the relation between $\vec{m}$ and $\vec{h}$ is not a scalar, but a tensor relation.

Equations (1.25 and 1.26) indicate that the amplitude of the magnetization response increases rapidly as the driving frequency $\omega$ approaches $\omega_0$, characteristic of a resonance process. In fact, the susceptibilities diverge at $\omega = \omega_0$, which is an unphysical situation, showing that relaxation cannot be neglected. Here we introduce the magnetic relaxation, or damping, in a phenomenological manner. This is often done by introducing an imaginary part in the resonance frequency that corresponds to multiplying the amplitudes in Eq. (1.13) by an exponential function that decays in time. Thus, replacing $\omega_0$ in Eqs. (1.25 and 1.26) by $\omega_0 - i\eta$, where $\eta$ is the magnetic relaxation rate, the components of the *rf* susceptibility tensor become approximately

$$\chi_{xx}(\omega) = \chi_{yy}(\omega) = \frac{\omega_M \omega_0}{\omega_0^2 - \omega^2 - i2\omega_0\eta}, \tag{1.27}$$

$$\chi_{yx}(\omega) = -\chi_{xy}(\omega) = i\frac{\omega_M \omega}{\omega_0^2 - \omega^2 - i2\omega_0\eta}, \tag{1.28}$$

where we have assumed that $\eta \ll \omega_0$, that means small damping. From these equations, we can readily obtain the real and imaginary parts of the susceptibility components, denoted, respectively, by $\chi'_{\mu\nu}$ and $\chi''_{\mu\nu}$. Since the susceptibility is large only in the vicinity of resonance, we consider $\omega \approx \omega_0$ so that $\omega_0^2 - \omega^2 \approx 2\omega_0(\omega_0 - \omega)$. Thus, the real and imaginary parts of the diagonal components of the susceptibility tensor become, approximately

$$\chi'_{xx} = \chi'_{yy} = \frac{\omega_M(\omega_0 - \omega)/2}{(\omega_0 - \omega)^2 + \eta^2}, \tag{1.29}$$

$$\chi''_{xx} = \chi''_{yy} = \frac{\omega_M \eta/2}{(\omega_0 - \omega)^2 + \eta^2}. \tag{1.30}$$

Assuming that the *rf* driving field has only a $h_x$ component, the time–average power absorbed by the material, per unit volume, is

$$P(\omega) = \frac{1}{2}\mu_0 \omega \chi''_{xx} h_x^2. \tag{1.31}$$

Figure 1.3 shows plots of the real and imaginary parts of the diagonal susceptibility as a function of the magnetic field intensity $H$, for a fixed driving frequency $\omega/2\pi = 9.8$ GHz. The reason for making the plot as a function of field, and not frequency, is that usually magnetic resonance experiments are carried out with a scanning field and a fixed frequency coinciding with that of a resonant structure,

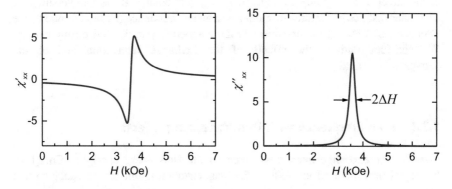

**Fig. 1.3** Real and imaginary parts of the diagonal susceptibility tensor calculated with $\omega/2\pi = 9.8$ GHz, $g = 2$, $\omega_M/\gamma = 3$ kOe, and $\eta/\omega = 0.04$

such as a microwave cavity. The other parameters used to make the plots in Fig. 1.3 are $g = 2$; $\omega_M/\gamma = 3$ kOe; and $\eta/\omega = 0.04$.

Note that the imaginary part has the shape of a Lorentzian function, with a maximum at the field for which $\omega_0 = \omega$, which is $H = 3.5$ kOe. Thus, due to Eq. (1.31), the field dependence of the microwave power absorbed by the sample also has a Lorentzian shape. As shown in Fig. 1.3, when the field is scanned and its value coincides with $\omega/\gamma$, the absorbed power is maximum, characterizing a magnetic resonance phenomenon. The difference between the two field intensities for which $\chi''_{xx}$ has a value half of the peak value is called *full linewidth*, while one half of which is called *half-width at half maximum* (HWHM), or simply *linewidth*, denoted by $\Delta H$. From Eq. (1.30), we can readily obtain a relation between the linewidth and the relaxation rate

$$\Delta H = \eta/\gamma. \tag{1.32}$$

Thus, the measurement of the linewidth in a magnetic resonance experiment gives information about the magnetic damping. The linewidth of the sample with susceptibility shown in Fig. 1.3 is $\Delta H = 140$ Oe, which is typical of polycrystalline ferrite materials. While the imaginary part of the susceptibility is proportional to the power absorption, the real part is proportional to the phase angle between the driving *rf* field and the magnetization response. For this reason, the imaginary part of the susceptibility is called absorptive component while the real part is called dispersive.

## 1.3    Ferromagnetic Resonance

The derivations presented in Sect. 1.2 are valid for the electronic spins in a paramagnetic material, in which there is no exchange interaction between neighboring spins and the magnetization is relatively small. In a ferromagnetic or ferrimagnetic material, the resonance frequency depends strongly on the sample shape, because of the effect of the *dipolar*, or *demagnetizing*, field created at the sample surfaces. The other distinctive features of the ferromagnetic resonance (FMR) are due to the effects of the exchange interaction and the magneto-crystalline anisotropy energy. We consider initially the effects of the exchange interaction and of the demagnetizing field.

### 1.3.1    FMR Frequency with Demagnetizing Effects

Consider the exchange interaction given by the Heisenberg energy in Eq. (1.3). In the uniform mode of precession, the magnetization is uniform in space so that all spins can be considered parallel. In this case, using the relation $\vec{S} = -\vec{M}S/M$, the energy of interaction of spin $\vec{S}_i$ with its $z$ nearest neighbors becomes

$U_i = 2zJ\vec{S}_i \cdot \vec{M}S/M$. Using Eqs. (1.1 and 1.4), we can write the exchange energy for the magnetic moment at site $i$ as

$$U_i = -\mu_0 \vec{\mu}_i \cdot \vec{H}_E, \tag{1.33}$$

where

$$\vec{H}_E = w\vec{M}, \qquad w = \frac{2zJS}{\mu_0 g\mu_B M}, \tag{1.34}$$

represents an effective magnetic field acting on any atomic magnetic moment by its neighbors due to the exchange interaction, and thus on the magnetization itself. Then, since the motion of the magnetization is governed by the Landau–Lifshitz equation (1.10), and $\vec{M} \times \vec{H}_E = 0$, we see that the presence of the exchange interaction does not change the frequency of the uniform FMR mode. As we will see in the next chapter, it does have an important contribution to the frequency of spin waves, for which the magnetization varies in space. Actually, the uniform mode is a spin wave with zero wave number, $k = 0$, because the phase of the precession does not vary in space. Since the quanta of spin waves are called *magnons*, the uniform mode is also called the $k = 0$ magnon.

In a ferromagnetic sample under an external magnetic field and in a state of single domain, the discontinuity of the magnetization at the surfaces and interfaces creates a demagnetizing field that has an important effect on the resonance frequency. If the sample has an arbitrary shape, this field is nonuniform and the resonance becomes complicated. However, in samples with ellipsoidal shape, the demagnetizing field is uniform and its effect can be treated quite nicely. Consider a sample in the form of an ellipsoid subject to an external uniform field $\vec{H}_0$, such that the magnetization $\vec{M}$ is saturated in an arbitrary direction. In this case, the internal field is also uniform and can be written as

$$\vec{H}_{int} = \vec{H}_0 - \bar{N} \cdot \vec{M}, \tag{1.35}$$

where $\bar{N}$ is the tensor of the demagnetizing factors. If we choose a Cartesian coordinate system with the $x$, $y$, $z$ directions along the symmetry axes of the sample, the tensor is diagonal and has components $N_x$, $N_y$, $N_z$. In this case, the diagonal components satisfy the relation

$$N_x + N_y + N_z = 1 \quad \text{(SI)}, \qquad N_x + N_y + N_z = 4\pi \quad \text{(CGS)}. \tag{1.36}$$

As we will show later, for samples with simple shapes, the demagnetization factors can be calculated using Eq. (1.36) and symmetry relations. Now consider that the external uniform field is applied along the $z$-axis. If the magnetization precesses about the field, it can be written as in Eq. (1.11), so that the internal field becomes

$$\vec{H}_{int} = -\hat{x}N_x m_x - \hat{y}N_y m_y + \hat{z}(H_0 - N_z M),\tag{1.37}$$

where we have assumed $M_z \approx M$. Thus, the magnetization precession creates time-varying demagnetizing fields in the $x$-, $y$-directions. Using Eqs. (1.11 and 1.37) and considering $m_x, m_y \ll M_z \approx M$, we obtain from Eq. (1.10) the equations of motion for the transverse magnetization components

$$\frac{dm_x}{dt} = -\gamma\mu_0 m_y \big[H_0 + (N_y - N_z)M\big],$$

$$\frac{dm_y}{dt} = \gamma\mu_0 m_x [H_0 + (N_x - N_z)M].$$

The solution of these equations has the same form as in Eq. (1.13), so that the resonance frequency becomes

$$\omega_0 = \gamma\mu_0 [H_0 + (N_x - N_z)M]^{1/2}[H_0 + (N_y - N_z)M]^{1/2}.\tag{1.38}$$

Equation (1.38) was derived by Charles Kittel [2] to explain some puzzling results of the first ferromagnetic resonance experiments [3], and is called Kittel equation. It is also called the frequency of the uniform mode, in distinction to the frequencies of modes with magnetization that varies in space as in spin waves, or magnons. Kittel equation shows that the FMR frequency depends on the sample shape and on the direction of the applied field. For this reason, the demagnetization is called shape effect, or shape anisotropy. Note that Eqs. (1.35–1.38) are equally valid for ferrimagnetic samples.

The demagnetization factors and the FMR frequency are readily obtained for samples with simple shapes. For a spherical sample, the three demagnetization factors must be the same, so that from Eq. (1.36) we have $N_x = N_y = N_z = 1/3$ in the SI, and $N_x = N_y = N_z = 4\pi/3$ in the CGS. Using this result in Eq. (1.38), we find that the FMR frequency is simply $\omega_0 = \gamma\mu_0 H_0$ in the SI and $\omega_0 = \gamma H_0$ in the CGS. For a very thin film, considering the $x$- and $z$-axes in the plane, and $y$ normal to the plane, $N_x = N_z = 0$ so that $N_y = 1$ in the SI and $N_y = 4\pi$ in the CGS. Thus, the FMR frequency for a thin film with the field applied in the plane is

$$\omega_0 = \gamma\mu_0 [H_0(H_0 + M)]^{1/2} \text{ (SI)}, \qquad \omega_0 = \gamma [H_0(H_0 + 4\pi M)]^{1/2} \text{ (CGS)}\tag{1.39}$$

Table 1.1 shows the values of the demagnetization factors and the equations for the FMR frequency for samples with simple forms and field configurations.

## 1.3.2   Ferromagnetic Resonance with Damping

The manner by which the magnetization relaxes toward equilibrium is governed by the atomic spin interactions and the detailed structure of the magnetic material. The

**Table 1.1** Demagnetization factors and equations for the FMR frequency (in the CGS) for samples with simple forms and field applied in the direction indicated by the arrow

| Sample shape and field direction | Demagnetizing factors | $\omega_0/\gamma$ |
|---|---|---|
| | $N_x = N_y = N_z = 4\pi/3$ | $H_0$ |
| | $N_x = N_y = 0$ <br> $N_z = 4\pi$ | $H_0 - 4\pi M$ |
| | $N_z = N_x = 0$ <br> $N_y = 4\pi$ | $[H_0(H_0 + 4\pi M)]^{1/2}$ |
| | $N_z = 0$ <br> $N_x = N_y = 2\pi$ | $H_0 + 2\pi M$ |
| | $N_y = 0$ <br> $N_x = N_z = 2\pi$ | $[H_0(H_0 - 2\pi M)]^{1/2}$ |

microscopic magnetic relaxation mechanisms have been investigated for many materials over the last decades since their understanding is important from the point of view of basic physics and for technological applications. Regardless of its detailed mechanism, the relaxation can be introduced phenomenologically in the study of the ferromagnetic resonance. This can be done in an ad hoc manner by introducing an imaginary part in the resonance frequency, so that the susceptibilities become complex, as shown in Sect. 1.2.2.

Another manner for introducing damping consists of adding a term in the equation of motion representing a torque that pulls the magnetization toward the equilibrium direction. One of the most widely used forms is that introduced by Gilbert [4] in Eq. (1.10), leading to the so-called Landau–Lifshitz–Gilbert (LLG) equation

$$\frac{d\vec{M}}{dt} = -\gamma\mu_0\vec{M} \times \vec{H}_{eff} + \frac{\alpha}{M}\vec{M} \times \frac{d\vec{M}}{dt}, \tag{1.40}$$

where $\vec{H}_{eff}$ represents the total effective field acting on the magnetization, produced by all spin interactions, and $\alpha$ is a dimensionless quantity called Gilbert damping parameter. Assuming, for simplicity, a uniform magnetization in an infinite sample, subject only to an applied field $\hat{z}H_0$ and considering solutions for the small-signal time-varying components of the magnetization $m_x$ and $m_y$ in the form $\exp(-i\omega t)$, Eq. (1.40) leads to the following linear equations,

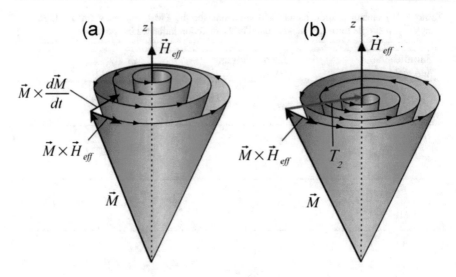

**Fig. 1.4** Illustration of relaxation processes of the magnetization precessing about an equilibrium direction. In (**a**) the magnetization relaxes with constant magnitude, characteristic of the Gilbert damping expressed by the LLG Eq. (1.40). (**b**) Illustrates a process that occurs in many insulating ferro- or ferrimagnetic materials. The transverse components of the magnetization relax rapidly to zero while the $z$-component stays constant, as described by the Bloch–Bloembergen relaxation terms in Eqs. (1.43 and 1.44) for $T_1 \gg T_2$

$$-i\omega m_x = (-\omega_0 + i\alpha\omega)m_y, \quad -i\omega m_y = (\omega_0 - i\alpha\omega)m_x, \tag{1.41}$$

where $\omega_0 = \gamma\mu_0 H_0$. Solution of Eq. (1.41) gives $m_x = -im_y$ and $\omega = \omega_0 - i\alpha\omega$. Thus, if the magnetization vector is deviated from the equilibrium direction, it precesses about the $z$-direction with transverse components that vary with exp $(-i\omega_0 t)$ exp $(-\alpha\omega t)$. Figure 1.4a illustrates the magnetization precession, with the Gilbert torque that pulls the magnetization toward the $z$-direction. Due to the damping, the transverse components of the magnetization decay exponentially in time with a relaxation rate given by $\eta = \alpha\omega$. From Eq. (1.40) it can be shown that the modulus of the magnetization vector is a constant of motion.

Thus, as the magnetization components $m_x$ and $m_y$ decay exponentially in time, $M_z$ increases in such a way that $M^2 = m_x^2 + m_y^2 + M_z^2$ remains constant. The Gilbert torque describes well the relaxation observed in bulk metallic ferromagnets, and for this reason Eq. (1.40) is used to describe the motion of the magnetization in these materials.

Similarly to what we did in Sect. 1.2.2, one can treat the ferromagnetic resonance with Gilbert damping by introducing in the effective field in Eq. (1.40) a microwave driving field with frequency $\omega$. In this case, one obtains for the components of the susceptibility tensor expressions identical to Eqs. (1.29 and 1.30), where $\eta = \alpha\omega$.

Thus, with Eq. (1.32), one obtains a relation between the FMR (half) linewidth in terms of the Gilbert damping parameter,

$$\Delta H = \alpha \omega_0 / \gamma. \tag{1.42}$$

Equation (1.42) is a central result of the Landau–Lifshitz–Gilbert phenomenology, the FMR linewidth scales linearly with the FMR resonance frequency.

Although the LLG equation is the most widely used form to describe the damped motion of the magnetization, it does not apply to insulating ferro- and ferrimagnetic materials with weak spin–orbit coupling. In these materials, the relaxation processes do not take place with constant magnetization as in bulk metallic ferromagnets. The process is better described by a relaxation form introduced by Bloch [5] for nuclear magnetic resonance and adapted by Bloembergen to paramagnetic or ferromagnetic relaxation [6]. The Bloch–Bloembergen (B–B) phenomenology considers that the longitudinal and transverse components of the magnetization have different relaxation rates, so that B–B equations of motion for the magnetization are written as

$$\frac{dm_{x,y}}{dt} = -\gamma \mu_0 (\vec{H}_{eff} \times \vec{M})_{x,y} - \frac{m_{x,y}}{T_2}, \tag{1.43}$$

$$\frac{dM_z}{dt} = -\gamma \mu_0 (\vec{H}_{eff} \times \vec{M})_z - \frac{M_z - M}{T_1}, \tag{1.44}$$

where $T_1$ and $T_2$ are respectively the longitudinal and transverse relaxation times and $1/T_1$ and $1/T_2$ are the corresponding relaxation rates. In insulating ferrimagnetic materials the longitudinal and transverse relaxations are governed by different physical processes so that $T_1$ and $T_2$ can be quite different. Usually $T_2$ is determined by the spin interactions that redistribute the energy of the precessing magnetization in the magnetic system, whereas $T_1$ is determined by processes that thermalize with the lattice. In many materials $T_1 \gg T_2$ so that the relaxation process occurs essentially in two steps, one in which the $z$-component of the magnetization remains constant while the tip of the magnetization vector spirals toward the $z$-axis with characteristic time $T_2$, as illustrated in Fig. 1.4b, followed by another step in which the length of magnetization increases to the saturation value in time $T_1$. This is essentially what happens in YIG, where $T_2$, caused by magnon–magnon processes that conserve $M_z$, is one order of magnitude smaller than $T_1$, which is long due to the weak spin–orbit coupling. Of course, the physical processes that are responsible for $T_1$ and $T_2$ occur simultaneously so that the two steps cannot actually be separated. In general, several mechanisms contribute to the relaxation, acting as independent channels through which the energy flows out of the excited magnetic state so that the total relaxation rate is the sum of the individual contributions and the total half-linewidth is

$$\Delta H = \sum_\lambda 1/(2^\upsilon \gamma T_\lambda), \tag{1.45}$$

where $T_\lambda$ is the relaxation time of mechanism $\lambda$, and $\upsilon$ is an exponent which is 1 for a mechanism contributing only to the longitudinal relaxation and 0 for transverse relaxation. In closing this section, we note that many authors call the source of the

FMR linewidth as Gilbert damping, regardless of its origin. As we have shown this is not strictly correct, some relaxation processes can be adequately described by the LLG equations but others require the B–B formulation. Only if a relaxation process has $T_2 = T_1/2$, the magnitude of $\vec{M}$ remains constant during the relaxation, and then the B–B and the LLG forms are essentially equivalent.

## 1.4    General Equation for the Ferromagnetic Resonance Frequency

In this section, we present the derivation of two equations that can be readily used to calculate the ferromagnetic resonance frequency for a material with various magnetic interactions, and for any direction of the applied magnetic field. The equations are used to calculate the FMR frequency including the effect of the crystalline anisotropy.

### 1.4.1    Effective Fields and FMR Frequency

Consider that the motion of the magnetization is governed by the Landau–Lifshitz equation (1.10), in which the field is the sum of the effective fields representing all magnetic interactions. Using Eqs. (1.4 and 1.9) one can show that the change in the total energy per unit volume $E$, containing the contributions of all magnetic interactions, due to a change in the magnetization $\delta\vec{M}$, is given by $\delta E = -\mu_0 \delta\vec{M} \cdot \vec{H}_{eff}$, where $\vec{H}_{eff}$ is the total effective field. Thus, the effective field is related to the energy density by

$$\vec{H}_{eff} = -\frac{1}{\mu_0} \nabla_{\vec{M}} E, \qquad (1.46)$$

where $\nabla_{\vec{M}}$ represents the gradient operator relative to the components of $\vec{M}$, that is, the $\alpha$ component of $\nabla_{\vec{M}}$ is $\partial E/\partial M_\alpha$. If we write the magnetization as $\vec{M} = \vec{m} + \hat{z} M_z$, where $z$ is the equilibrium direction, since the modulus of $\vec{M}$ is constant, the dependence of the energy on the magnetization can be written in terms of the transverse components $E(m_x, m_y)$. Thus, the effective field in Eq. (1.46) has components in the $x$-$y$ plane only. Considering for $\vec{m}$ the time dependence exp$(-i\omega_0 t)$ and $m \ll M$, Eq. (1.10) gives

$$-i\omega_0 m_x = -\gamma M [\nabla_{\vec{M}} E]_y, \qquad (1.47)$$

$$-i\omega_0 m_y = \gamma M [\nabla_{\vec{M}} E]_x. \qquad (1.48)$$

Now we expand the energy $E(\vec{m})$ in power series of $m_\alpha$, where $\alpha$, $\beta$ denote either $x$ or $y$

$$E(\vec{m}) \cong E_0 + \frac{1}{2}\sum_{\alpha,\beta}\frac{\partial^2 E}{\partial m_\alpha \partial m_\beta}\bigg|_0 m_\alpha m_\beta, \tag{1.49}$$

where we have considered $|\partial E/\partial m_\alpha|_0 = 0$ at the equilibrium position. Since $m \ll M$, in Cartesian coordinates we can write $m_x = M\phi_x$ and $m_y = M\phi_y$, where $\phi_x$ and $\phi_y$ represent the deviation angles of the magnetization vector from the equilibrium direction. Using the expansion in Eq. (1.49) is Eqs. (1.47 and 1.48) we obtain

$$i\frac{\omega_0 M}{\gamma}\phi_x = E_{xy}\phi_x + E_{yy}\phi_y,$$

$$-i\frac{\omega_0 M}{\gamma}\phi_y = E_{xx}\phi_x + E_{yx}\phi_y,$$

where $E_{\alpha\beta} \equiv \partial^2 E/\partial m_\alpha \partial m_\beta$ is calculated at the equilibrium position. These equations can be written in matrix form

$$\begin{bmatrix} E_{xy} - i\omega_0 M/\gamma & E_{yy} \\ E_{xx} & E_{yx} + i\omega_0 M/\gamma \end{bmatrix}\begin{pmatrix} \phi_x \\ \phi_y \end{pmatrix} = 0. \tag{1.50}$$

The zero of the main determinant of the system gives the FMR frequency

$$\omega_0 = \frac{\gamma}{M}(E_{xx}E_{yy} - E_{xy}^2)^{1/2}, \tag{1.51}$$

where we have used the fact that $E_{xy} = E_{yx}$. For systems with the equilibrium magnetization in an arbitrary direction, it is useful to have another expression for the FMR frequency in terms of the polar and azimuthal angles of a spherical coordinate system $(r, \theta, \varphi)$. This can be obtained with

$$\vec{M} = \hat{e}_r M_r + \hat{e}_\theta m_\theta + \hat{e}_\varphi m_\varphi, \tag{1.52}$$

where $m_\theta$, $m_\varphi \ll M_r \approx M$, so that we can write $m_\theta = M\delta\theta$ and $m_\varphi = M\sin\theta\,\delta\varphi$, where $\delta\theta$, $\delta\varphi$ represent the deviations of the polar and azimuthal angles from the equilibrium direction, denoted by $\theta_0$, $\varphi_0$. Using Eqs. (1.46) in (1.10) in spherical coordinates and considering the time dependence $\exp(-i\omega_0 t)$ for $m_\theta$, $m_\varphi$ we obtain

$$\left(-i\frac{\omega_0}{\gamma}M\sin\theta_0 + E_{\varphi\theta}\right)m_\theta + \frac{1}{\sin\theta_0}E_{\varphi\varphi}m_\varphi = 0,$$

$$E_{\theta\theta}m_\theta + \left( \frac{1}{\sin\theta_0} E_{\theta\varphi} - i\frac{\omega_0}{\gamma} M \right) m_\varphi = 0,$$

where the angles of the equilibrium direction are calculated with $\partial E/\partial\theta = \partial E/\partial\varphi = 0$ and $E_{\alpha\beta}$ denotes the second derivative $\partial E^2/\partial\alpha\partial\beta$ calculated at the angles of equilibrium. The zero of the main determinant of the system gives the FMR frequency

$$\omega_0 = \frac{\gamma}{M\sin\theta_0} \left( E_{\theta\theta} E_{\varphi\varphi} - E_{\theta\varphi} E_{\varphi\theta} \right)^{1/2}, \qquad (1.53)$$

Equation (1.53), first derived by Suhl [7] and independently by Smit and Beljers [8], is very useful for calculating the FMR frequency in a variety of material configurations, as will be shown in the remainder of this chapter. Notice that the coordinate system should be chosen such that the polar angle of equilibrium is not 0 or $\pi$.

## 1.4.2    Effect of Crystalline Anisotropy on the FMR Frequency

In a crystalline magnetic material, there is a magnetic interaction not considered previously that tends to orient the magnetization along certain crystal axes of symmetry. This interaction can be represented by a magneto-crystalline energy, also called anisotropy energy. The form of the energy and its magnitude depend on the crystal symmetry and the chemical composition of the material.

The microscopic origin of the anisotropy lies in the fact that the electron spin interacts with the orbital moment by means of the spin–orbit coupling, and the electronic orbitals interact with the electric field created by the ionic charges of the lattice. The effect of the anisotropy on the magnetization dynamics can be studied using a phenomenological expression for the anisotropy energy. The form of the energy depends on the direction of the magnetization relative to the crystalline axes and also on the crystal symmetry. In a crystal with uniaxial symmetry, the anisotropy energy density can be expressed by a power series expansion of the form

$$E_u = -K_{u1}\cos^2\theta - K_{u2}\cos^4\theta, \qquad (1.54)$$

where $\theta$ is the angle between the magnetization vector and the direction of symmetry, and $K_{u1}$ and $K_{u2}$ are, respectively, the anisotropy constants of first and second order. There are no terms in odd powers of $\cos\theta$ because the energy does not change with the inversion of $\vec{M}$. If both $K_{u1}$ and $K_{u2}$ are positive, the energy is minimum for $\theta = 0, \pi$. Thus, this is called the **easy axis** of the crystal. On the other hand, if both $K_{u1}$ and $K_{u2}$ are negative, the energy is minimum for $\theta = \pm \pi/2$, and the plane perpendicular to the symmetry ($\theta = 0$) is called **easy plane** of the magnetization. Cobalt is a metallic ferromagnet that crystallizes with a hexagonal structure, has uniaxial anisotropy with energy well described by Eq. (1.54), with anisotropy constants $K_{u1} = 4.1 \times 10^6 \, erg/cm^3$ and $K_{u2} = 1.0 \times 10^6 \, erg/cm^3$ at room

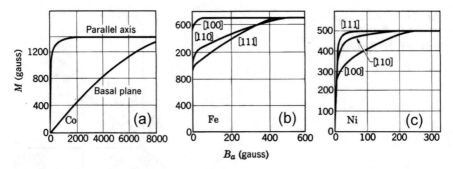

**Fig. 1.5**  Magnetization curves for Co, Fe, and Ni [9]

temperature. Since they are both positive, the easy axis of Co is the symmetry direction. For this reason, the magnetization of Co rises much faster with increasing field if this is applied along the parallel axis instead of in the basal plane, as can be seen in Fig. 1.5a.

In magnetic crystals with cubic structure, such as Fe, Ni, and YIG, there are several equivalent directions, so that the expression for the anisotropy energy requires the use of several angles. One can use the direction cosines $\alpha_1$, $\alpha_2$, $\alpha_3$ to characterize the direction of $\vec{M}$ relative to the crystal axes [100], [010], and [001], respectively. As in the case of uniaxial symmetry, the anisotropy energy should contain only even powers of the direction cosines because of inversion symmetry. Also, the energy should be invariant under interchanges of $\alpha_i$ among themselves due to cubic symmetry. Finally, it does not contain the lowest term $\alpha_1^2 + \alpha_2^2 + \alpha_3^2$ because it is a constant. Thus, the cubic anisotropy energy density can be represented by

$$E_c \cong K_{c1}(\alpha_1^2\alpha_2^2 + \alpha_2^2\alpha_3^2 + \alpha_3^2\alpha_1^2) + K_{c2}\alpha_1^2\alpha_2^2\alpha_3^2, \qquad (1.55)$$

where $K_{c1}$ and $K_{c2}$ are, respectively, the first- and second-order cubic anisotropy constants. By minimizing the energy in Eq. (1.55) one can show that the easy and hard directions of magnetization are along one of the principal symmetry axes, <100>, <110>, or <111>, depending on the relative values of the anisotropy constants. For $K_{c1} > 0$ and $K_{c2} \geq -9\,K_{c1}$ the easy axis is along <100> while for $K_{c1} < 0$ and $K_{c2} < -9\,K_{c1}$ the easy axis is along <111>. The constants for Fe at room temperature are $K_{c1} = 4.5 \times 10^5$ erg/cm$^3$ and $K_{c2} = 1.5 \times 10^5$ erg/cm$^3$ and the easy axes are along <100>, as shown in Fig. 1.5b. At room temperature, in Ni, $K_{c1} = -5 \times 10^4$ erg/cm$^3$ and in YIG $K_{c1} = -5 \times 10^3$ erg/cm$^3$, while $K_{c2}$ is negligible, so that in both materials the easy axes are <111>, while <100> are hard axes.

In order to calculate the effect of the crystalline anisotropy on the FMR frequency, initially we consider a spherical sample in which, as we previously saw, the demagnetizing effects cancel out. Then we consider only two contributions to the magnetic energy, the interaction of the magnetization with the applied field that magnetizes the sample to saturation, and the anisotropy energy. The interaction with the field is represented by the Zeeman energy per unit volume

**Fig. 1.6** Coordinates used to express the field and magnetization vectors

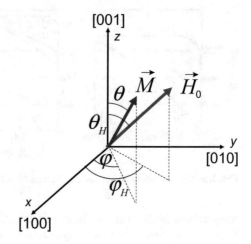

$$E_Z = -\mu_0 \vec{M} \cdot \vec{H}_0. \tag{1.56}$$

With the magnetization vector and the applied field in arbitrary directions, characterized by the polar and azimuthal angles in a coordinate system with the axes along the $<100>$ axes of a cubic lattice, as shown in Fig. 1.6, the Zeeman energy becomes

$$E_Z = -\mu_0 M H_0 [\sin \theta \sin \theta_H \cos (\varphi - \varphi_H) + \cos \theta \cos \theta_H], \tag{1.57}$$

where $\theta_H$, $\varphi_H$ are the angles of the applied field and $\theta$, $\varphi$ are the angles of the magnetization. Consider in the cubic anisotropy energy density given by Eq. (1.55) only the first-order constant and write it in terms of the magnetization components

$$E_A = \frac{K_{c1}}{M^4} (M_x^2 M_y^2 + M_y^2 M_z^2 + M_z^2 M_x^2). \tag{1.58}$$

This can be expressed in terms of the polar and azimuthal angles of the magnetization vector

$$E_A = \frac{K_{c1}}{4} (\sin^4 \theta \sin^2 2\varphi + \sin^2 2\theta). \tag{1.59}$$

The angles of the equilibrium magnetization are given by $\partial E/\partial \theta = \partial E/\partial \varphi = 0$, where $E = E_Z + E_A$ is the total energy. For a general direction of the magnetic field one has to solve the equations for equilibrium for both $\theta$ and $\varphi$. In order to obtain simpler equations, we consider that the field is applied in the (001) plane. In this case, with $\theta_H = \pi/2$, $\partial E/\partial \theta = 0$ gives for the equilibrium direction $\theta_0 = \pi/2$, meaning that the magnetization is also in the (001) plane. The equation $\partial E/\partial \varphi = 0$ gives for the azimuthal angle of equilibrium

$$2\mu_0 M H_0 \sin(\varphi_H - \varphi_0) = K_{c1} \sin 4\varphi_0. \tag{1.60}$$

This equation has analytical solutions only for $\varphi_H = 0$, $\pi/4$, $\pi/2$ and corresponding angles in the other quadrants, in which case $\varphi_0 = \varphi_H$, meaning that the equilibrium magnetization is aligned with the field. For a general field angle, Eq. (1.60) has to be solved numerically. However, for Fe and YIG, the anisotropy energy is much smaller than the Zeeman energy, $K_{c1} \ll 2\mu_0 M H_0$, so that $\varphi_H - \varphi_0 \cong K_{c1} \sin 4\varphi_0/(2\mu_0 M H_0) \ll 1$.

In order to calculate the FMR frequency with Eq. (1.53) we need the second derivatives with respect to the angles. Using Eqs. (1.57 and 1.59) one can show that for $\theta_0 = \pi/2$ we have $E_{\theta\varphi} = E_{\varphi\theta} = 0$ and

$$E_{\theta\theta} = \mu_0 M H_0 \cos(\varphi_H - \varphi_0) + K_{c1}(2 - \sin^2 2\varphi_0),$$

$$E_{\varphi\varphi} = \mu_0 M H_0 \cos(\varphi_H - \varphi_0) + 2K_{c1} \cos 4\varphi_0.$$

Using these equations in Eq. (1.53), with the approximation $\varphi_H - \varphi_0 \ll 1$, we obtain for the FMR frequency

$$\omega_0 = \gamma\mu_0 \left[ H_0 + \frac{K_{c1}}{\mu_0 M}(2 - \sin^2 2\varphi_H) \right]^{1/2} \left[ H_0 + \frac{2K_{c1}}{\mu_0 M} \cos 4\varphi_H \right]^{1/2}. \tag{1.61}$$

For the field applied along one of the <100> axes, with $\varphi_H = 0$, $\pi/2$ this gives

$$\omega_0 = \gamma\mu_0(H_0 + H_{A100}), \tag{1.62}$$

where $H_{A100} = 2K_{c1}/(\mu_0 M)$ represents an effective anisotropy field. For the field applied along one of the <110> axes, with $\varphi_H = \pi/4$, Eq. (1.61) gives

$$\omega_0 = \gamma\mu_0 \left( H_0 + \frac{K_{c1}}{\mu_0 M} \right)^{1/2} \left( H_0 - \frac{2K_{c1}}{\mu_0 M} \right)^{1/2}, \tag{1.63}$$

so that the anisotropy cannot be represented by a simple effective field. Note that when the field is aligned along the <100> or <110> axes, the relations $\theta_0 = \theta_H$ and $\varphi_0 = \varphi_H$ are exact, so that Eqs. (1.62 and 1.63) are also exact.

The FMR frequency can also be calculated exactly if the field is applied along one of the <111> axes. If we set $\theta_H = \arccos(\sqrt{3}/2)$ and $\varphi_H = \pi/4$, one can show that the angles of the magnetization are given by the relations $\theta_0 = \theta_H$ and $\varphi_0 = \varphi_H$, and the FMR frequency calculated with Eq. (1.53) becomes

$$\omega_0 = \gamma\mu_0(H_0 + H_{A111}), \tag{1.64}$$

where $H_{A111} = -4K_{c1}/(3\mu_0 M)$ represents an effective anisotropy field for the <111> direction.

Using for Fe, at room temperature, $K_{c1} = 4.2 \times 10^5$ erg/cm$^3$ and $M = 1707$ G, we find $H_{A100} = 492$ Oe and $H_{A111} = -328$ Oe. A positive $H_{A100}$ means that the anisotropy field adds to the external field, an indication that $<100>$ is an easy axis of anisotropy. On the other hand, $<111>$ is a hard axis. Using for YIG, at room temperature, $K_{c1} = -5 \times 10^3 \, \text{erg/cm}^3$ and $M = 140 \, \text{G}$, we find $H_{A100} = -71 \, \text{Oe}$ and $H_{A111} = 48 \, \text{Oe}$, showing that the easy axes are $<111>$, while $<100>$ are hard axes.

## 1.5  Ferromagnetic Resonance in Thin Films

### 1.5.1  Experimental Measurement of Ferromagnetic Resonance

A ferromagnetic resonance experiment consists in exciting a ferro- or ferrimagnetic sample with an *rf* magnetic field of a microwave radiation applied perpendicularly to the static field and measuring the response of the magnetization dynamics. Since the variation of the frequency produces changes in the electromagnetic response of the microwave circuit, usually the FMR experiments are done with fixed frequency and scanning static field intensity. The magnetization dynamics is excited when the field intensity approaches the value for which the resonance frequency coincides with the excitation frequency.

A schematic diagram of the experimental setup for measuring FMR is shown in Fig. 1.7a. The main components of the setup are microwave generator; circulator; microwave cavity or some kind of fixture to excite the sample; electromagnet; microwave detector; and electronic analog or digital system to process the signal from the detector. The radiation from the microwave generator is directed to the circulator by a transmission line, usually either a coaxial cable or a hollow

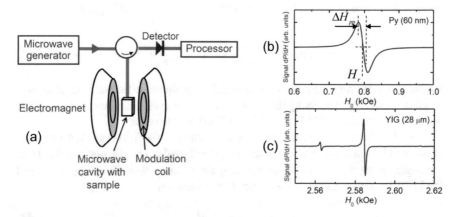

**Fig. 1.7** (**a**) Schematic apparatus used for ferromagnetic resonance (FMR) experiments. (**b**) Field-derivative FMR absorption spectrum for a permalloy film with thickness 60 nm measured with frequency 8.6 GHz [10]. Reproduced from A. Azevedo et al., Appl. Phys. Lett. **104**, 052402 (2014), with the permission of AIP Publishing. (**c**) FMR absorption spectrum for a 28-um thick YIG film measured at 9.4 GHz [11]. Reproduced from S. M. Rezende et al., Appl. Phys. Lett. **102**, 012402 (2013), with the permission of AIP Publishing

waveguide. The circulator is a nonreciprocal device that employs the gyromagnetic properties of ferrites. Power entering in a certain port leaves only from the adjacent port, following the sense of the arrow. Thus, the power from the generator is directed by the circulator to the arm that leads to the sample, while the signal returning from the sample is directed to the detector. The detector consists of a high-frequency diode that rectifies the microwave field producing a low-frequency signal corresponding to the variation of the peak value of the microwave. The signal from the detector is processed by analog and/or digital electronic equipment.

The FMR signal is proportional to the power absorbed by the sample. Hence, according to Eq. (1.31), it is proportional to the microwave *rf* field, to the sample volume, and to the absorptive susceptibility given by Eq. (1.30). For samples with small volume and small susceptibility, two schemes are used to increase the signal. The first consists in using a resonant cavity, usually made with a hollow rectangular or cylindrical waveguide shorted at the ends, and placing the sample in a position of maximum *rf* magnetic field. The second consists in using a pair of Helmholtz coils to create an alternating field that is superimposed to the field of the electromagnet, as illustrated in Fig. 1.7a. The modulation field has a frequency of a few kilohertz and amplitude that can be varied from a fraction of oersted to several oersteds. A modulation field with amplitude much smaller than the linewidth results in an FMR signal that is proportional to the derivative of the absorptive susceptibility relative to the magnetic field. This signal is then amplified with a phase-sensitive (or phase lock-in) amplifier increasing the signal-to-noise ratio by orders of magnitude. Since the imaginary part of the susceptibility varies with field as a Lorentzian function, when the static field of the electromagnet is swept and superimposed with the modulation field, the resulting FMR signal has the shape of a Lorentzian derivative.

A typical FMR spectrum is shown in Fig. 1.7b for a 60-nm thick film of $Ni_{81}Fe_{19}$, called permalloy, an important material for scientific investigations and for practical applications. The measurements were made using a microwave rectangular waveguide cavity with $Q = 2000$ at a frequency of 8.6 GHz [10]. Two quantities are readily measured in the derivative absorption spectrum, indicated in Fig. 1.7b, the resonance field $H_r$ and the peak-to-peak linewidth $\Delta H_{pp}$. $H_r$ is the field value for which the absorption derivative changes sign, that corresponds to maximum absorption, and $\Delta H_{pp}$ is the separation between the maximum and the minimum. Using Eqs. (1.30 and 1.32) one can show that this is related to the FMR linewidth by

$$\Delta H = \frac{\sqrt{3}}{2} \Delta H_{pp}. \tag{1.65}$$

From the value measured in Fig. 1.7b, $\Delta H_{pp} = 27$ Oe, we obtain with Eq. (1.65) the linewidth of the Py film, $\Delta H = 23.4$ Oe. This is smaller than the values measured in other simple metallic films such as Co, Fe, Ni, and CoFe. The spectrum shown in Fig. 1.7c was measured with a 28-µm thick crystalline YIG film at 9.4 GHz [11]. Since the sample volume is orders of magnitude larger than in the Py film, the measurement was made with a shorted waveguide, not a resonant cavity. In fact, microwave cavities are not adequate to study thick YIG films because the strong FMR interferes with the cavity resonance. The peak-to-peak linewidth measured in Fig. 1.7c is $\Delta H_{pp} = 0.7$ Oe, corresponding to a linewidth of the YIG film $\Delta H = 0.6$ Oe. This is almost two orders of magnitude smaller than in Py. The fact that YIG has the smallest room temperature damping of all known materials makes it the best material for studies of many magnonic phenomena [12].

## 1.5.2  FMR in Crystalline Films with Cubic Anisotropy

In very thin magnetic films there is another important contribution to the energy not considered previously, namely, the surface anisotropy, also called out-of-plane anisotropy. This originates from the fact that there is a broken translation symmetry in the crystalline electric field at the surface or interface that reflects in the magnetic energy due to the spin–orbit coupling. The surface anisotropy can be represented phenomenologically by an energy per unit area of the form $-(K_S/M^2)(\widehat{n} \cdot \vec{M})^2$, where $K_S$ is the surface anisotropy constant and $\widehat{n}$ denotes the unit vector normal to the film plane [13]. Thus, the surface anisotropy energy per unit volume is

$$E_S = -\frac{K_S}{M^2 t}\left(n \cdot \vec{M}\right)^2, \tag{1.66}$$

where $t$ is the film thickness. Clearly, for $K_S > 0$ the surface anisotropy tends to pull the magnetization out of the plane, while for $K_S < 0$ it favors an in-plane magnetization. The effective surface anisotropy field obtained from (1.66) has intensity $H_S = 2K_S/(\mu_0 M t)$. We can find the FMR frequency for a magnetic film with cubic volume anisotropy under an external field applied in the film plane, using for the total energy per unit volume

$$E = -\mu_0 \vec{H}_0 \cdot \vec{M} + \frac{K_{c1}}{M^4}\left(M_x^2 M_y^2 + M_y^2 M_z^2 + M_z^2 M_x^2\right) + \frac{1}{2}\mu_0\left(n \cdot \vec{M}\right)^2$$
$$-\frac{K_S}{M^2 t}\left(n \cdot \vec{M}\right)^2, \tag{1.67}$$

where the first term represents the Zeeman energy of the interaction with the external field, the second represents the cubic anisotropy energy, the third is the demagnetizing energy obtained by integrating the effective field in Eq. (1.37) with respect to the magnetization with demagnetizing factors $N_n = 1$ and the two others null, and the last one is the surface anisotropy. Using the angular coordinates of Fig. 1.6, the first two terms are identical to Eqs. (1.57 and 1.59), and the last two become $E_{dem} + E_S = M^2(\mu_0/2 - K_s/M^2 t)\cos^2\theta$. The angles of the equilibrium magnetization are given by $\partial E/\partial\theta = \partial E/\partial\varphi = 0$, where $E = E_Z + E_A + E_{dem} + E_S$ is the total energy. For a general direction of the magnetic field one has to solve the equations for equilibrium for both $\theta$ and $\varphi$.

Consider a crystalline film grown in the (001) plane with an external field applied in the plane. In this case, with $\theta_H = \pi/2$, assuming that the surface anisotropy does not pull the magnetization out of the plane, $\partial E/\partial \theta = 0$ gives for the equilibrium direction $\theta_0 = \pi/2$. The equation $\partial E/\partial \varphi = 0$ gives for the azimuthal angle of equilibrium the same result as in Eq. (1.60). Using in Eq. (1.53) the second derivatives of the energy calculated as previously, we obtain for the FMR frequency

$$\omega_0 = \gamma\mu_0 \left[ H_0 \cos(\varphi_H - \varphi_0) + M_{eff} + \frac{K_{c1}}{\mu_0 M}(2 - \sin^2 2\varphi_0) \right]^{1/2}$$
$$\times \left[ H_0 \cos(\varphi_H - \varphi_0) + \frac{2K_{c1}}{\mu_0 M}\cos 4\varphi_0 \right]^{1/2}, \quad (1.68)$$

where we have introduced the effective magnetization $M_{eff} = M - H_S$. In the CGS this is replaced by $4\pi M_{eff} = 4\pi M - H_S$ and $\mu_0 = 1$. For the field applied along one of the <100> axes, with $\varphi_0 = \varphi_H = 0, \pi/2$ in Eq. (1.68), we obtain the FMR frequency for a crystalline film

$$\omega_0 = \gamma\mu_0(H_0 + H_{A100})^{1/2}(H_0 + H_{A100} + M_{eff})^{1/2}, \quad \text{(SI)} \quad (1.69a)$$

$$\omega_0 = \gamma(H_0 + H_{A100})^{1/2}(H_0 + H_{A100} + 4\pi M_{eff})^{1/2}. \quad \text{(CGS)} \quad (1.69b)$$

One of the most used techniques for measuring the various material parameters of a magnetic film is angular-dependent ferromagnetic resonance. Figure 1.8 shows the scheme and results of FMR experiments carried out with a 10.2 nm thick film of (001) Fe, grown by DC magnetron sputtering on a (001) MgO substrate [14]. The sample is glued to the tip of an insulating rod and introduced through a small hole in the back wall of a rectangular microwave cavity with resonance frequency 11 GHz and quality factor $Q = 2000$, in a position of maximum microwave field $h$ and zero electric field, as illustrated in Fig. 1.8a. The waveguide is placed between the poles of an electromagnet so that the static magnetic field and the microwave field are in the film plane and kept perpendicular to each other as the sample is rotated. This configuration allows measurements of the angular dependence of the FMR field $H_r$ and the linewidth.

## 1.5.3 FMR in Ferromagnetic Films with Exchange Bias

When a ferromagnetic (FM) film is grown in contact with an antiferromagnetic (AF) layer, its magnetization becomes subject to another type of anisotropy, called unidirectional anisotropy. This phenomenon, called exchange bias, arises from the exchange interaction between the spins of the FM and the spins of the AF layer through the FM/AF interface. This effect was first observed in nanoparticles of FM materials embedded in AF matrices in the 1950s [15] and only in the last few

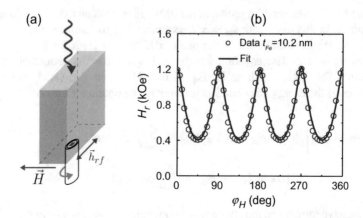

**Fig. 1.8** (a) Illustration of the setup for FMR experiments with the scheme for introducing the sample through a hole in the back wall of a microwave waveguide cavity and for rotating the film in the plane. (b) Angular dependence of the FMR field measured at 11 GHz in an Fe (10.2 nm) film in the (001) plane, exhibiting the fourfold symmetry typical of cubic anisotropy. The symbols represent the data and the solid line represents a fit with the FMR field calculated with Eq. (1.68) using the azimuthal angle obtained numerically with Eq. (1.60) [14]. Reproduced from J. R. Fermin et al., J. Appl. Phys. **85**, 7316 (1999), with the permission of AIP Publishing

decades became subject of investigations and applications in very thin magnetic films [16].

A characteristic feature of the exchange bias phenomenon is the shift of the hysteresis curve in the magnetic field axis when this is applied in the direction of the unidirectional anisotropy. Figure 1.9a shows the hysteresis curve of a bilayer made of a permalloy film ($Ni_{81}Fe_{19}$) with thickness 7.5 nm, grown on a 35-nm thick layer of the metallic antiferromagnet $Ir_{20}Mn_{80}$. The field shift measured in the hysteresis loop of Fig. 1.9a is 146 Oe [17]. Permalloy is a convenient material for studying various phenomena and for some applications because it has very small crystalline anisotropy, and thus very narrow hysteresis loop. This is due to the fact that the anisotropy constant $K_{c1}$ is negative in Ni and positive in Fe, so that with the composition 81–19 the net anisotropy is very small.

The measurement shown in Fig. 1.9 was made with a magneto-optical Kerr effect (MOKE) setup, that is widely used to measure the magnetic properties of very thin films. The MOKE technique employs a low power laser with polarization modulated by an electro-optical device that undergoes a Kerr effect in the film proportional to the magnetization [18]. Thus, the change in the polarization of the light reflected from a magnetic film gives a signal proportional to the magnetization. The MOKE technique is very convenient for studies of very thin films because they have small volumes and thus small net magnetic moment to be detected by direct magnetometry methods.

The shift in field of the hysteresis curve is due to the fact that the sublattice spins of the AF layer closest to the interface interact with the FM spins at the interface by means of the exchange interaction. Since the AF sublattice spins are essentially

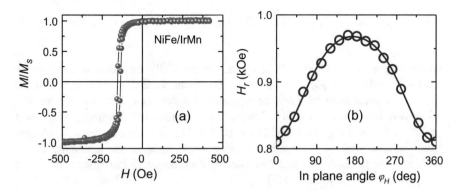

**Fig. 1.9** (a) MOKE hysteresis loop of a $Ni_{81}Fe_{19}$ (7.5 nm)/$Ir_{20}Mn_{80}$ (35 nm) bilayer measured with field applied parallel to the direction of the exchange bias. The measured field shift is 146 Oe. (b) In-plane angular dependence of the FMR field measured at 9.4 GHz. The symbols represent the data and the solid line represents a numerical fit that gives the value $H_{eb} = 95$ Oe [17]. Reproduced from R. L. Rodríguez-Suárez et al., J. Appl. Phys. **123**, 043901 (2018), with the permission of AIP Publishing

frozen, an external field applied to the FM/AF bilayer acts on the FM magnetization without disturbing the AF spin arrangement. Thus, in a very simplified description, the effect of the exchange bias can be represented by an energy per unit volume of the form $-\mu_0 \vec{H}_{eb} \cdot \vec{M}$, where $\vec{H}_{eb}$ is the exchange bias field. This field has a direction defined by the AF spin arrangement and has an amplitude that depends on the composition of the bilayer and the details of the FM/AF interface. Since it originates from an interface phenomenon, the exchange bias field varies inversely with the FM film thickness, $H_{eb} \sim 1/t$.

In order to calculate the FMR frequency of a film with exchange bias, consider for the energy per unit volume

$$E = -\mu_0 \vec{H}_0 \cdot \vec{M} + \frac{1}{2}\mu_0 (\hat{n} \cdot \vec{M})^2 - \frac{K_S}{M^2 t}(\hat{n} \cdot \vec{M})^2 - \mu_0 \vec{H}_{eb} \cdot \vec{M}, \qquad (1.70)$$

where the various terms represent, in order from the left, Zeeman, demagnetizing, surface anisotropy and exchange bias contributions. Consider that the film is in the $x$–$y$ of the coordinate system in Fig. 1.6, with the external field in the film plane at an angle $\varphi_H$ with the $x$-axis, and that the exchange bias field is along $\varphi = 0$. The energy in Eq. (1.70) can be written in terms of the angle coordinates and the direction of equilibrium magnetization is given by $\partial E/\partial\theta = \partial E/\partial\varphi = 0$. For $\theta_H = \pi/2$ one finds $\theta_0 = \pi/2$, while the azimuthal angle $\varphi_0$ is given by

$$H_{eb} \sin\varphi_0 = H_0 \sin(\varphi_H - \varphi_0). \qquad (1.71)$$

Using in Eq. (1.53) the second derivatives of the energy in Eq. (1.70) expressed in terms of the angular coordinates, we obtain for the FMR frequency of the film with exchange bias

$$\omega_0 = \gamma\mu_0[H_0\cos(\varphi_H - \varphi_0) + M_{eff} + H_{eb}\cos\varphi_0]^{1/2}[H_0\cos(\varphi_H - \varphi_0) + H_{eb}\cos\varphi_0]^{1/2},$$

$$(1.72)$$

where $M_{eff} = M - 2K_S/\mu_0 M^2 t$ in the SI. In the CGS, this term is replaced by $4\pi M_{eff} = 4\pi M - 2K_S/M^2 t$ and $\mu_0 = 1$. The symbols in Fig. 1.9b represent the angular dependence of the FMR field measured in the same film of Fig. 1.9b, at a microwave frequency of 9.4 GHz. The solid line represents a fit with Eq. (1.72) that yields for the exchange bias field $H_{eb} = 95$ Oe. The fact that this value is different from the field shift of the hysteresis curve is attributed to the sensitivity of the exchange bias phenomenon to the measurement technique, an intriguing and still not completely resolved question [19].

## Problems

1.1 Consider a magnetization vector $\vec{M}$ under an external magnetic field with its motion governed by Eq. (1.40) with the Gilbert damping. Show that if $\vec{M}$ is pulled out of equilibrium by some external perturbation, it describes a spiraling motion toward the equilibrium direction with constant magnitude. Hint, use the dot products of both terms on the right-hand side of Eq. (1.40) with $\vec{M}$.

1.2 Find the equations for the ferromagnetic resonance frequency of a sample with the shape of a thin cylinder, with magnetization $M$, no crystalline anisotropy, with an external field $\vec{H}_0$ applied either parallel or perpendicular to the cylinder axis.

1.3 Using the expression for the power absorbed by a sample under FMR, show that the linewidth (half-width at half maximum) is related to peak-to-peak linewidth by $\Delta H = \sqrt{3}\Delta H_{pp}/2$.

1.4 A spherical ferromagnetic sample with saturation magnetization $4\pi M$ has crystalline uniaxial anisotropy with energy per unit volume $E_A = -K_u\cos^2\theta$, where $\theta$ is the angle between the magnetization and the symmetry axis. Show that the FMR frequency with a static magnetic field $H_0$ applied along the symmetry axis is given by $\omega_0 = \gamma(H_0 + H_A)$, where $H_A$ is an effective anisotropy field.

1.5 Consider a spherical ferromagnetic sample with crystalline cubic anisotropy with energy per unit volume as in Eq. (1.55). Assume that the magnetization is saturated.

(a) Find the directions of the magnetization relative to the principal symmetry axes for which the anisotropy energy is maximum or minimum.

(b) Determine the conditions that the anisotropy constants $K_{c1}$ and $K_{c2}$ must obey for the energy to be minimum either along the <100> or <111> axes.

(c) Using the values for $K_{c1}$ and $K_{c2}$ given in the text, determine which are the easy and hard axes in Fe, Ni, and YIG.

1.6 Consider a thin ferromagnetic film with crystalline cubic and surface anisotropies with energy per unit volume as in Eq. (1.67). Considering the film magnetized in the plane by an external field $H_0$, calculate the second derivatives of the energy and find the equation for the FMR frequency using Eq. (1.53).

1.7 A ferromagnetic resonance experiment is carried out with various magnetic films magnetized in the plane, using a microwave spectrometer operating with frequency 10 GHz. Neglecting all types of anisotropy, calculate the field for resonance for films made of Fe, Permalloy, and YIG, that have saturation magnetization $4\pi M$, respectively, 21.45 kG, 11.5 kG, and 1.76 kG.

# References

1. Landau, L.D., Lifshitz, E.M.: On the theory of the dispersion of magnetic permeability in ferromagnetic bodies. Phys. Z. Sowjetunion. **8**, 153 (1935)
2. Kittel, C.: Interpretation of Anomalous Larmor frequencies in ferromagnetic resonance experiment. Phys. Rev. **71**, 270 (1947)
3. Griffiths, J.H.E.: Anomalous high-frequency resistance of ferromagnetic metals. Nature. **158**, 670 (1946)
4. Gilbert, T.L.: Lagrangian formulation of the gyromagnetic equation of the magnetization field. Phys. Rev. **100**, 1243 (1955)
5. Bloch, F.: Nuclear induction. Phys. Rev. **70**, 460 (1946)
6. Bloembergen, N.: On the ferromagnetic resonance in nickel and supermalloy. Phys. Rev. **78**, 572 (1950)
7. Suhl, H.: Ferromagnetic resonance in nickel ferrite between one and two kilomegacycles. Phys. Rev. **97**, 555 (1955)
8. Smit, J., Beljers, G.: Ferromagnetic resonance absorption in $BaFe_{12}O_{19}$, a highly anisotropic crystal. Philips Res. Rep. **10**, 113 (1955)
9. Honda, K., Kaya, S.: On magnetization of single crystals of iron. Sci. Rept. Tohoku Imp. Univ. **15**, 721 (1926)
10. Azevedo, A., Alves Santos, O., Fonseca Guerra, G.A., Cunha, R.O., Rodríguez-Suárez, R., Rezende, S.M.: Competing spin pumping effects in magnetic hybrid structures. Appl. Phys. Lett. **104**, 052402 (2014)
11. Rezende, S.M., Rodríguez-Suárez, R.L., Soares, M.M., Vilela-Leão, L.H., Ley Domínguez, D., Azevedo, A.: Enhanced spin pumping damping in yttrium iron garnet/Pt bilayers. Appl. Phys. Lett. **102**, 012402 (2013)
12. Serga, A.A., Chumak, A.V., Hillebrands, B.: YIG Magnonics. J. Phys. D. Appl. Phys. **43**, 264002 (2010)
13. Rado, G.T.: Theory of ferromagnetic resonance and static magnetization in ultrathin crystals. Phys. Rev. B. **26**, 295 (1982)
14. Fermin, J.R., Azevedo, A., de Aguiar, F.M., Li, B., Rezende, S.M.: Ferromagnetic resonance linewidth and anisotropy dispersions in thin Fe films. J. Appl. Phys. **85**, 7316 (1999)
15. Meiklejohn, W.H., Bean, C.P.: New magnetic anisotropy. Phys. Rev. **102**, 1413 (1956)
16. Nogués, J., Schuller, I.K.: Exchange bias. J. Magn. Magn. Mat. **192**, 203 (1999)
17. Rodríguez-Suárez, R.L., Oliveira, A.B., Estrada, F., Maior, D.S., Arana, M., Alves Santos, O., Azevedo, A., Rezende, S.M.: Rotatable anisotropy on ferromagnetic/antiferromagnetic bilayer investigated by Brillouin light scattering. J. Appl. Phys. **123**, 043901 (2018)
18. Qiu, Z.Q., Bader, S.D.: Surface magneto-optic Kerr effect. Rev. Sci. Instrum. **71**, 1243 (2000)
19. Xi, H., White, R.M., Rezende, S.M.: Irreversible and reversible measurements of exchange anisotropy. Phys. Rev. B. **60**, 14637 (1999)

# Further Reading

Abragam, A., Bleaney, B. (eds.): Electron Paramagnetic Resonance of Transition Ions. Clarendon, Oxford (1970)

Blundell, S.: Magnetism in Condensed Matter. Oxford University Press, Oxford (2001)

Chikazumi, S.: Physics of Magnetism. Wiley, New York (1964)

Coey, J.M.D.: Magnetism and Magnetic Materials. Cambridge University Press, Cambridge (2009)

Guimarães, A.P.: Magnetism and Magnetic Resonance in Solids. Wiley, New York (1998)

Guimarães, A.P.: Principles of Nanomagnetism, 2nd edn. Springer, Cham (2017)

Heinrich, B., Bland, J.A.C. (eds.): Ultrathin Magnetic Structures II. Springer, Heidelberg (1994)

Kittel, C.: Introduction to Solid State Physics, 8th edn. Wiley, New York (2004)

Lax, B., Button, K.: Microwave Ferrites and Ferrimagnetics. McGraw-Hill, New York (1962)

Morrish, A.H.: The Physical Principles of Magnetism. IEEE Press, New York (2001)

Reis, M.: Fundamentals of Magnetism. Elsevier, Amsterdam (2013)

Slichter, C.P.: Principles of Magnetic Resonance. Springer, Berlin (1980)

Smit, J., Wijn, H.P.J.: Ferrites. Wiley, New York (1959)

Sohoo, R.F.: Microwave Magnetics. Harper and Row, New York (1985)

Vonsovskii, S.V.: Ferromagnetic Resonance. Pergamon, New York (1966)

White, R.M.: Quantum Theory of Magnetism, 3rd edn. Springer, Heidelberg (2007)

# Spin Waves in Ferromagnets: Semiclassical Approach

In Chap. 1, we considered only the case of the magnetization dynamics that is uniform in space. The more general situation consists of a magnetization that varies in space, with elementary excitations that are called spin waves, the quanta of which are magnons. In this chapter, we shall treat spin waves considering that the spins are classical vectors with motion governed by the classical equation for the torque. We begin with a study of spin waves in an one-dimensional chain of classical spins. Then we present a macroscopic view of long-wavelength spin waves in a 3-dimensional ferromagnet, with an approach based on the Landau–Lifshitz equation of motion for the magnetization. Then we consider that the magnetization interacts with the lattice vibrations giving origin to coupled spin and elastic waves, called magnetoelastic waves, and discuss the conservation laws involved. We also present the concept of coupled spins and electromagnetic waves, called magnetic polaritons. Finally, we present two of the most important experimental techniques to study magnetoelastic waves, microwave excitation, and Brillouin light scattering.

## 2.1 Spin Waves in a Linear Ferromagnetic Chain

In order to have a quick view of the main properties of spin waves, we treat initially a simple system, a linear chain of classical spins illustrated in Fig. 2.1. Consider that the chain has $N$ spins with magnitude $S$, coupled with nearest-neighbor exchange interaction, uniformly spaced, and separated by a distance $a$. Initially, we consider only the Heisenberg interaction as in Eq. (1.1), so that the energy is given by

$$U_{exc} = -2J \sum_i \vec{S}_i \cdot \vec{S}_{i+1}, \qquad (2.1)$$

© Springer Nature Switzerland AG 2020
S. M. Rezende, *Fundamentals of Magnonics*, Lecture Notes in Physics 969,
https://doi.org/10.1007/978-3-030-41317-0_2

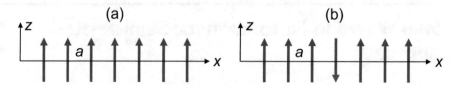

**Fig. 2.1** (a) Linear chain of classical spins in the ground state. (b) Linear chain of spins in an excited state with one spin reversed

where $J$ is the nearest neighbor exchange parameter and $\vec{S}_i$ denotes the classical spin at the coordinate $x_i = ia$. In the ground state all spins are parallel, as in Fig. 2.1a, so that with $\vec{S}_i \cdot \vec{S}_{i+1} = S^2$ the exchange energy of the system is $U_0 = -2JNS^2$.

In the early studies of the magnetic properties of matter, Pierre Weiss considered that the first excited state consisted of one localized spin reversed, as in Fig. 2.1b. In this case, the exchange energy of the linear chain becomes $U_1 = U_0 + 8JS^2$. Although the Weiss molecular field model explained well the overall temperature dependence of the magnetization in ferromagnets, in 1930 Felix Bloch showed that the low-lying excitations of the spin system consisted of nonlocalized, collective spin deviations. Bloch called these excitations *spin waves*, and showed that they dominate the magnetic thermodynamics at low temperatures [1].

We treat spin waves in a linear chain of spins that have motion governed by the classical mechanical equation for the torque. The torque acting on the spin $\vec{S}_i$ can be found by considering that it has an associated magnetic moment $\vec{\mu}_i = -g\mu_B \vec{S}_i$, and that the torque acting on the moment has the form $\vec{\tau} = \vec{\mu}_i \times \vec{B}_T$, where $\vec{B}_T$ is an effective field (induction) representing all interactions on the spin $\vec{S}_i$. This field can be found considering that the energy of the moment has the form $U_i = -\vec{\mu}_i \cdot \vec{B}_T$. Comparison with Eq. (2.1) shows that in the linear spin chain the effective field arising from the exchange interaction is

$$\vec{H}_{exc}^{eff} = -\frac{2J}{g\mu_B\mu_0}(\vec{S}_{i-1} + \vec{S}_{i+1}). \qquad (2.2)$$

We consider that in addition to the exchange interactions with their neighbors, the spins are subject to an applied static magnetic field $\vec{H}$, so that the total field acting on the spins is $\vec{B}_T = \mu_0(\vec{H} + \vec{H}_{exc}^{eff})$. Using the torque equation $d\hbar\vec{S}/dt = \vec{\tau}$ we obtain the equation of motion for the spin $\vec{S}_i$

$$\frac{d\vec{S}_i}{dt} = -\gamma\mu_0\,\vec{S}_i \times (\vec{H} + \vec{H}_{exc}^{eff}), \qquad (2.3)$$

where $\gamma = g\mu_B/\hbar$ is the gyromagnetic ratio. Since the magnetic moment has a direction opposite to the spin, to be consistent with the ground state depicted in

Fig. 2.1a, we consider that the external field is applied in the $-\hat{z}$ direction, so that $\vec{H} = -\hat{z}H$. Then the equation for the spin component $S_i^x$ becomes

$$\frac{dS_i^x}{dt} = \gamma\mu_0 S_i^y \left[ H + \frac{2J}{g\mu_B\mu_0} \left( S_{i-1}^z + S_{i+1}^z \right) \right] - \gamma\mu_0 \frac{2J}{g\mu_B\mu_0} \left( S_{i-1}^y + S_{i+1}^y \right) S_i^z.$$

Considering that the amplitude of the spin excitation is small, we linearize this equation using $S_i^x, S_i^y << S_i^z \approx S$. Then the equations for the two transverse spin components become

$$\frac{dS_i^x}{dt} = \gamma\mu_0 H S_i^y + \frac{2JS}{\hbar} \left( 2S_i^y - S_{i-1}^y - S_{i+1}^y \right), \tag{2.4}$$

$$\frac{dS_i^y}{dt} = -\gamma\mu_0 H S_i^x - \frac{2JS}{\hbar} \left( 2S_i^x - S_{i-1}^x - S_{i+1}^x \right). \tag{2.5}$$

Equations (2.4 and 2.5) show that the motion of the spin in any site is coupled to the motions of the neighboring spins, indicating that their solutions must be collective excitations. Consider for possible solutions excitations in the form of harmonic travelling waves

$$S_i^x = A_x \, e^{i(k x_i - \omega t)}, \quad S_i^y = A_y \, e^{i(k x_i - \omega t)}, \tag{2.6}$$

where $\omega$ is the angular frequency and $k$ is the wave number. Substitution of (2.6) into Eq. (2.4) leads to

$$-i\omega A_x = A_y \left[ \gamma\mu_0 H + \frac{2JS}{\hbar} \left( 2 - e^{-ika} - e^{ika} \right) \right], \tag{2.7}$$

where we have cancelled the term $i(kx_i - \omega t)$ on both sides. Equation (2.7) can be written as

$$-i\omega A_x = A_y \left[ \gamma\mu_0 H + \frac{4JS}{\hbar} \left( 1 - \cos ka \right) \right]. \tag{2.8}$$

Similarly, we obtain from Eq. (2.5)

$$-i\omega A_y = -A_x \left[ \gamma H + \frac{4JS}{\hbar} \left( 1 - \cos ka \right) \right]. \tag{2.9}$$

Equations (2.8 and 2.9) can be written in matrix form

$$\begin{bmatrix} i\omega & \left[ \gamma\mu_0 H + \frac{4JS}{\hbar} \left( 1 - \cos ka \right) \right] \\ -\left[ \gamma\mu_0 H + \frac{4JS}{\hbar} \left( 1 - \cos ka \right) \right] & i\omega \end{bmatrix} \begin{pmatrix} A_x \\ A_y \end{pmatrix} = 0. \tag{2.10}$$

The solution of Eq. (2.10) is obtained by equating the main determinant to zero, which gives for the frequency

$$\omega_k = \gamma \mu_0 H + \frac{4JS}{\hbar}(1 - \cos ka). \tag{2.11}$$

This equation represents the dependence of the spin wave frequency on the wave number and is called *dispersion relation*. Substitution of Eq. (2.11) in either (2.8) or (2.9) gives the relation between the amplitudes of the spin components $A_y = -iA_x \equiv -iA_0$. Thus, the real parts of the transverse spin components become

$$S_i^x = A_0 \cos (kx_i - \omega_k t), \quad S_i^y = A_0 \sin (kx_i - \omega_k t), \tag{2.12}$$

while the longitudinal component is $S_i^z \approx S$. These equations show that the classical picture of a spin wave in one dimension consists of spins precessing circularly about the equilibrium direction, as illustrated in Fig. 2.2 at a certain instant of time. In a travelling wave propagating in the $+x$ direction, the spin precession has the same amplitude along the chain and has a phase that varies with the position as $\phi_i = kx_i$. The shortest distance between two spins that precess with the same phase corresponds to the wavelength, related to the wave number by $\lambda = 2\pi/k$.

As will be presented in Chap. 3, a quantum formulation shows that the spin excitations are quantized. The quanta of spin waves are called magnons. For the one-dimensional chain just studied the energy of one magnon is

$$\hbar \omega_k = \hbar \gamma \mu_0 H + 4JS(1 - \cos ka). \tag{2.13}$$

This equation shows that at $k = 0$, the magnon energy is determined only by the magnetic field intensity $\hbar \omega_0 = \hbar \gamma \mu_0 H$. As we saw in Chap. 1, this is due to the fact that if all spins precess in phase there is no contribution from the exchange energy. As the wave number increases, the phase difference of precession for neighboring

(a)

(b)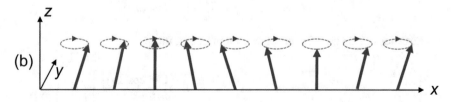

**Fig. 2.2** Illustration of a spin wave in a linear chain of classical spins propagating in the $+x$-direction. The distance between the two spins at the ends corresponds to one wavelength. (**a**) Top view of the spins (**b**) Side view

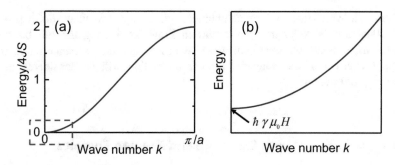

**Fig. 2.3** Dispersion relation for magnons in a linear chain. (**a**) View of the full Brillouin zone. (**b**) Zoom near the origin ($ka << 1$)

spins increases and so does the exchange energy. Figure 2.3a shows the magnon dispersion relation over the positive side of the first Brillouin zone. For magnetic field intensities typical of laboratories, the Zeeman contribution to the energy is many orders of magnitude smaller than the exchange, so that the energy gap at $k = 0$ cannot be seen in the plot. For $ka = \pi$ the exchange energy is maximum and has a value $8JS$, which is precisely the energy of the linear chain with one spin reversed, as in Fig. 2.1b. This is so because $ka = \pi$ corresponds precisely to neighboring spins with opposite phases. The important insight of Bloch was that there is a wealth of collective modes with smaller wave numbers that dominate the thermodynamics at low temperatures.

Spin waves with small wave numbers are important for many magnonic phenomena. Using the binomial expansion of the exchange term in Eq. (2.13) for $ka \ll 1$ we obtain the dispersion relation for low-wave number magnons

$$\hbar \omega_k \approx \hbar \gamma \mu_0 H + 2JS a^2 k^2, \tag{2.14}$$

which is a quadratic dispersion relation with an energy gap, as shown in Fig. 2.3b. Note that the energy scale in Fig. 2.3b is several orders of magnitude smaller than in (a). While the frequencies of spin waves with $ka \ll 1$ are in the microwave range, for $k \sim 1/a$ they are in the terahertz range.

## 2.2  Macroscopic Treatment of Spin Waves in 3D

In this section, we use a semiclassical approach to treat spin waves in a 3-dimensional ferromagnet [2]. The approach is based on the Landau–Lifshitz equation to describe the time and spatial evolution of the magnetization and is valid only for wave numbers satisfying two conditions: (a) $kL \gg 1$, where $L$ is the dimension of the sample in the direction of propagation, so that the wavelength is much smaller than the sample dimension and the boundary conditions can be ignored; (b) $ka \ll 1$, where $a$ is the lattice parameter, so that the discreteness of the crystal lattice is ignored and the medium can be considered a continuum. These conditions are typically satisfied by wave numbers in the range of $10^3$ cm$^{-1}$ to

$10^6$ cm$^{-1}$, in which many important magnonic phenomena take place. In Chap. 4, we will study long-wavelength spin waves in films considering the full boundary conditions. Initially, we treat spin waves with Zeeman and exchange interactions only. The effect of the magnetic dipolar interaction will be considered in the following subsection.

## 2.2.1   Spin Waves with Exchange and Zeeman Energies

In order to describe the magnetization motion with the Landau–Lifshitz equation, we need an expression for the effective exchange field in the case of a nonuniform magnetization in a 3D ferromagnet. For this, we assume that the spin is a continuous function of space so that it has derivatives with respect to the coordinates. In a spin wave with $ka \ll 1$ the spin varies slowly in space and we can use a Taylor series expansion to express the spin at a site $\vec{r}_i + \vec{\delta}$ in terms of the spin at site $\vec{r}_i$, where $\vec{\delta}$ denotes the position vector of a nearest neighbor relative to $\vec{r}_i$. Consider a ferromagnet with cubic crystal structure with lattice parameter $a$ and use a Cartesian coordinate system with the axes $x$, $y$, $z$ along the <100> directions. The relations between two nearest neighbor spins along the $x$-axis relative to $\vec{r}_i$ become

$$\vec{S}(\vec{r}_i - \hat{x}a) \approx \vec{S}(\vec{r}_i) - a\frac{\partial \vec{S}}{\partial x}\bigg|_{\vec{r}_i} + \frac{a^2}{2}\frac{\partial^2 \vec{S}}{\partial x^2}\bigg|_{\vec{r}_i}, \tag{2.15}$$

and

$$\vec{S}(\vec{r}_i + \hat{x}a) \approx \vec{S}(\vec{r}_i) + a\frac{\partial \vec{S}}{\partial x}\bigg|_{\vec{r}_i} + \frac{a^2}{2}\frac{\partial^2 \vec{S}}{\partial x^2}\bigg|_{\vec{r}_i}. \tag{2.16}$$

Similar expressions are obtained for the nearest neighbor spins along the $y$- and $z$-axes. With Eq. (2.1) and following the same procedure used to obtain Eq. (2.2) for the linear chain, we obtain the effective exchange field for the simple cubic lattice

$$\vec{H}_{exc}^{eff} = -\frac{2J}{g\mu_B\mu_0}\sum_{\vec{\delta}}\vec{S}(\vec{r}_i + \vec{\delta}) = -\frac{2J}{g\mu_B\mu_0}[6\vec{S}(\vec{r}_i) + a^2\nabla^2\vec{S}(\vec{r}_i)], \tag{2.17}$$

where $\nabla^2\vec{S}(\vec{r}_i)$ denotes the Laplacian operator of the spin function evaluated at site $\vec{r}_i$. Using $\vec{M}(\vec{r}_i) = -(g\mu_B/a^3)\vec{S}(\vec{r}_i)$ to express the relation between the magnetization and the spin, we obtain the effective exchange field in terms of the magnetization

$$\vec{H}_{exc}^{eff} = w\vec{M} + \frac{D}{M}\nabla^2\vec{M}, \tag{2.18}$$

where the two new parameters are

$$w = 6\frac{2JS}{g\mu_B\mu_0 M}, \quad D = \frac{2JSa^2}{g\mu_B\mu_0}. \tag{2.19}$$

The parameter $D$ in (2.18) and (2.19) is called *exchange parameter*. Some authors also use $A = DM$, where $A$ is called *exchange stiffness*. Consider that in addition to the exchange field, the magnetization is subject to an applied static magnetic field in the $z$-direction that magnetizes the sample, so that the internal field is $\vec{H}_{int} = \hat{z}H_z = \hat{z}(H_0 + H_A - N_z M)$, where $H_A$ is the anisotropy field (assuming that the field is applied in a direction of crystal symmetry). Writing the magnetization as $\vec{M}(\vec{r},t) = \hat{x}m_x(\vec{r},t) + \hat{y}m_y(\vec{r},t) + \hat{z}M_z$, where the small letters denote small-signal quantities that vary in space and time, and using for the total magnetic field $\vec{H}_T = \hat{z}H_z + \vec{H}_{exc}^{eff}$, the Landau–Lifshitz equation (1.10) gives for the transverse magnetization components

$$\frac{\partial m_x}{\partial t} = -\gamma\mu_0 m_y H_z + \gamma\mu_0 M_z \frac{D}{M}\nabla^2 m_y, \tag{2.20}$$

$$\frac{\partial m_y}{\partial t} = -\gamma\mu_0 M_z \frac{D}{M}\nabla^2 m_x + \gamma\mu_0 m_x H_z. \tag{2.21}$$

Introducing the circularly polarized magnetization $m^- = m_x - im_y$ and considering $M_z \approx M$, we obtain from Eqs. (2.20 and 2.21)

$$i\frac{\partial m^-}{\partial t} = -\gamma\mu_0 D\nabla^2 m^- + \gamma\mu_0 H_z m^-. \tag{2.22}$$

This is a wave equation which, for a uniform magnetic field $H_z$, has solution in the form of harmonic travelling waves

$$m^-(\vec{r},t) = m_0 e^{i(\vec{k}\cdot\vec{r} - \omega_k t)} \tag{2.23}$$

where $\vec{k}$ is the wave vector and $\omega_k$ the spin wave frequency, given by

$$\omega_k = \gamma\mu_0\left(H_z + Dk^2\right). \tag{2.24}$$

Equation (2.23) represents a magnetization that precesses about the $z$-direction with frequency $\omega_k$, with an amplitude $m_0$ that does not vary in space, and with a phase $\vec{k} \cdot \vec{r}$ that changes linearly with the travelled distance, as in Fig. 2.2. The spin wave frequency is represented by a quadratic dispersion relation with the same form as the one for the linear spin chain in the region $ka \ll 1$, shown in Fig. 2.3b.

Note that Eq. (2.22) that governs the evolution of the small-signal magnetization has the same form as the Schrödinger equation for the wave function of a particle. This has an interesting practical consequence. If magnons are excited with $k \approx 0$ at some position in a nonuniform magnetic field $H_z(z)$, they behave like electrons in a nonuniform potential. They accelerate toward regions of lower magnetic field with increasing wave number $k$ and thus increasing linear momentum. This property was important in the early days of experiments with spin wave propagation carried out with bulk YIG samples in the form of rods or slabs. The nonuniform internal fields of the bulk samples were very convenient for studying many properties of spin waves [3–5].

Let us now see the effect of damping on the spin wave properties. Substitution of $\vec{M}(\vec{r}, t) = \hat{x} m_x(\vec{r}, t) + \vec{y} m_y(\vec{r}, t) + \hat{z} M_z$ and $\vec{H}_T = \hat{z} H_z + \vec{H}_{exc}^{\,\text{eff}}$ in the Landau–Lifshitz–Gilbert equation (1.40) gives for the circularly polarized magnetization an equation that is the same as (2.22) with an additional term

$$i \frac{\partial m^-}{\partial t} - \alpha \frac{\partial m^-}{\partial t} = -\gamma \mu_0 D \nabla^2 m^- + \gamma \mu_0 H_z m^-. \tag{2.25}$$

Considering that $\alpha \ll 1$, the solution for harmonic travelling waves in a uniform magnetic field now has the form

$$m^-(\vec{r}, t) = m_0 \, e^{i(\vec{k} \cdot \vec{r} - \omega_k t)} \, e^{-\alpha \, \omega_k t}, \tag{2.26}$$

showing that the wave propagates with an amplitude that decays exponentially in time with a relaxation rate $\eta_k = \alpha \omega_k$.

## 2.2.2   Effect of the Volume Magnetic Dipolar Fields on Spin Waves

The magnetic dipolar interaction was considered in the previous subsection only as the source of the static demagnetizing field in the direction of the applied field. For spin waves with $kL \gg 1$ the rf demagnetizing fields are negligible, because the variations at the sample surfaces of the transverse rf magnetization with wavelength $\lambda \ll L$ result in vanishing surface dipolar fields.

However, the properties of spin waves are influenced by the presence of volume magnetic dipolar fields. This effect actually vanishes for waves propagating in the direction of the static field, so that the dispersion relation in Eq. (2.24) is exact. But the effect is important for waves propagating in other directions. Since the dipolar interaction is anisotropic and of long range, its treatment is difficult for a general nonuniform magnetization. However, the treatment of the volume dipolar field is relatively simple for harmonic travelling spin waves, as we shall see here.

The volume dipolar magnetic field is created by spatial variations in the magnetization, so it can be calculated with Maxwell's equations

$$\nabla \cdot (\vec{H} + \vec{M}) = 0, \quad \nabla \times \vec{H} = \varepsilon \frac{\partial \vec{E}}{\partial t}, \tag{2.27}$$

where $\vec{E}$ is the electric field and $\varepsilon$ is the electric permittivity. For spin waves with $kL \gg 1$, the wave number is much larger than the one for an electromagnetic wave with the same frequency, so that the role of the electric field on the propagation is negligible. Thus, the second equation in (2.27) becomes $\nabla \times \vec{H} \approx 0$, in what is called the magnetostatic limit. Considering for the magnetization and the magnetic field the forms $\vec{M}(\vec{r},t) = \hat{z}M_z + \vec{m}(\vec{r},t)$ and $\vec{H}(\vec{r},t) = zH_z + \vec{h}_{dip}(\vec{r},t)$, where the spatial and time variations are contained in the transverse $rf$ components $\vec{m}(\vec{r},t)$ and $\vec{h}_{dip}(\vec{r},t)$, the dipolar field can be calculated with

$$\nabla \cdot (\vec{h}_{dip} + \vec{m}) = 0, \quad \nabla \times \vec{h}_{dip} = 0. \tag{2.28}$$

For a harmonic travelling wave propagating in an arbitrary direction,

$$\vec{m}(\vec{r},t) = \vec{m}\, e^{i(\vec{k}\cdot\vec{r} - \omega_k t)} \tag{2.29}$$

where $\vec{m}$ is a complex vector with constant amplitude. Substitution of Eq. (2.29) into (2.28) leads to

$$\vec{k} \cdot (\vec{h}_{dip} + \vec{m}) = 0, \quad \vec{k} \times \vec{h}_{dip} = 0. \tag{2.30}$$

With the vector product by the wave vector $\vec{k}$ to the left of the second equation in (2.30), and using a vector identity we have

$$\vec{k} \times \vec{k} \times \vec{h}_{dip} = \vec{k}(\vec{k} \cdot \vec{h}_{dip}) - \vec{h}_{dip} k^2 = 0.$$

Finally, combining this result with the first equation in (2.30) we obtain the volume dipolar field created by the $rf$ magnetization in a travelling spin wave

$$\vec{h}_{dip} = -\frac{1}{k^2} \vec{k}(\vec{k} \cdot \vec{m}). \tag{2.31}$$

With a Cartesian coordinate system such that the wave vector is in the $x$–$z$ plane, as illustrated in Fig. 2.4, since $\vec{m}$ has only $x$ and $y$ components, Eq. (2.31) gives for the dipolar field

$$\vec{h}_{dip} = \hat{x} h_{dip}^x = -\hat{x}\frac{k_x^2}{k^2} m_x = -\hat{x}\sin^2\theta_k\, m_x. \tag{2.32}$$

**Fig. 2.4** Coordinates used to express the wave vector, magnetization, and magnetic dipolar field

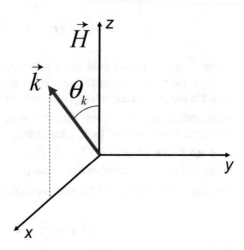

Using in the Landau–Lifshitz equation (1.10) the Zeeman, exchange and dipolar fields, we obtain the equations for the transverse magnetization components of a wave described by Eq. (2.29)

$$i\omega_k m_x = \gamma\mu_0(H_z + Dk^2)m_y, \tag{2.33}$$

$$-i\omega_k m_y = \gamma\mu_0(H_z + Dk^2 + M\sin^2\theta_k)m_x. \tag{2.34}$$

Multiplying these equations and crossing out the term $m_x\, m_y$, we obtain the spin wave dispersion relation for an arbitrary direction of the wave vector relative to the magnetic field,

$$\omega_k = \gamma\mu_0\left[(H_z + Dk^2)(H_z + Dk^2 + M\sin^2\theta_k)\right]^{1/2}, \text{(SI)} \tag{2.35}$$

$$\omega_k = \gamma[(H_z + Dk^2)(H_z + Dk^2 + 4\pi M\sin^2\theta_k)]^{1/2}. \text{(CGS)} \tag{2.36}$$

Note that for $\theta_k = 0$ Eq. (2.35) coincides with (2.24), which means that spin waves propagating in the direction of the static field have no dipolar field. This is due to the fact that, in this case, the magnetic dipoles precess about the static field maintaining the tips (monopoles) in planes parallel to the phase plane, as illustrated in Fig. 2.5a. In this case, the magnetic dipole density is uniform along the direction of propagation, and so is the magnetization, so that no dipolar field is created. On the other hand, for $\theta_k > 0$ the planes of precession are not parallel to the phase planes, so that the magnetic dipole density is not uniform and a dipolar field is created. Figure 2.5b illustrates the precession of the dipoles for the case $\theta_k = \pi/2$, showing clearly the change in the dipole density along the direction of propagation.

Figure 2.6 shows the spin wave dispersion relations for YIG with $\theta_k = 0$ and $\theta_k = \pi/2$, calculated with Eq. (2.36) for an internal magnetic field $H_z = 1.5$ kOe with

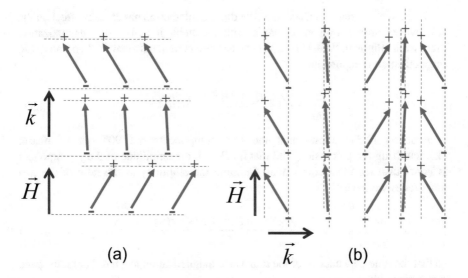

**Fig. 2.5** Illustration of the behavior of the magnetic dipole densities for spin waves propagating parallel (**a**) and perpendicular (**b**) to the static magnetic field

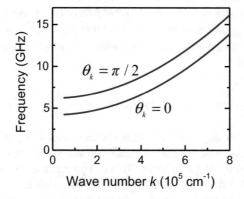

**Fig. 2.6** Dispersion relations for spin waves in YIG with $H_z = 1.5$ kOe

the following parameters for YIG: $\gamma = 2\pi \times 2.8$ GHz/kOe, $4\pi M = 1.76$ kG, and $D = 5.4 \times 10^{-9}$ Oe cm$^{-2}$. The dispersions for intermediate angles lie between the two curves, so that we actually have a continuum manifold of curves. Note that the curves do not extend to small wave numbers because they are valid for $kL \gg 1$. For $k = 0$ the magnon frequency is given by the expression for the ferromagnetic resonance as discussed in Chap. 1. As we shall see in Chap. 4, the detailed curves in the region $k \approx 0$ depend on the shape of the sample and the direction of propagation relative to the static field.

Besides the increase in frequency, the dipolar interaction has another effect on the spin waves propagating at an angle with the field, it makes the magnetization precession elliptical. With Eq. (2.34), we obtain the ratio between the two transverse magnetization components

$$\frac{m_y}{m_x} = i \frac{\gamma \mu_0 (H_z + Dk^2 + M \sin^2 \theta_k)}{\omega_k}. \tag{2.37}$$

Equation (2.37) shows that the two components are 90° out of phase, characterizing the precession, and that for $\theta_k > 0$ the amplitude of the $y$ component is larger than for the $x$ component. Defining the ellipticity as the ratio of the two amplitudes we have

$$e = \frac{\gamma \mu_0 (H_z + Dk^2 + M \sin^2 \theta_k)}{\omega_k}, \tag{2.38}$$

so that the time dependence of the transverse magnetization at a fixed point in space is described by

$$\vec{m}(t) = m_0 (\hat{x} \cos \omega_k t + \hat{y} \, e \sin \omega_k t). \tag{2.39}$$

This result has an important consequence, the $z$ component of the magnetization is not constant, it varies in time with frequency $2\omega_k$. To see this use the fact that the modulus of the magnetization is invariant so that $M_z = [M^2 - m_x^2 - m_y^2]^{1/2}$. Using Eq. (2.39) and considering that $m_0 \ll M$, with a binomial expansion of the expression for $M_z$ we find

$$M_z \approx M - \frac{m_0^2}{2M} \left( \cos^2 \omega_k t + e^2 \sin^2 \omega_k t \right) = M - \frac{m_0^2}{4M} (1 - e^2) \cos 2\omega_k t. \tag{2.40}$$

This result has an important practical consequence. It shows that if a driving microwave magnetic field is applied parallel to the static field, it can excite spin waves with half the driving frequency. This is a parametric process, called *parallel pumping*, in which a photon generates two magnons with half frequency, so as to conserve energy. In Chap. 6, we will study this and other parametric processes used to generate spin waves in several applications.

## 2.3  Magnetoelastic Waves

Due to the spin–orbit interactions, the elastic deformations in a magnetic material couple to the spin excitations, giving rise to magnetostrictive, or magnetoelastic, properties [6, 7]. This is what ultimately relaxes the magnetization dynamics in any material. Here we are interested in the coupling between spin waves and elastic waves, that manifests in the region where they have similar frequencies and wave vectors, where they form hybrid excitations, called magnetoelastic waves, or magnon–phonon excitations. Magnetoelastic waves with frequency of a few GHz were extensively studied in bulk YIG crystals in the 1960s, motivated by scientific interest and potential application in controllable delay lines [3–5, 8, 9]. Recently there was a revival of interest in these excitations in the context of new magnonic phenomena.

Initially, we briefly treat some elastic properties of materials in order to introduce the concept of elastic waves. Let us consider that the material is a continuous solid, elastically isotropic, with average mass density $\rho$. The elastic deformations of the solid are expressed in terms of the displacement vector $\vec{u} = \vec{r} - \vec{r}'$, where $\vec{r}$ is the initial position of a volume element, and $\vec{r}'$ is the position after deformation. The deformation of the solid can be expressed by the symmetric *strain tensor*

$$e_{ij} = \frac{1}{2}\left(\frac{\partial u_i}{\partial x_j} + \frac{\partial u_j}{\partial x_i}\right), \tag{2.41}$$

where $i, j = 1, 2, 3$ denote the $x$, $y$, $z$ coordinates (not to be confused with the lattice sites). The deformation of a small volume is caused by forces that act on it, due both to the neighboring elements and to external agents. We consider only the first ones, that are called elastic forces. The elastic force per unit volume can be written as

$$\vec{f}_{el} = \sum_{i=1}^{3} \hat{x}_i \sum_{j=1}^{3} \frac{\partial \sigma_{ij}}{\partial x_j}, \tag{2.42}$$

where $\sigma_{ij}$ is the symmetric *stress tensor*, that represents the $i$ component of the force per unit area acting along a normal vector $\hat{x}_j$. According to Hooke's law, for a small deformation the displacement is proportional to the stress, so that one can write

$$\sigma_{ij} = \sum_{l=1}^{3} \sum_{m=1}^{3} c_{ijkl}\, e_{kl}, \tag{2.43}$$

where $c_{ijkl}$ are the components of a fourth-rank tensor of elastic stiffness constants, also called moduli of elasticity, or simply *elastic constants*. It can be shown that due to the symmetry of the strain and stress tensors, there are only 21 independent elastic constants. Also, by means of symmetry operations, it can be shown that the number of nonzero elastic constants decreases as the symmetry of the lattice structure increases.

For cubic crystals, considering the coordinate axes along <100> directions, there are only three nonzero elastic constants, and the notation can be simplified as: $c_{iiii} = c_{11}$, $c_{iijj} = c_{12}$, $c_{ijij} = c_{44}$. Furthermore, it can be shown that for an isotropic solid the constants are related by $c_{11} - c_{12} = 2c_{44}$, so that an isotropic cubic crystal has only two nonzero independent elastic constants. This condition holds very approximately for YIG, which has elastic constants at room temperature $c_{11} = 2.69 \times 10^{11}$ N/m$^2$ and $c_{44} = 7.64 \times 10^{10}$ N/m$^2$ in the SI, and $c_{11} = 2.69 \times 10^{12}$ dyn/cm$^2$ and $c_{44} = 7.64 \times 10^{11}$ dyn/cm$^2$ in the CGS.

In the context of Hooke's law, the elastic energy in a cubic crystal takes the form

$$U_{el} = \frac{1}{2}c_{11}\left(e_{xx}^2 + e_{yy}^2 + e_{zz}^2\right) + c_{12}\left(e_{xx}e_{yy} + e_{yy}e_{zz} + e_{zz}e_{xx}\right) + 2c_{44}\left(e_{xy}^2 + e_{yz}^2 + e_{zx}^2\right).$$
$$(2.44)$$

The equation of motion for the displacement in an elastically strained solid is given by Newton's force law $\partial^2 \vec{u}/\partial t^2 = \vec{f}_{el}$, so that with Eqs. (2.42 and 2.43) we have

$$\rho \frac{\partial^2 u_i}{\partial t^2} = \frac{\partial}{\partial x_j}\left(c_{ijkl}\frac{\partial u_k}{\partial x_l}\right), \qquad (2.45)$$

where, in order to simplify the equations, we have used the notation by which repeated subscripts imply summation over all three values. With the relations between the elastic constants for a cubic crystal, Eq. (2.45) gives

$$\rho \frac{\partial^2 u_i}{\partial t^2} = c_{11}\frac{\partial^2 u_i}{\partial x_i^2} + c_{12}\frac{\partial^2 u_i}{\partial x_i \partial x_j} + c_{44}\frac{\partial^2 u_i}{\partial x_j^2}. \qquad (2.46)$$

This equation has solution of the type $\vec{u}(\vec{r},t) = \vec{u}\, e^{i(\vec{k}\cdot\vec{r}-\omega t)}$ that represents an elastic wave. Consider a cubic crystal with the coordinate axes $x$, $y$, $z$ along <100> directions and assume that the displacements vary only along the $z$-axis, which means that $\vec{k} = \hat{z}k$. Equation (2.46) gives for the three components of the displacement

$$\rho \frac{\partial^2 u_x}{\partial t^2} = c_{44}\frac{\partial^2 u_x}{\partial z^2}, \quad \rho \frac{\partial^2 u_y}{\partial t^2} = c_{44}\frac{\partial^2 u_y}{\partial z^2}, \qquad (2.47)$$

$$\rho \frac{\partial^2 u_z}{\partial t^2} = c_{11}\frac{\partial^2 u_z}{\partial z^2}. \qquad (2.48)$$

Equations (2.47 and 2.48) are wave equations for three independent normal modes with linear dispersion relations given by $\omega_\mu = v_\mu k$, where $v_\mu$ is the elastic wave velocity. Equation (2.48) implies that the displacement is along the direction of propagation, or $\vec{u}//\vec{k}$, characterizing a *longitudinal elastic wave*, with velocity

$$v_l = \omega_l/k = \sqrt{c_{11}/\rho}. \qquad (2.49)$$

On the other hand, Eqs. (2.47) apply to two independent *transverse, or shear, elastic waves*, with velocity

$$v_t = \omega_t/k = \sqrt{c_{44}/\rho}. \qquad (2.50)$$

The elastic waves are quantized, and their quanta are called *phonons*. Each phonon of a mode $\mu$ has energy $\hbar\omega_\mu$. Note that for propagation along an arbitrary direction in a cubic crystal, the velocities are given by expressions different from Eqs. (2.49 and 2.50), and the two transverse modes have different velocities. In YIG, using the elastic constants given before and the mass density $\rho = 5.17$ g/cm$^3$ at room temperature, we find for waves propagating along <100> the velocities $v_l = 7.21 \times 10^3$ m/s and $v_t = 3.84 \times 10^3$ m/s.

The coupling between spin and elastic waves is due to the magnetoelastic interaction [6, 7]. To first order in the strain components $e_{ij}$ and to second order in the magnetization components (the first order ones are zero due to inversion symmetry), the magnetoelastic energy is written as $E_{me} = b_{ijkl} M_i M_j e_{kl}$, where $b_{ijkl}$ are the components of a fourth-rank tensor of *magnetoelastic constants*. For a cubic crystal with the coordinate axes along <100> directions, there are only two nonzero constants, $b_{iiii} = b_1/M^2$ and $b_{ijij} = b_2/M^2$. Thus, for a cubic crystal the magnetoelastic energy density becomes

$$E_{me} = \frac{b_1}{M^2}\left(m_x^2 e_{xx} + m_y^2 e_{yy} + M_z^2 e_{zz}\right) + \frac{2b_2}{M^2}\left(m_x m_y e_{xy} + m_y M_z e_{yz} + M_z m_x e_{zx}\right)$$

(2.51)

Considering only terms in first order in the small-signal magnetization components, Eq. (2.51) reduces to

$$E_{me} = \frac{2b_2}{M^2}\left(m_y M_z e_{yz} + M_z m_x e_{zx}\right).$$  (2.52)

The effect of the magnetoelastic interaction enters in the equation of motion for the spins by means of an effective field introduced in the Landau–Lifshitz equation. On the other hand, it enters in the equation of motion for the elastic waves by means of an effective force introduced in Eq. (2.45). Using the expression (2.52) for the energy in Eq. (1.46), and considering $M_z \approx M$, we obtain the small-signal transverse component of the effective magnetoelastic field

$$\vec{h}_{me} = -\frac{2b_2}{\mu_0 M}\left(\hat{x}\,e_{zx} + \hat{y}\,e_{yz}\right).$$  (2.53)

It can be shown [6, 9–11] that the force per unit volume that acts on the elastic displacements by the magnetoelastic interaction has component $i$ given by

$$f_i^{me} = \frac{\partial}{\partial x_j}\left[\frac{\partial E_{me}}{\partial(\partial u_i/\partial x_j)}\right],$$  (2.54)

where again we use repeated subscripts to denote summation. Writing the magnetization as $\vec{M} = \hat{z}M_z + \vec{m}$ in the Landau–Lifshitz equation (1.10), in which the total field includes the applied magnetic field, demagnetization, anisotropy, exchange, dipolar and magnetoelastic contributions, considering $M_z \approx M$, we obtain the linearized equation of motion for the small-signal magnetization

$$\frac{\partial \vec{m}}{\partial t} = -\gamma \mu_0 M \hat{z} \left( -\frac{H_z}{M} \vec{m} + \vec{h}_{exc} + \vec{h}_{dip} + \vec{h}_{me} \right).$$ (2.55)

Adding in Eq. (2.45) the force density (2.54) due to the magnetoelastic energy, we obtain the equation of motion for the elastic displacement

$$\rho \frac{\partial^2 u_i}{\partial t^2} = \frac{\partial}{\partial x_j} \left( c_{ijkl} \frac{\partial u_k}{\partial x_l} \right) + \frac{\partial}{\partial x_j} \left[ \frac{\partial E_{me}}{\partial (\partial u_i / \partial x_j)} \right].$$ (2.56)

From now on we shall consider the case of a cubic crystal, with coordinates $x$, $y$, $z$ along the <100> axes. With Eqs. (2.55 and 2.56), we obtain the equations of motion for the small-signal magnetization and elastic displacement components

$$\frac{\partial m_x}{\partial t} = \gamma \mu_0 (-H_z + D \nabla^2) m_y + \gamma \mu_0 M h_y^{dip} - \gamma b_2 \left( \frac{\partial u_y}{\partial z} + \frac{\partial u_z}{\partial y} \right),$$ (2.57)

$$\frac{\partial m_y}{\partial t} = \gamma \mu_0 (H_z - D \nabla^2) m_x - \gamma \mu_0 M h_x^{dip} + \gamma b_2 \left( \frac{\partial u_z}{\partial x} + \frac{\partial u_x}{\partial z} \right),$$ (2.58)

$$\rho \frac{\partial^2 u_x}{\partial t^2} = c_{44} \nabla^2 u_x + c_{11} \frac{\partial}{\partial x} \nabla \cdot \vec{u} + \frac{b_2}{M} \frac{\partial m_x}{\partial z},$$ (2.59)

$$\rho \frac{\partial^2 u_y}{\partial t^2} = c_{44} \nabla^2 u_y + c_{11} \frac{\partial}{\partial y} \nabla \cdot \vec{u} + \frac{b_2}{M} \frac{\partial m_y}{\partial z},$$ (2.60)

$$\rho \frac{\partial^2 u_z}{\partial t^2} = c_{44} \nabla^2 u_z + c_{11} \frac{\partial}{\partial z} \nabla \cdot \vec{u} + \frac{b_2}{M} \left( \frac{\partial m_x}{\partial x} + \frac{\partial m_y}{\partial y} \right).$$ (2.61)

Equations (2.57–2.61) have wave-like solutions that represent the coupled spin–elastic waves. As we shall see, actually the coupling is important only in the vicinity of the crossover of the spin and elastic wave dispersions. The wave solutions of Eqs. (2.57–2.61) for an arbitrary direction of propagation involve all five small-signal variables, indicating that the spin wave couples with the three elastic wave modes.

For simplicity, we shall work the solution only for the case of waves propagating along the direction of the applied magnetic field. In this case $\partial/\partial x = \partial/\partial y = 0$ and Eq. (2.61) show that the displacement $u_z$ does not depend on the magnetization, which means that spin waves do not couple with longitudinal elastic waves. Considering in Eqs. (2.57–2.60) wave solutions $\vec{u}(\vec{r}, t) = \vec{u} \, e^{i(kz - \omega t)}$ and $\vec{m}(\vec{r}, t) = \vec{m} \, e^{i(kz - \omega t)}$ and taking into account that for $\vec{k} = \hat{z} k$ the dipolar field vanishes, we obtain for the small-signal magnetization and displacement components the following coupled equations

$$-i\omega m_x = -\gamma\mu_0(H_z + Dk^2)m_y - i\gamma b_2 k u_y, \tag{2.62}$$

$$-i\omega m_y = \gamma\mu_0(H_z + Dk^2)m_x + i\gamma b_2 k u_x, \tag{2.63}$$

$$-\rho\omega^2 u_x = -c_{44}k^2 u_x + i\frac{b_2}{M}k m_x, \tag{2.64}$$

$$-\rho\omega^2 u_y = -c_{44}k^2 u_y + i\frac{b_2}{M}k m_y. \tag{2.65}$$

Introducing the circularly polarized quantities $m^- = m_x - i m_y$, $u^- = u_x - i u_y$, and using the magnon and phonon dispersion relations in Eqs. (2.24 and 2.50), Eqs. (2.62–2.65) lead to

$$(\omega - \omega_k)m^- = i\gamma b_2 k u^-, \tag{2.66}$$

$$(\omega^2 - \omega_t^2)u^- = -i\frac{b_2}{M\rho}k m^-. \tag{2.67}$$

These equations readily give for the magnetoelastic wave dispersion

$$(\omega - \omega_k)(\omega^2 - \omega_t^2) - \frac{1}{2}\omega_t \sigma_k^2 = 0, \tag{2.68}$$

where

$$\sigma_k = b_2 \left(\frac{2\gamma k}{\rho v_t M}\right)^{1/2} \tag{2.69}$$

is a parameter that expresses the coupling between magnons and phonons. Equation (2.68) has three roots, because the magnetoelastic excitation involves one spin wave and two independent transverse elastic waves. Note that if the magnetoelastic interaction vanishes, $b_2 = 0$ and $\sigma_k = 0$, and the three roots of Eq. (2.68) are $\omega_k$ and $\pm\omega_t$, corresponding to unperturbed spin and elastic waves. The two signs in $\omega_t$ correspond to (+) and (−) circularly polarized phonons. With nonzero magnetoelastic interaction, Eq. (2.68) describes three coupled spin and elastic waves, called *magnetoelastic waves*. However, since the magnon frequency is always positive, the phonon with negative frequency has negligible coupling with magnons. Thus, we can approximately eliminate the negative root $\omega \approx -\omega_t$ in Eq. (2.68), so that it reduces to an equation of second degree

$$(\omega - \omega_k)(\omega - \omega_t) - \sigma_k^2/4\omega_t = 0, \tag{2.70}$$

that has two solutions

$$\omega_a = \frac{\omega_t + \omega_k}{2} + \frac{1}{2}[(\omega_t - \omega_k)^2 + \sigma_k^2]^{1/2}, \tag{2.71}$$

$$\omega_b = \frac{\omega_t + \omega_k}{2} - \frac{1}{2}[(\omega_t - \omega_k)^2 + \sigma_k^2]^{1/2}, \tag{2.72}$$

where $\omega_a$ and $\omega_b$ denote, respectively, the frequencies of the upper and lower dispersion curves. Figure 2.7 shows the magnetoelastic dispersion curves for YIG with an internal field $H_z = 2.0\,\text{kOe}$, calculated with the unperturbed magnon and phonon frequencies given by Eqs. (2.24 and 2.50), using the following parameters: $4\pi M = 1.76\,\text{kG}$, $\gamma = 2.8 \times 2\pi \times 10^6 \ \text{s}^{-1}\text{Oe}^{-1}$, $D = 5.4 \times 10^{-9}\,\text{Oe} \ \text{cm}^2$, $b_2 = 7.0 \times 10^6 \ \text{erg/cm}^2$, $\rho = 5.2 \ \text{g/cm}^3$, and $v_t = 3.84 \times 10^5$ cm/s. In the region around the point where the magnon and phonon curves cross, called *crossover region*, the upper and lower curves are separated. As one can see in Fig. 2.7, and also with an analysis of Eqs. (2.71 and 2.72), excitations with frequency and wave number far from the crossover have almost pure magnon or phonon character. However, in the crossover region, the normal modes are mixtures of coupled magnetic and elastic excitations, or magnetoelastic waves, or magnon–phonon hybrid excitations. The zoom of the crossover region in Fig. 2.7b shows that for $H = 2$ kOe the magnon and phonon curves cross at a frequency 5.73 GHz and wave number $k_{\text{cross}} = 0.938 \times 10^5 \ \text{cm}^{-1}$. It also shows that the maximum separation between the two curves, called frequency splitting, is only 0.12 GHz, which is quite small compared to the magnon and phonon frequencies.

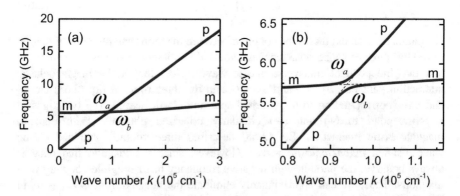

**Fig. 2.7** Dispersion relations for magnetoelastic waves in YIG propagating along the magnetic field $H_z = 2.0$ kOe, applied parallel to a < 100> axis. (**a**) The upper and lower branches are represented, respectively, by wine and navy colors. (**b**) Zoom of the crossover region

## 2.4 Energy and Momentum Conservation Relations for Magnetoelastic Waves

Waves transport energy and momentum, and since these quantities are conserved each type of wave has appropriate conservation relations involving power flow, energy density, and other quantities. In electromagnetism, the Poynting's theorem establishes a simple relation between the power flow per unit area and the time rate of change of the energy density. In this section, we study the conservation relations for a magnetoelastic ferromagnet considering only time-varying small-signal quantities, since these are the ones that matter for magnetoelastic waves.

The total energy in a magnetoelastic medium is the sum of magnetic, elastic, and magnetoelastic energies. The magnetic energy has three contributions, Zeeman, exchange, and magnetic dipolar. Using the binomial expansion $M_z \approx M - |\vec{m}|^2/2M$, and considering a field $H_z = H_0 + H_A - N_z M$, that includes anisotropy and demagnetizing contributions, we obtain with Eq. (1.56) the small-signal Zeeman energy density

$$E_Z = \frac{1}{2}\mu_0 \frac{H_z}{M} |\vec{m}|^2. \tag{2.73}$$

To find the exchange energy density, we use the small-signal exchange field calculated with Eq. (2.18), $\vec{h}_{exc} = (D/M)\nabla^2\vec{m}$, and obtain for the total small-signal exchange energy

$$U_{exc} = -\frac{1}{2}\mu_0 \int \vec{h}_{exc} \cdot \vec{m}\, d\vec{r} = -\frac{1}{2}\mu_0 \frac{D}{M} \int \nabla^2\vec{m} \cdot \vec{m}\, d\vec{r}, \tag{2.74}$$

which can be written as

$$U_{exc} = -\frac{1}{2}\mu_0 \frac{D}{M} \int m_i \frac{\partial}{\partial x_j} \frac{\partial m_i}{\partial x_j}\, d\vec{r}, \tag{2.75}$$

where the subscript $i$ runs only over two values $i = 1, 2$, corresponding to the coordinates $x$, $y$, while subscript $j$ can have all three values $j = 1, 2, 3$ corresponding to $x$, $y$, $z$. Integrating Eq. (2.75) by parts we have

$$U_{exc} = -\frac{1}{2}\mu_0 \frac{D}{M} \left[ \int m_i \frac{\partial m_i}{\partial x_j}\, d\vec{r} - \int \frac{\partial m_i}{\partial x_j} \frac{\partial m_i}{\partial x_j}\, d\vec{r} \right].$$

The first integral in this equation vanishes, so that the exchange energy becomes

$$U_{exc} = \frac{1}{2}\mu_0 \frac{D}{M} \int \frac{\partial m_i}{\partial x_j} \frac{\partial m_i}{\partial x_j}\, d\vec{r}.$$

Hence, the small-signal exchange energy density can be written as

$$E_{exc} = \frac{1}{2}\mu_0 \frac{D}{M}\left[|\nabla m_x|^2 + |\nabla m_y|^2\right]. \tag{2.76}$$

The energy stored in the magnetic dipolar field is the one known in electromagnetism, $\mu_0|\vec{h}|^2/2$, where we have omitted the subscript dip. Thus, the total small-signal magnetic energy density with Zeeman, exchange, and dipolar contributions is

$$E_m = \frac{1}{2}\mu_0 \frac{H_z}{M}|\vec{m}|^2 + \frac{1}{2}\mu_0 \frac{D}{M}\left[|\nabla m_x|^2 + |\nabla m_y|^2\right] + \frac{1}{2}\mu_0|\vec{h}|^2, \tag{2.77}$$

where the magnetic anisotropy and surface dipolar components are hidden in the Zeeman term. Equation (2.44) gives the potential elastic energy density for a cubic crystal. Here we will consider for phonons a general expression containing kinetic and potential energies

$$E_p = \frac{1}{2}\rho\left|\frac{\partial \vec{u}}{\partial t^2}\right|^2 + \frac{1}{2}c_{ijkl}\frac{\partial u_i}{\partial x_j}\frac{\partial u_k}{\partial x_l}. \tag{2.78}$$

For the magnetoelastic energy, we use the general expression $E_{me} = b_{ijkl}M_iM_je_{kl}$ and consider that terms linear in the small-signal magnetization must have either $M_i = M_z$ or $M_j = M_z$, so that only three subscripts are necessary. Thus, introducing new magnetoelastic constants related to the original ones by $b_{ikl} = b_{i3kl} = b_{3jkl} = b_{ijkl}M$, the small-signal magnetoelastic energy density becomes

$$E_{me} = b_{ijk}m_i\frac{\partial u_j}{\partial x_k}, \tag{2.79}$$

Note that for a cubic crystal with the coordinate axes along <100> directions the constants become

$$b_{113} = b_{131} = b_{223} = b_{232} = \frac{b_2}{M}, \tag{2.80}$$

and all others are equal to zero, so that Eq. (2.79) reduces to (2.52). The total small-signal energy density is then

$$E_T = E_m + E_p + E_{me}. \tag{2.81}$$

The law of energy conservation for waves can be expressed with differential operators in the same form as in Poynting's theorem

$$\nabla \cdot \vec{S}_T + \frac{\partial E_T}{\partial t} = -p_T, \tag{2.82}$$

where $\vec{S}_T$ is the power flow vector (energy per area per time), which in electromagnetism is called Poynting vector, $\partial E_T/\partial t$ is the rate of change of the total energy density stored in the system, and $-p_T$ is the total power per unit volume dissipated in the magnetoelastic medium.

Here we expect that $\vec{S}_T = \vec{S}_{em} + \vec{S}_m + \vec{S}_p + \vec{S}_{me}$, where $\vec{S}_{em} = \vec{e} \times \vec{h}$ is the electromagnetic Poynting vector and the other three terms are, in order, magnetic, elastic (phonon), and magnetoelastic contributions to the power flow. Notice that the small-signal electric field $\vec{e}$ is of higher order in $1/k$ so that the electric energy $\varepsilon|\vec{e}|^2/2$ is negligible compared to the terms in Eq. (2.81). However, in the Poynting vector the electric field is of first order and must be taken into account. In order to find the components of the power flow, we use the energy in Eq. (2.81) with the equations for its three components, calculate $\partial E_T/\partial t$ and cast the result in the form of Eq. (2.82), using the equations of motion (2.55) and (2.56) for the magnetization and the elastic displacement. One can then show that [10–12]

$$S_j = (\vec{e} \times \vec{h})_j - \mu_0 \frac{D}{M} \frac{\partial m_i}{\partial x_j} \frac{\partial m_i}{\partial t} - c_{ijkl} \frac{\partial u_k}{\partial x_l} \frac{\partial u_k}{\partial t} - b_{ikj} m_i \frac{\partial u_k}{\partial t}, \tag{2.83}$$

where the four terms represent, in order, electromagnetic, exchange, elastic, and magnetoelastic contributions to the power flow. With this result and with Eq. (2.81), one can show that the total small-signal energy density and the power flow vector satisfy the conservation law

$$\nabla \cdot \vec{S}_T + \frac{\partial E_T}{\partial t} = \frac{1}{2} \mu_0 \frac{|\vec{m}|^2}{M} \frac{\partial H_z}{\partial t} - p_T, \tag{2.84}$$

where the first term on the right-hand side is the time rate of change of the magnetic energy due to time changes in the internal magnetic field. This result is important in connection with experiments in which the biasing field varies in time, as will be presented in Sect. 2.6.

Another small-signal conservation law is the invariance of the momentum density of magnetoelastic waves in spatially uniform fields. The total momentum density has four contributions, electromagnetic, magnetic, elastic, and magnetoelastic momenta. The first is well known, and being of higher order is neglected here. The magnetic momentum density has been shown [5, 10] to have components $i$

$$g_{mi} = \frac{1}{2\gamma M} \left( \vec{m} \times \frac{\partial \vec{m}}{\partial x_i} \right) \cdot \hat{z}. \tag{2.85}$$

The elastic contribution to the momentum density is [5, 10, 11]

$$g_{pi} = \frac{1}{2} \rho \left( \frac{\partial^2 \vec{u}}{\partial x_i \partial t} \cdot \vec{u} - \frac{\partial \vec{u}}{\partial t} \cdot \frac{\partial \vec{u}}{\partial x_i} \right). \tag{2.86}$$

If all parameters are spatially invariant, it can be shown [11] that the total momentum density is conserved

$$\frac{\partial}{\partial t}(\vec{g}_m + \vec{g}_p) = 0. \tag{2.87}$$

Interestingly, no magnetoelastic component enters into the momentum conservation law. As will be shown in the Sect. 2.6, if a spin wave propagates in a ferromagnetic sample subject to a magnetic field that changes in time in a step-like manner, the spin wave can be converted into an elastic wave. In this process, if the field is uniform in the region of propagation, the momentum of the spin wave is totally converted into elastic momentum.

As will be shown in Chap. 3, the magnetic and elastic excitations are quantized, and, as long as all parameters are static and spatially uniform, harmonic plane waves are solutions of the equations of motion, the following relations are valid

$$E_T = E_m + E_p + E_{me} = n\hbar\omega, \tag{2.88}$$

$$\vec{g}_T = \vec{g}_m + \vec{g}_p = n\hbar\vec{k}, \tag{2.89}$$

$$\vec{S}_T = E_T\vec{v}_g, \tag{2.90}$$

where $n$ is the number of quasiparticles per unit volume and $\vec{v}_g$ is the group velocity.

## 2.5    Coupled Spin–Electromagnetic Waves: Magnetic Polariton

Spin waves can also couple to electromagnetic waves by means of the direct interaction between the spins and the *rf* magnetic field of the radiation. Similar to the case of magnetoelastic waves, the coupling is strongest in the region around the crossing of the magnon and photon dispersion relations. Thus, for spin waves in ferromagnets, that have frequencies in the low microwave range, the crossing occurs in the region $k \approx 1 - 10$ cm$^{-1}$. The hybrid magnon–photon excitations that form in this region are called magnetic polaritons, and have similarities with the phonon polariton extensively studied in dielectrics and semiconductors, where the lattice vibration couples with the electric field of the radiation [13]. The properties of electromagnetic waves in ferromagnetic materials were extensively studied in the 1950s and 1960s, motivated by their applications in microwave ferrite devices (See Lax and Button; Stancil and Prabhakar). More recently the subject gained renewed interest in connection with the use of the coupling between the strong ferromagnetic resonance of YIG with microwave cavity photons for possible use in solid-state quantum information systems [14–16].

The starting point to study the magnetic polariton is the propagation of electromagnetic radiation in a magnetic material, that is governed by Maxwell's equations. In CGS units they are

$$\nabla \times \vec{h} = \frac{\varepsilon}{c} \frac{\partial \vec{e}}{\partial t}, \quad \nabla \times \vec{e} = -\frac{1}{c} \frac{\partial}{\partial t} (\bar{\mu} \cdot \vec{h}), \tag{2.91}$$

where $\vec{e}$, $\varepsilon$, and $\bar{\mu}$ are, respectively, the *rf* electric field, dielectric constant (considered to be real and isotropic), and the *rf* magnetic permeability tensor. Considering harmonic plane-wave solutions, $\exp(i\vec{k} \cdot \vec{r} - i\omega t)$, propagating with wave vector $\vec{k}$, Eqs. (2.91) lead to the following equations

$$\vec{k} \times \vec{h} = -\frac{\omega \varepsilon}{c} \vec{e}, \quad \vec{k} \times \vec{e} = \frac{\omega}{c} (\bar{\mu} \cdot \vec{h}), \tag{2.92}$$

that lead to an expression for the wave vector

$$\vec{k} \times (\vec{k} \times \vec{h}) = -(\varepsilon \omega^2 / c^2) \bar{\mu} \cdot \vec{h}. \tag{2.93}$$

For simplicity, we shall consider only waves propagating in the $z$-direction of the static field $\vec{H}_0$, with circularly polarized magnetic field $\vec{h}^{\pm} = h(\hat{x} \pm i\hat{y})$ and small-signal magnetization $\vec{m}^{\pm} = m(\hat{x} \pm i\hat{y})$. In this case, the permeability tensor has only diagonal elements, $\mu^{\pm} = 1 + 4\pi\chi^{\pm}$, where $\chi^{\pm} = m^{\pm}/h^{\pm}$ is the scalar *rf* magnetic susceptibility for circular polarization. From Eqs. (1.20 and 1.21) it follows that $\chi^{\pm} = \gamma M/(\omega_0 \mp \omega)$, where $\omega_0 = \gamma H_0$ is the $k = 0$ spin wave frequency. Using this result in Eq. (2.93), we obtain for the wave numbers of the circularly polarized waves

$$k^{\pm} = \frac{\omega \varepsilon^{1/2}}{c} \left( \frac{\omega_M + \omega_0 \mp \omega}{\omega_0 \mp \omega} \right)^{1/2}, \tag{2.94}$$

where $\omega_M = \gamma 4\pi M$. This equation shows that there are two solutions for the wave number, one for each sense of circular polarization. For frequencies far from resonance, $\omega \gg \omega_0, \omega_M$, the term in the square root is approximately unit, and the dispersion approaches $k^{\pm} = \omega \varepsilon^{1/2}/c$, characteristic of photons. For frequencies close to $\omega_0$, the wave number $k^+$ diverges, characterizing a flat dispersion, appropriate for magnons with negligible exchange. In the region near the crossing of the two dispersions, the coupling of photons and magnons is strong and the excitation is a magnetic polariton. In the frequency range between $\omega_0$ and $\omega_0 + \omega_M$, the term in the square root is negative and there is no real solution for the wave number. This is a forbidden frequency band, or frequency gap, for which the radiation does not propagate in the magnetic material in the direction of the static magnetic field. Similarly to the magnetoelastic waves, photons with $(-)$ circular polarization have negligible coupling with magnons. Figure 2.8 shows the magnetic polariton dispersion curves calculated for Galium doped YIG, with $4\pi M = 0.3$ kG, $H_0 = 2.0$ kOe, and $\varepsilon = 8$. The dashed lines represent the pure photon and pure magnon dispersions. As in the case of magnetoelastic waves, far from the crossover region, the dispersions approach the photon or the magnon curves.

**Fig. 2.8** Magnetic polariton dispersion relations in Ga–YIG calculated with Eq. (2.94) using $4\pi M = 0.3\,\text{kG}$, $\varepsilon = 8$, and $H_0 = 2.0\,\text{kOe}$. The dashed lines represent the pure photon and pure magnon dispersions

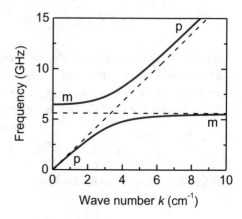

## 2.6    Experimental Techniques to Study Long-Wavelength Spin Waves

### 2.6.1   Microwave Techniques

The techniques most widely used to study long wavelength spin waves in ferromagnets employ microwave radiation. One advantage of the microwave techniques is that the *rf* magnetic field of the radiation couples directly with the magnetic moments. For a monochromatic driving field with frequency $\omega$, the average microwave power absorbed by a spin wave with frequency $\omega_k = \omega$ is

$$P = \frac{1}{2}\mu_0 \omega \, \text{Re} \left[ \int \vec{h} \cdot \vec{m}^* \, d\vec{r} \right], \tag{2.95}$$

where $\vec{h}$ and $\vec{m}$ are the complex vectors representing the driving field and the small-signal magnetization with frequency $\omega$, and the integral is carried out in the sample volume. Consider, for simplicity, that the driving field is along the $x$-direction. The power absorbed by a spin wave with magnetization $m_x(\vec{k}) \exp(i\vec{k} \cdot \vec{r})$ from a driving field that has a spatial variation with Fourier component $h_x \exp(i\vec{q} \cdot \vec{r})$ is

$$P = \frac{1}{2}\mu_0 \omega V \, \text{Re} \left[ h_x(\vec{q}) m_x^*(\vec{k}) \right] \delta_{\vec{q},\vec{k}}, \tag{2.96}$$

where $\delta_{\vec{q},\vec{k}}$ is the Kronecker delta that has value 1 for $\vec{q} = \vec{k}$ and zero otherwise. Using the definition of the susceptibility in (1.22), this result leads to

$$P = \frac{1}{2}\mu_0 \omega V \chi_{xx}'' \left| h_x(\vec{k}) \right|^2. \tag{2.97}$$

This expression shows that in order to excite a spin wave with a certain wave vector $\vec{k}$ directly, the driving magnetic field must have a Fourier component at the same wave vector. This result can be seen as a manifestation of momentum conservation in the conversion of photons into magnons. Consider a driving field with frequency 10 GHz in a microwave waveguide, that has wavelength $\lambda \sim 3$ cm and wave number $q = (2\pi/\lambda) \sim 2$ cm$^{-1}$. In this case, in bulk samples with uniform internal magnetizing and driving fields, one can excite only the $k \sim 0$ magnon, that is, the ferromagnetic resonance, as studied in Chap. 1. Spin waves with wave numbers up to $k \sim 10^2$ can be excited with fine wire antennas.

Other schemes can be used to excite magnons with higher $k$ in films by microwave radiation. One is to use a metallic multielement antenna deposited on the surface of the magnetic film by lithographic processes, so that the microwave field varies periodically in space with a wavelength much shorter than in waveguides [17]. Another scheme consists in using the fact that a driving field applied in the plane of a metallic film is internally nonuniform because it has to satisfy the boundary conditions at the film surfaces. In this case it can drive spin waves with wavelength comparable to the film thickness. This is the method that was used in the first observation [18] of spin wave resonances in a thick film of permalloy, with the static field perpendicular to the film. In this case, plane spin waves with wave vector perpendicular to the film plane form standing waves between the two film surfaces, with wavelength $d = n(\lambda/2)$, where $d$ is the film thickness and $n$ is an integer. As the field is scanned, the wave number varies and the spin wave resonances produce absorption peaks, as shown in Fig. 2.9. Each peak corresponds to a different wavelength, with index number $n$ as indicated in the upper horizontal axis.

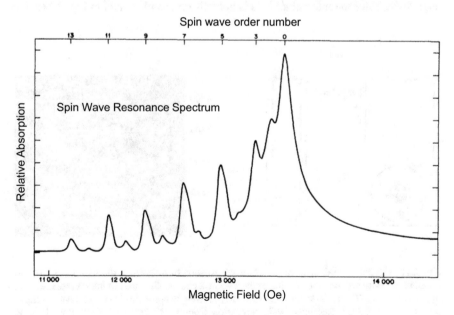

**Fig. 2.9** Spin wave resonance spectrum in a 560-nm thick film of permalloy, driven by microwave radiation with frequency 8.89 GHz [18]. Reprinted with permission from M. H. Seavey, Jr. and P. E. Tannenwald, Phys. Rev. Lett. **1**, 168 (1958). Copyright (1958) by the American Physical Society

In the early days of spin wave studies with YIG, in the 1960s, crystalline films were not available, so experiments were done with bulk single crystal samples of several shapes. In a YIG sample in the form of a sphere, if the applied field is uniform the internal field is also uniform. In this case, one can excite magnetostatic spin wave resonances with small wave numbers, but not spin waves with large wave numbers. However, in samples with the shape of disks, rods, or slabs, an applied uniform internal field produces a nonuniform internal field, due to demagnetizing effects at the surfaces. In this case, a spin wave generated with $k{\sim}0$ at some position in the sample, propagates in a direction of decreasing field, as discussed in Sect. 2.2.1. If the field is static, the frequency of the wave does not change, but the wave number varies since momentum is not conserved. Thus, the wave number increases as to satisfy the dispersion relation at each position of the sample.

Figure 2.10 shows the sample arrangement and some results of the first experiments to demonstrate spin wave propagation [19, 20]. The experiments used a single crystal YIG disk, with diameter 3.19 mm and thickness 0.33 mm, placed in a cylindrical microwave cavity fed by a waveguide with pulsed microwave radiation of frequency 9.4 GHz. The disk is magnetized to saturation by a static field applied perpendicularly to its plane, as illustrated in Fig. 2.10a. The cavity is of the reentrant type, with a post to concentrate the microwave magnetic field in the disk and perpendicular to the static field. The first broad pulse in Fig. 2.10b corresponds to the microwave reflection from the cavity. The other pulses are echoes of the magnetoelastic propagation in the disk, which have time delays that vary with the field.

The origin of the echoes in the oscilloscope traces in Fig. 2.10b can be understood as follows. Since the internal *dc* field is nonuniform, as illustrated in Fig. 2.11c, the

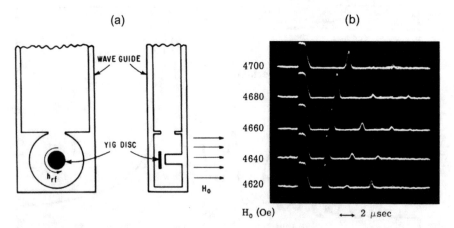

**Fig. 2.10** Experimental demonstration of the propagation of magnetoelastic waves [19, 20]. (**a**) Microwave cavity showing the position of the YIG disk and the directions of the static and driving microwave fields. (**b**) Oscilloscope traces (2 μs/div) showing magnetoelastic wave echoes at various magnetic field values. Reprinted with permission from J. R. Eshbach, Phys. Rev. Lett. **8**, 357 (1962). Copyright (1962) by the American Physical Society

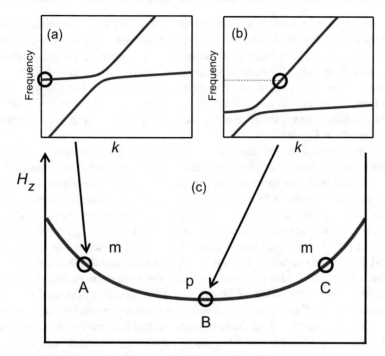

**Fig. 2.11** (**a**), (**b**) Magnetoelastic dispersion relations corresponding to the magnetic field values in a YIG disk magnetized perpendicularly to the plane with the internal field profile in (**c**)

microwave pulse drives a spin wave only in the small region where the field has a value for which the spin wave has wave number $k \sim 0$, point A in the figure. Figure 2.11a shows the dispersion relation corresponding to the field at point A, with the circle indicating the operating point. The wave packet, with a circular wave front, propagates toward the center of the disk with constant frequency. Since the field decreases, the wave number increases and approaches the crossover region.

If the field were uniform, the two branches of the magnetoelastic dispersion would correspond to truly normal modes, so that the excitation in one branch would not undergo branch hopping. However, since the field varies in space, there is a finite probability for the excitation to jump to the other branch. This probability depends on the value of the magnetic field gradient at the position where the magnon and phonon wave numbers are equal. If the field gradient is much smaller than a certain critical gradient, the probability for branch hopping is negligible [21]. This is the case of the experiments of Eshbach [19, 20], so that the initial spin wave packet is almost completely converted into an elastic wave, in other words, the magnon–phonon conversion efficiency is nearly 100%, as illustrated by the circle in Fig. 2.11b.

Note that after reaching the center of the disk as an elastic wave packet, the wave front radius increases again and at some point it is converted back into a spin wave. As illustrated in Fig. 2.11c, at point C the spin wave has $k \sim 0$, so that it radiates an electromagnetic pulse, and then propagates back toward the center of the disk. The process is repeated several times until the excitation dies away due to the magnetic and elastic damping. Each time the wave packet reaches the positions of points A and C in Fig. 2.11c, it radiates a microwave pulse, resulting in the traces shown in Fig. 2.10b. Note that as the static field increases, the field profile in Fig. 2.11c moves upward and the radius of the initial wave front decreases, so that the wave packet stays longer with spin wave character. Since the spin wave velocity is smaller than that of the elastic wave, the time delay increases, as observed in the experimental results shown in Fig. 2.10b.

One disadvantage of the disk-shaped YIG sample for applications and for the studies of spin wave propagation is that the waves have circular wave fronts. A few years after the initial experiments with disks [19], researchers began to study spin wave propagation using YIG cylinders, or rods. YIG rods were extensively used to investigate magnetostatic waves, that propagate from one end of the rod to the other [22, 23], and also magnetoelastic waves, that are confined to a region near one of the end faces [24, 25]. Magnetostatic waves, which will be studied in Chap. 4, are excited at a field range below the one needed to excite magnetoelastic waves.

With the external field applied along the axis, the demagnetizing field is larger near the end faces than in the middle of the rod, so that the internal field along the axis has a profile as shown in Fig. 2.12a. Thus, a spin wave with $k \sim 0$ excited at the point where $H_z = \omega/\gamma\mu_0$, propagates toward the nearest end face with increasing wave number. At some point, the spin wave is converted into an elastic wave that reaches the end face. If the surface is optically polished, the elastic wave is reflected and returns toward the $k \sim 0$ point, called the *turning point*. Then it returns back toward the end surface, and the process repeats until the excitation is damped out. Figure 2.12b shows the oscilloscope traces of a magnetoelastic echo train measured in a YIG rod, produced by the process just described, in experiments with microwave pulsed driving of frequency 1.3 GHz and applied field $H_0 = 1.15$ kOe [25, 26].

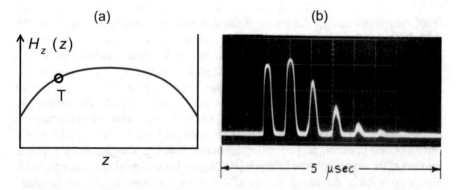

(a)                                    (b)

$H_z(z)$

T

z

5 μsec

**Fig. 2.12** (a) Profile of the internal magnetic field in a YIG rod magnetized along the axis. T indicates the turning point where spin waves are excited and return after being reflected at the end face. (b) Magnetoelastic echo train measured in a YIG rod excited by a fine wire antenna near the end face, with frequency 1.3 GHz and external applied field of 1.15 kOe [25, 26]. Reproduced from S. M. Rezende and F. R. Morgenthaler, J. Appl. Phys. **40**, 537 (1969), with the permission of AIP Publishing

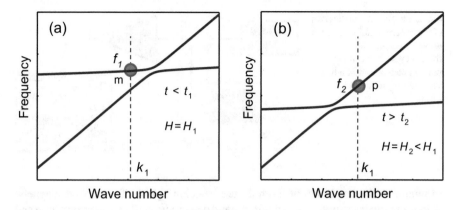

**Fig. 2.13** Illustration of the conversion of an elastic excitation (phonon) into a spin wave (magnon) in a time-varying magnetic field. (**a**) At $t < t_1$ a spin wave propagates in the medium with frequency $f_1$ and wave number $k_1$. (**b**) At $t > t_2$ the field is at a lower value and the excitation is converted into an elastic wave of frequency $f_2$ and the same wave number $k_1$

Another manner to convert magnons into phonons, or vice versa, consists in using an applied magnetic field that varies in time but not in space. In this case, the wave number of the spin wave remains constant but the frequency changes. For simplicity consider a ferromagnetic material under a uniform field that varies in time in a step-like manner. Let us assume that prior to an instant of time $t_1$ the internal field $H_1$ is constant, between $t_1$ and $t_2$ it decreases monotonically in time, and after $t_2$ it remains constant at a value $H_2 < H_1$. We assume also that, prior to $t_1$, a magnetoelastic wave pulse with essentially pure spin wave character is propagating in the material. Figure 2.13a illustrates the magnetoelastic dispersion at $t < t_1$ showing that the magnon has frequency $f_1$ and wave number $k_1 < k_{\text{cross}}$ in the upper branch. If the variation of the magnetic field is slow enough so that the excitation stays in the same eigenmode, at $t > t_2$ it should remain in the upper branch [5]. Assuming that during the transient the field stays uniform in space, the wave number of the excitation remains constant during the process due to momentum conservation. Thus, as the field varies, the frequency of the spin wave changes and reaches the crossover region, where the excitation becomes an essentially elastic wave with wave number $k_1$ and frequency $f_2$, as shown in Fig. 2.13b. Conversely, if the initial state is a phonon excitation, as in Fig. 2.12b, and the field increases in time, it is converted into a magnon excitation.

Spin wave frequency shift and magnon–phonon conversion in pulsed magnetic fields were observed experimentally some time ago, both with long-wavelength magnetostatic modes and exchange-dominated spin waves [25, 26]. The experiments were carried out at room temperature with an axially magnetized (100) YIG rod, with length 1 cm, diameter 0.3 cm, and optically polished end faces. Figure 2.14 shows a schematic diagram of the experimental setup. Spin wave packets were generated and detected by shorted fine wire antennas placed close to one of the end faces. The excitation and detection antennas are crossed so as

**Fig. 2.14** Schematic diagram of the experimental setup used to observe spin wave frequency shift and magnon–phonon conversion in a time-varying magnetic field

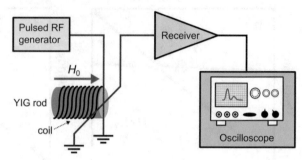

to minimize the amplitude of the reflected launching pulse. The applied magnetic field is the superposition of two fields, a static one produced by an electromagnet and a time-varying one produced by a coil wound around the YIG rod and driven by a pulsed current source. The rise time of the pulsed field had a time gradient much smaller than the critical gradient in order to avoid branch hopping [5].

The microwave apparatus consists basically of the microwave source, that produces *rf* bursts with frequency in the range 1–2 GHz and duration 0.2–0.5 μs, and the receiving system. The latter can be a broadband microwave amplifier, followed by a diode detector, or a narrowband super heterodyne system, composed of a local oscillator, mixer, IF amplifier, and diode detector. In the narrowband receiver, by varying the local oscillator frequency one can tune the amplifier and measure the pulse frequency. The detected pulses are displayed on an oscilloscope screen, together with the waveform of the current pulses that produce the time-varying field. The latter can be positioned in time with respect to the *rf* bursts by means of delayed synchronization pulses.

Figure 2.15 shows oscilloscope traces that demonstrate the frequency conversion in experiments with magnetoelastic waves excited in the YIG rod by a fine wire antenna, fed by microwave with frequency 1.1 GHz and pulse width 0.4 μs [25, 26]. The upper traces correspond to current steps applied to the coil and the lower traces are due to the video-detected microwave pulses. The broadening and rounding of the pulses are due to the narrow frequency band of the receiver used in the experiments. The initial pulse corresponds to the launching burst and the delayed one is generated by the radiation of the spin wave at the turning point. The magnetic field of the electromagnet is kept fixed at 648 Oe, corresponding to a magnetoelastic delay of 3 μs, and the field step has amplitude 59 Oe. In Fig. 2.15a, the field step is applied after the magnetoelastic echo is detected, so that there is no frequency shift. Thus, the two pulses have the same frequency and both are picked up by the receiver tuned to 1.1 GHz. In (b) the field step is applied 2.3 μs after the launching pulse, when the wave is in a magnon state, producing a frequency shift. Thus, the magnetoelastic echo is not seen because the receiver is tuned to 1.1 GHz. As shown in (c), if the receiver is tuned to 1.266 GHz, only the frequency shifted magnetoelastic echo is detected. The ratio between the frequency shift and the field step is 2.86 MHz/Oe, which is very close to the gyromagnetic ratio in YIG. Finally, Fig. 2.15d shows the traces with the field step applied 1.0 μs after the launching pulse, when the wave

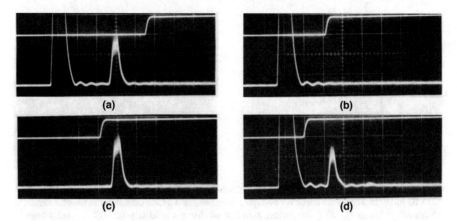

**Fig. 2.15** Experimental demonstration of spin wave frequency shift in a YIG rod under a time-varying magnetic field. Input frequency 1.1 GHz, static field 648 Oe, field step 59 Oe, time scale 1 µs/div. In (**a**), (**b**), and (**d**), the receiver is tuned to 1.1 GHz. In (**c**) it is tuned to 1.266 GHz [25, 26]. Reproduced from S. M. Rezende and F. R. Morgenthaler, J. Appl. Phys. **40**, 537 (1969), with the permission of AIP Publishing

packet is an elastic wave. In this case there is no frequency shift because phonons are not sensitive to field changes, so both pulses have the original frequency.

Magnon–phonon conversion in a time-varying field was also experimentally demonstrated in a YIG rod using the same setup of Fig. 2.14. The magnetoelastic waves were injected and detected in the sample as either magnons or shear phonons. The coupling of microwave radiation to magnons is made with a fine wire antenna, while the coupling with phonons employs a piezoelectric transducer, made of a CdS film and electrodes deposited in one of the end faces, as illustrated in Fig. 2.16a.

Figure 2.16b shows the oscilloscope traces that demonstrate the conversion of magnons into phonons. The pulses are sharp and narrow because the receiver used here is the broadband one. Magnons are excited at the turning point by pulsed microwave radiation with frequency 1.1 GHz and propagate toward the end face, as explained earlier. The external field is 600 Oe, smaller than the one used in the experiments of Fig. 2.15, so that the turning point is closer to the middle of the rod, where the field is more uniform. In this case, the delay of the first magnetoelastic pulse is 5.0 µs. At a time 2.0 µs after the launching pulse, when the magnon wave packet is travelling toward the end face, a negative field step with amplitude 77 Oe is applied, as shown in the upper trace in Fig. 2.16b. Then magnons are converted into phonons by the time-varying field, as illustrated in Fig. 2.13. The field step down also has the effect of removing the turning point, since the maximum field in the middle of the rod axis becomes smaller than $\omega/\gamma\mu_0$. Thus, the phonons are not converted back into magnons by the nonuniform field, so that the phonon wave packet travels back and forth between the two end faces and produces the pulse train detected by the CdS transducer, shown in Fig. 2.16b. Note that the round-trip time of the phonons in the whole length of the rod is smaller than the delay time of the magnetoelastic echo in a shorter distance because the velocity of phonons is larger than of magnons.

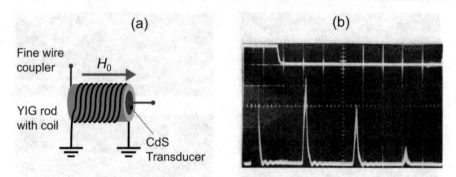

**Fig. 2.16** Experimental demonstration of magnon–phonon conversion by a pulsed magnetic field. (a) YIG rod with a coil with a fine wire coupler in one end and a CdS transducer in the other end. (b) Oscilloscope traces showing the current step that produces a field step of −77 Oe and a train of elastic wave pulses detected by the CdS transducer. Input frequency 1.1 GHz, static field 600 Oe, time scale 2 µs/div [25, 26]. Reproduced from S. M. Rezende and F. R. Morgenthaler, J. Appl. Phys. **40**, 537 (1969), with the permission of AIP Publishing

### 2.6.2 Brillouin Light Scattering Techniques

Inelastic light scattering techniques are among the most important ones to study elementary excitations in solids. The techniques are based on the fact that when photons of a certain frequency interact with excitations in a material, they may be inelastically scattered with a different frequency due to the creation or annihilation of the quanta of the excitations. Measurements of the scattered photon frequencies give information on the excitation frequencies and the strength of the interaction mechanisms.

Figure 2.17 illustrates the two simplest processes involving magnons. Photons in a laser beam with frequency $\omega_L$ and wave vector $\vec{k}_L$ interact in a magnetic material with magnons of frequency $\omega_m$ and wave vector $\vec{k}_m$. Energy and momentum are conserved, so that in the Stokes process, illustrated in Fig. 2.17(a), where one magnon is created, the scattered photons have frequency $\omega_S = \omega_L - \omega_m$ and wave vector $\vec{k}_S = \vec{k}_L - \vec{k}_m$. On the other hand, in the Anti-Stokes process, illustrated in Fig. 2.17b, one magnon is destroyed, so that the scattered photons have frequency $\omega_S = \omega_L + \omega_m$ and wave vector $\vec{k}_S = \vec{k}_L + \vec{k}_m$. The mechanisms of interaction between photons and magnons will be studied in Chap. 3.

The apparatus used in a light scattering experiment consists basically of three components: a laser source that produces the light beam to impinge on the sample; a spectrometer to analyze the frequencies of the scattered photons; and a very low noise photodetector. The main challenge of the experimental setup is to discriminate the signal of the very weak inelastically scattered light from the strong elastically scattered component, called Rayleigh light. The type of spectrometer depends on the magnitude of the frequency shift, that is equal to the magnon frequency. As will be seen in Chap. 5, antiferromagnets with high anisotropy have magnon frequencies in

**Fig. 2.17** Diagrammatic representation of the Stokes and anti-Stokes processes in inelastic light scattering by magnons. The equations on the right indicate the energy and momentum of the scattered photons

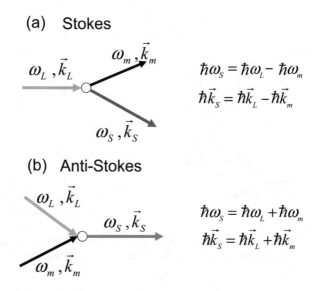

(a)  Stokes

$$\hbar\omega_s = \hbar\omega_L - \hbar\omega_m$$
$$\hbar\vec{k}_s = \hbar\vec{k}_L - \hbar\vec{k}_m$$

(b)  Anti-Stokes

$$\hbar\omega_s = \hbar\omega_L + \hbar\omega_m$$
$$\hbar\vec{k}_s = \hbar\vec{k}_L + \hbar\vec{k}_m$$

the terahertz range. In this case one uses the techniques of Raman light scattering, that employ grating spectrometers. In ferromagnets, magnons with wave numbers comparable to that of visible light, $k_L \sim 10^5$ cm$^{-1}$, have frequency in the range of several gigahertz. In this case one uses the technique of Brillouin light scattering, in which the spectrometer is a Fabry–Perot interferometer.

A simple Fabry–Perot interferometer, illustrated in Fig. 2.18a, consists of two parallel fused silica plates with inner surfaces that are very flat, highly polished, and coated to have reflectivity on the order of 90% [27]. Collimated light incident perpendicularly to the plates undergoes interferences between the two inner surfaces, so that the transmission has peaks at the optical resonance condition $d = m\lambda/2$, where $d$ is the distance between the two mirrors, $\lambda$ is the wavelength, and $m$ is an integer. If one of the plates is mounted on piezoelectric transducers, the mirror distance $d$ can be scanned so that the transmitted light is frequency analyzed. Figure 2.18b showing the instrument transmission demonstrates how the interferometer analyzes the light wavelength, and thus its frequency $f = c/\lambda$, by scanning the mirror distance.

The frequency difference between two neighboring transmission peaks is called the free spectral range (FSR) of the instrument, and is given by FSR $= c/2d$. For a mirror distance of $d = 5$ mm, $FSR = 30$ GHz, which is a convenient range to study low-frequency magnons. Note that with $d = 5$ mm, collimated visible light with $\lambda = 500$ nm has a transmission peak with $m = 10^4$. Thus, in order to scan a range of 30 GHz, the mirror distance has to vary only 0.5 µm, which can be comfortably achieved with PZT transducers. Current instruments use a repetitive scanning with a frequency of few scans per second, and employ active electronic stabilization to optimize the parallelism of the mirrors. Stabilization control is essential for signals on the order of few photons per second that require several hours of photon counting data accumulation to overcome the noise.

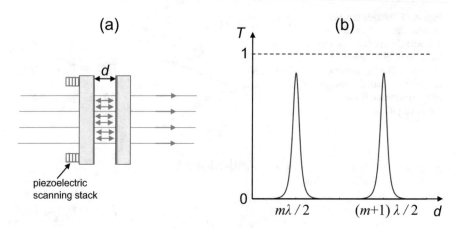

**Fig. 2.18** (a) Illustration of the Fabry–Perot interferometer with a piezoelectric scan. (b) Interferometer transmission as a function of the distance $d$ between the mirrors

A major improvement in the Brillouin light scattering (BLS) technique was introduced in the early 1970s by John Sandercock [27, 28]. By using two corner cube prisms to reflect the light back and forth through the interferometer, he managed to operate a Fabry–Perot interferometer in a multipass mode. This increases the ratio between the maximum and minimum in the transmission by many orders of magnitude. Multipass interferometers made possible the use of BLS to study excitations in opaque materials, because it has to be done in the backscattering configuration, that produces very intense Rayleigh scattered light.

The 5-pass Fabry–Perot interferometer was employed by Sandercock in backscattering BLS experiments to observe thermally excited acoustic magnons and phonons in a YIG sphere with diameter 2 mm. Figure 2.19 shows a spectrum measured with the sphere under a magnetic field of 3.5 kOe, using a He–Ne laser with wavelength $\lambda = 632.8$ nm and power 10–20 mW [28]. The magnon (M) and longitudinal acoustic phonon (LA) peaks have frequency shifts of approximately 14 GHz and 53 GHz. These values are consistent with the frequencies calculated for magnon and phonon in YIG, with the wave number appropriate for the backscattering configuration

$$\left| \vec{k}_S - \vec{k}_L \right| = 2n\frac{2\pi}{\lambda}, \tag{2.98}$$

where $n$ is the index of refraction and $\lambda$ the wavelength of light.

Note that since the interferometer scan produces a periodic train of transmission peaks, the spectrum in Fig. 2.19 shows three Rayleigh lines and the corresponding shifted Stokes and anti-Stokes lines. As described in Sandercock and Wettling [28], the light scattered by magnons is circularly polarized, while the light scattered by phonons is linearly polarized in the direction of the incident light. By using an analyzer with the axis at right angle to incident polarization, the intensity of the Rayleigh decreases sharply making possible an accurate measurement of the magnon frequency. In the spectrum of Fig. 2.19, measured without the analyzer to

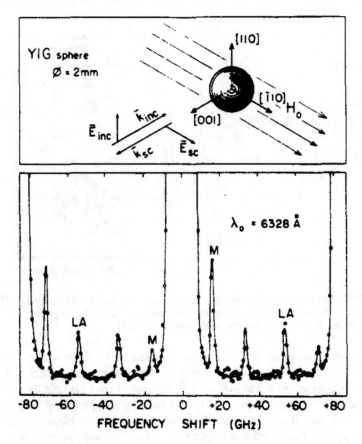

**Fig. 2.19** Brillouin light scattering spectrum measured in a YIG sphere with diameter 2 mm under a magnetic field of 3.5 kOe [28]. Reproduced with kind permission from Elsevier Inc.

allow the observation of the magnon and phonon peaks simultaneously, the magnon line is at the shoulder of the Rayleigh line and seems to have a higher frequency.

Another very important improvement in interferometer design was the introduction of the multipass-tandem operation by John Sandercock in the 1980s [29]. The new instrument consists of two Fabry–Perot interferometers operating in tandem, having a corner cube reflector and a mirror positioned so that the light passes three times in each interferometer, as illustrated in Fig. 2.20. The scanning mirrors of both interferometers are mounted on a translation stage that is scanned by means of a piezoelectric transducer, in such a way that the two mirrors move together. However, since the axes of the two interferometers make an angle $\theta$, when the translation stage is scanned in a distance $\delta L$, the mirror spacing of FP1 changes by $\delta L$ while the spacing in FP2 changes by $\delta L \cos \theta$. The angle is such that the ratio between the two is about 0.95. Thus, if the two interferometers are aligned so that their transmission peaks coincide at some wavelength, as the scan proceeds the next order transmission peaks do not quite coincide. This scheme avoids having the train of Rayleigh lines in

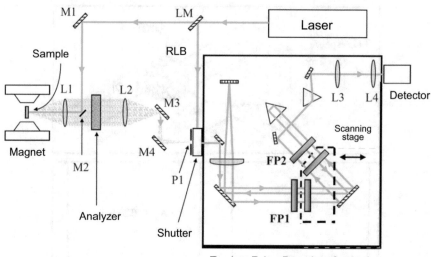

**Fig. 2.20** Schematic diagram of the apparatus for Brillouin light scattering studies. The main component is the tandem multipass interferometer used to analyze the frequencies of the light scattered by the sample in the backscattering configuration [30]

the BLS spectrum as in Fig. 2.20. The optimization of the mirror distance and parallelism in each Fabry–Perot is made by piezoelectric transducers that act on the nonscanning mirrors. The active feedback control uses the light of a reference laser beam (RLB) shown in Fig. 2.19. When the scan is approaching the position of the Rayleigh line, that means zero frequency shift, a shutter blocks the incoming scattered light so that the reference light with constant intensity is used for the signal optimization.

The commercialization of the Sandercock multipass tandem Fabry–Perot interferometer greatly contributed to make Brillouin light scattering techniques more widely used for the investigation of materials [29]. One important example is in the studies of magnetic multilayers, made of metallic ferromagnetic (FM) thin films intercalated with thin layers of a nonmagnetic metal (NM). In FM/NM/FM trilayers, there are two spin wave modes with frequencies that depend on the coupling between the FM layers, as will be studied in Chap. 4. Using BLS techniques to measure the frequencies of spin waves in trilayers, Peter Grünberg discovered in 1986 that for some materials and interlayer thickness, the exchange coupling between the FM films can be antiferromagnetic [31]. Figure 2.21 shows the BLS spectra of spin waves in trilayers of Fe, demonstrating that the frequencies depend on the material and thickness of the interlayer [32]. This development led to the discovery in 1989 of the giant magnetoresistance (GMR) effect in antiferro-magnetically coupled multilayers by Albert Fert [33] and independently by Grünberg [34]. Fert and Grünberg received the Physics Nobel Prize in 2007 for their discovery of the GMR that gave birth to the field of Spintronics.

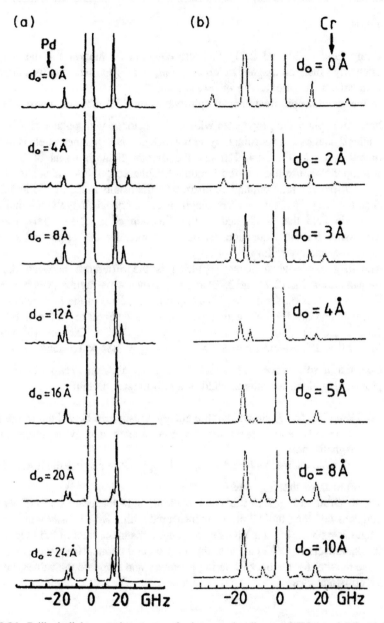

**Fig. 2.21** Brillouin light scattering spectra of spin waves in trilayers of FM materials intercalated by a very thin metallic layer of varying thickness. (**a**) Py/Pd/Py. (**b**) Py/Cr/Co [31]. Reprinted from P. Grünberg et al., Phys. Rev. Lett. **57**, 2442 (1986). Copyright (1986) by the American Physical Society

## Problems

2.1 Using Eqs. (2.20 and 2.21) find the equation of motion for the negative circularly polarized magnetization $m^- = m_x - i m_y$ and interpret its solution in comparison with the one for $m^+ = m_x + i m_y$.

2.2 In the macroscopic treatment of spin waves in a 3D material, consider that a harmonic spin wave propagates with wave vector $\vec{k}$ with polar and azimuthal angles $\theta_k$ and $\varphi_k$ in a coordinate system where the static magnetic field is in the $z$-direction, as in Fig. 2.4. Express the dipolar field in terms of its $x$ and $y$ components, use them in the Landau–Lifshitz equation to calculate the spin wave dispersion relation, and compare with the results in Eqs. (2.35 and 2.36).

2.3 A spin wave propagates in a ferromagnetic cubic crystal along the internal static magnetic field that is aligned in the direction of a $<100>$ axis. Find an expression for the dispersion relation that includes the effects of Zeeman, exchange, and crystalline anisotropy.

2.4 The magnetoelastic frequency splitting is the difference between the two frequencies in Eqs. (2.71 and 2.72) at the crossover wave number, where $\omega_t = \omega_k$. Find an expression for the frequency splitting in terms of the parameter $\sigma_k$ and calculate its value for YIG, with the parameters used to make the plot in Fig. 2.7: $H_z = 2.0$ kOe, $4\pi M = 1.76$ kG, $\gamma = 2.8 \times 2\pi \times 10^6$ s$^{-1}$Oe$^{-1}$, $b_2 = 7.0 \times 10^6$ erg/cm$^3$, $\rho = 5.2$ g/cm$^3$, and $v_t = 3.84 \times 10^5$ cm/s.

2.5 Consider a spin wave with small-signal magnetization $\vec{m} \exp(i \vec{k} \cdot \vec{r} - i\omega t)$ propagating along the internal field of a ferromagnetic sample.

(a) Using Eq. (2.77) shows that the energy density can be written in the form $E_m = n\hbar\omega$ and express the magnon number density in terms of the magnetization.

(b) Calculate the momentum density of the spin wave and use the result of item (a) to show that $\vec{g}_m = n\hbar\vec{k}$.

2.6 The spectrum in Fig. 2.19 shows the Brillouin light scattering peaks of acoustic magnons and longitudinal phonons measured with a He–Ne laser with a wavelength $\lambda = 632.8$ nm in a YIG sphere under a magnetic field of 3.5 kOe. Using the parameters for YIG calculate the magnon and phonon frequencies with the wave number for the backscattering geometry and compare the values with the experimental data.

## References

1. Bloch, F.: Zur Theorie des Ferromagnetismus. Z. Physik. **61**, 206 (1930)
2. Herring, C., Kittel, C.: On the theory of spin waves in ferromagnetic media. Phys. Rev. **81**, 869 (1951)
3. Eshbach, J.R.: Spin-wave propagation and the magnetoelastic interaction in yttrium iron garnet. Phys. Rev. Lett. **8**, 357 (1962)

4. Strauss, W.: Magnetoelastic waves in yttrium Iron garnet. J. Appl. Phys. **36**, 118 (1965)
5. Rezende, S.M., Morgenthaler, F.R.: Magnetoelastic waves in time-varying magnetic fields. I. Theory. J. Appl. Phys. **40**, 524 (1969)
6. Kittel, C.: Interaction of spin waves and ultrasonic waves in ferromagnetic crystals. Phys. Rev. **110**, 836 (1958)
7. Akhiezer, A. I., Bar'yakhtar, V. G., Peletminskii, S. V.: Coupled magnetoelastic waves in ferromagnetic media and ferroacoustical resonance, Zh. Eksperim. i Teor. Fiz. **35**, 228 (1958) [English transl.: Soviet Phys.-JETP **8**, 157 (1959)]
8. Auld, B.A.: Magnetostatic and magnetoelastic wave propagation in solids. In: Wolfe, R. (ed.) Applied Solid State Science, vol. 2. Academic Press, New York (1971)
9. Schlömann, E.: Generation of phonons in high-power ferromagnetic resonance experiments. J. Appl. Phys. **31**, 1647 (1960)
10. Morgenthaler, F.R.: Exchange energy, stress, and momentum in a rigid ferrimagnet. J. Appl. Phys. **38**, 1069 (1967)
11. Morgenthaler, F. R.: Small Signal Power and Momentum Theorems for a Magnetoelastic Ferromagnet. Microwave and Quantum Magnetics Group Technical Report 14, Massachusetts Institute of Technology (1967)
12. Camley, R.E., Maradudin, A.A.: Power flow in magnetoelastic media. Phys. Rev. B. **24**, 1255 (1981)
13. Mills, D.L., Burstein, E.: Polaritons: the electromagnetic modes of media. Rep. Prog. Phys. **37**, 817 (1974)
14. Huebl, H., Zollitsch, C.W., Lotze, J., Hocke, F., Greifenstein, M., Marx, A., Gross, R., Goennenwein, S.T.B.: High cooperativity in coupled microwave resonator ferrimagnetic insulator hybrids. Phys. Rev. Lett. **111**, 127003 (2013)
15. Bai, L., Harder, M., Chen, Y.P., Fan, X., Xiao, J.Q., Hu, C.-M.: Spin pumping in electrodynamically coupled magnon-photon systems. Phys. Rev. Lett. **114**, 227201 (2015)
16. Hyde, P., Bai, L., Harder, M., Dyck, C., Hu, C.-M.: Linking magnon-cavity strong coupling to magnon-polaritons through effective permeability. Phys. Rev. B. **95**, 094416 (2017)
17. Lim, J., Bang, W., Trossman, J., Kreisel, A., Jungfleisch, M.B., Hoffmann, A., Tsai, C.C., Ketterson, J.B.: Direct detection of multiple backward volume modes yttrium iron garnet at micron scale wavelengths. Phys. Rev. B. **99**, 014435 (2019)
18. Seavey Jr., M.H., Tannenwald, P.E.: Direct observation of spin wave resonance. Phys. Rev. Lett. **1**, 168 (1958)
19. Eshbach, J.R.: Spin wave propagation and the magnetoelastic interaction in yttrium iron garnet. Phys. Rev. Lett. **8**, 357 (1962)
20. Eshbach, J.R.: Spin wave propagation and the magnetoelastic interaction in yttrium iron garnet. J. Appl. Phys. **34**, 1298 (1963)
21. Schlömann, E., Joseph, R.I.: Generation of spin waves in nonuniform magnetic fields. III. Magnetoelastic interaction. J. Appl. Phys. **35**, 2382 (1964)
22. Olson, F.A., Yaeger, J.R.: Microwave delay techniques using YIG. IEEE Trans. Microw. Theory Tech. **13**, 63 (1965)
23. Damon, R.W., van de Vaart, H.: Propagation of magnetostatic spin waves at microwave frequencies, II. Rods. J. Appl. Phys. **37**, 2445 (1966)
24. Strauss, W.: Magnetoelastic waves in yttrium iron garnet. J. Appl. Phys. **36**, 118 (1965)
25. Rezende, S. M.: Magnetoelastic and magnetostatic waves in time-varying magnetic fields. PhD Thesis Presented to the Massachusetts Institute of Technology (1967); Microwave and Quantum Magnetics Group Technical Report **19**, MIT (1967)
26. Rezende, S.M., Morgenthaler, F.R.: Magnetoelastic waves in time-varying magnetic fields. II. Experiments. J. Appl. Phys. **40**, 537 (1969)
27. Sandercock, J.R.: Trends in brillouin scattering: studies of opaque materials, supported films, and central modes. In: Cardona, M., Guntherodt, G. (eds.) Topics in Applied Physics: Light Scattering in Solids III, vol. 51. Spinger, Heidelberg (1982)

28. Sandercock, J.R., Wettling, W.: Light scattering from thermal acoustic magnons in yttrium iron garnet. Solid State Comm. **13**, 1729 (1973)
29. Patton, C.: Magnetic excitations in solids. Phys. Rep. **103**, 251 (1984)
30. Rodríguez-Suárez, R. L.: Magnetoelectronic Phenomena in Metallic Interfaces. PhD Thesis presented to the Physics Department, Universidade Federal de Pernambuco, Recife, Brazil (2006)
31. Grünberg, P., Schreiber, R., Pang, Y., Brodsky, M.B., Sowers, H.: Layered magnetic structures: evidence for antiferromagnetic coupling of Fe layers across Cr interlayers. Phys. Rev. Lett. **57**, 2442 (1986)
32. Vohl, M., Barnás, J., Grünberg, P.: Effect of interlayer exchange coupling on spin wave spectra in magnetic double layers: theory and experiments. Phys. Rev. B. **39**, 12003 (1989)
33. Baibich, M.N., Broto, J.M., Fert, A., Nguyen Van Dau, F., Petroff, F., Etienne, P., Creuzet, G., Friederich, A., Chazelas, J.: Giant magnetoresistance of (001)Fe/(001)Cr magnetic superlattices. Phys. Rev. Lett. **61**, 2472 (1988)
34. Binasch, G., Grünberg, P., Saurenbach, F., Zinn, W.: Enhanced magnetoresistance in layered magnetic structures with antiferromagnetic interlayer exchange. Phys. Rev. B. **39**, 4828 (1989)

## Further Reading

Akhiezer, A.I., Bar'yakhtar, V.G., Peletminskii, S.V.: Spin Waves. North-Holland, Amsterdam (1968)
Cottam, M.G., Lockwood, D.J.: Light Scattering in Magnetic Solids. Wiley, New York (1986)
Gurevich, A.G., Melkov, G.A.: Magnetization Oscillations and Waves. CRC Press, Boca Raton, FL (1994)
Hayes, W., Loudon, R.: Scattering of Light by Crystals. Wiley, New York (1978)
Kabos, P., Stalmachov, V.S.: Magnetostatic Waves and their Applications. Chapman and Hall, London (1994)
Keffer, F.: Spin Waves. In: Flugge, S. (ed.) Handbuch der Physik, vol. XVIII/B. Springer, Heidelberg (1966)
Kittel, C.: Introduction to Solid State Physics, 8th edn. Wiley, New York (2004)
Lax, B., Button, K.: Microwave Ferrites and Ferrimagnetics. McGraw-Hill, New York (1962)
Landau, L.D., Lifshitz, E.M.: Theory of Elasticity. Pergamon, New York (1970)
Sparks, M.: Ferromagnetic Relaxation. Mc Graw-Hill, New York (1964)
Stancil, D.D., Prabhakar, A.: Spin Waves: Theory and Applications. Springer Science, New York (2009)
White, R.M.: Quantum Theory of Magnetism, 3rd edn. Springer, Heidelberg (2007)

# Quantum Theory of Spin Waves: Magnons

3

The classical treatment of spin waves presented in the previous chapter provides only a limited view of their properties. Spin waves are quantum objects and several phenomena involving them require that they are treated as such. In this chapter, we present a quantum approach of spin waves based on the Holstein–Primakoff transformation from the spin operators into magnon operators. This formalism is not restricted to low temperatures and, as we will show in the rest of the book, it can be used to explain quantitatively several magnonic phenomena observed experimentally. Next, we present the properties of magnons, and show that coherent magnon states are the quantum states that describe classical spin waves. The three- and four-magnon interactions, that are conveniently treated with the Holstein–Primakoff formalism, are used to calculate magnon relaxation mechanisms and energy renormalization. The quantum treatments of magnetoelastic waves, magnetic polaritons, and inelastic light scattering are also presented. The last section is devoted to details of magnons in yttrium iron garnet and the calculation of their contributions to the thermodynamic properties.

## 3.1 Ferromagnetic Magnons

### 3.1.1 Bloch Spin Waves

The quantization of spin waves is a natural consequence of the fact that the atomic spin angular momentum is quantized. We begin the quantum treatment with a short review of the properties of the spin operators and their eigenstates. The Cartesian components of the spin operator $\vec{S}_i$, at site $i$, and in units of $\hbar$, satisfy the commutation relations

© Springer Nature Switzerland AG 2020                                    71
S. M. Rezende, *Fundamentals of Magnonics*, Lecture Notes in Physics 969,
https://doi.org/10.1007/978-3-030-41317-0_3

$$[S_i^\alpha, S_j^\beta] = i\varepsilon_{\alpha\beta\gamma}\,\delta_{ij}\,S_i^\gamma, \tag{3.1}$$

where $\alpha$, $\beta$, $\gamma$ stand for the coordinates $x$, $y$, $z$, and $\varepsilon_{\alpha\beta\gamma} = \pm 1$ is the Levi-Civita antisymmetric tensor. Its elements are $+1$ for $\alpha \neq \beta \neq \gamma$ in the order $x$, $y$, $z$ or its cyclic permutation, $-1$ for $\alpha \neq \beta \neq \gamma$ in the reverse order, and zero otherwise. The $z$ component of $\vec{S}_i$ has eigenstates $|S_i^z\rangle$, where $S_i^z$ has integer values from $-S$ to $+S$, that obey the eigenvalue equations

$$S_i^z|S_i^z\rangle = S_i^z\,|S_i^z\rangle, \qquad (\vec{S}_i \cdot \vec{S}_i)|S_i^z\rangle = S(S+1)\,|S_i^z\rangle. \tag{3.2}$$

Introduce the raising (+) and lowering (−) spin operators

$$S_i^\pm = S_i^x \pm iS_i^y, \tag{3.3}$$

which satisfy the commutation relations

$$[S_i^\pm, S_j^z] = \mp S_i^\pm \delta_{ij}, \qquad [S_i^+, S_j^-] = 2S_i^z \delta_{ij}. \tag{3.4}$$

Using Eqs. (3.3 and 3.4), it can be readily shown that the operators $S_i^+$ act on the eigenstates to give

$$S_i^\pm|S_i^z\rangle = [S(S+1) - S_i^z(S_i^z \pm 1)]^{1/2}\,|S_i^z \pm 1\rangle. \tag{3.5}$$

As we will see in the next subsection, this is an essential relation for the introduction of the magnon operators. Consider now a simple ferromagnet with Zeeman and exchange energies only. The exchange interaction is further assumed to be isotropic and nonzero only for nearest neighbors. From the energies in Eqs. (1.4 and 2.1) we can write the spin Hamiltonian as

$$\mathrm{H} = -g\mu_B \sum_i H_z S_i^z - J \sum_{i,\delta} \vec{S}_i \cdot \vec{S}_{i+\delta}, \tag{3.6}$$

where $\vec{S}_i$ is spin angular momentum operators at site $i$, $H_z$ is the internal static magnetic field in the $z$-direction of a Cartesian coordinate system, and $\vec{\delta}$ is the vector connecting site $i$ with its nearest neighbors. Notice in the sign of the Zeeman energy that we have assumed the spin to have the same direction of the magnetic moment to avoid the awkward situation of Fig. 2.1, where the spins point in a direction opposite to the field. Since the negative charge of the electron is a convention, the choice of a positive charge does not alter the basic results. We have written the Zeeman energy in the CGS to avoid carrying the vacuum permeability $\mu_0$ in the equations. Notice also that the factor 2 in the exchange energy does not appear explicitly because each pair of spins is counted twice in the sum over lattice sites. Introducing in Eq. (3.6) the raising and lowering spin operators, the Hamiltonian becomes

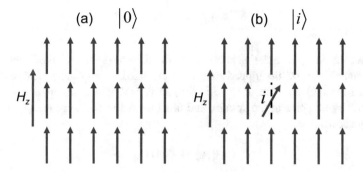

**Fig. 3.1** (a) Illustration of the ground state of a ferromagnetic spin system with Zeeman and exchange interactions. (b) State with one localized spin deviation at site $i$

$$H = -g\mu_B \sum_i H_z S_i^z - \frac{J}{2} \sum_{i,\delta} (S_i^+ S_{i+\delta}^- + S_i^- S_{i+\delta}^+ + 2S_i^z S_{i+\delta}^z), \quad (3.7)$$

which, with the commutation relations in Eq. (3.4), gives

$$H = -g\mu_B \sum_i H_z S_i^z - J \sum_{i,\delta} (S_i^- S_{i+\delta}^+ + S_i^z S_{i+\delta}^z), \quad (3.8)$$

We denote by $|0\rangle = |S_1^z = S, S_2^z = S, \ldots \ldots S_N^z = S\rangle$ the state pictured in Fig. 3.1a. Clearly, this is an eigenstate of the Hamiltonian, i.e.,

$$H|0\rangle = E_0|0\rangle, \quad E_0 = -g\mu_B H_z NS - 2zJNS^2, \quad (3.9)$$

where $z$ is the number of nearest neighbors and $N$ the number of spins in the system. Obviously, $|0\rangle$ is also the state of minimum energy. In an attempt to find the first excited state of the system, one may think on the state illustrated in Fig. 3.1b, that has one spin deviation at site $i$, represented by

$$|i\rangle = |S, S, S, \ldots \ldots S_i^z = S - 1, \ldots \ldots S, S, S\rangle = \frac{1}{\sqrt{2S}} S_i^- |0\rangle. \quad (3.10)$$

However, it is easy to see that this is not an eigenstate of the Hamiltonian in Eq. (3.7), because the term $S_i^- S_{i+\delta}^+$ flips the spin at site $i$ down and its nearest neighbor up. The actual solution for the first excited eigenstate is obtained with a linear combination of states of the type in Eq. (3.10). Due to the translation symmetry of the system this state turns out to be

$$|1k\rangle = \frac{1}{\sqrt{2SN}} \sum_i e^{i\vec{k} \cdot \vec{r}} S_i^- |0\rangle. \quad (3.11)$$

In order to understand the physical meaning of the state in Eq. (3.11), use the operator $S_i^z$ to find the number of spin deviations. We have

$$\sum_i S_i^z |1k\rangle = (NS - 1)|1k\rangle. \tag{3.12}$$

Thus, the net spin deviation is one unit. Although the spin deviation is the same as in the state $|i\rangle$, in $|1k\rangle$ the flipped spin is not localized, it propagates. The probability that in state $|1k\rangle$ the spin at site $i$ is flipped is $1/N$, which means that the flip is "shared" by all spins in the system, and this lowers the exchange energy. It can be readily shown that

$$H|1k\rangle = E_{1k}|1k\rangle, \tag{3.13a}$$

where the energy is

$$E_{1k} = E_0 + g\mu_B H_z + 2zJS(1 - \gamma_k), \tag{3.13b}$$

and

$$\gamma_k = \frac{1}{z} \sum_i e^{i\vec{k}\cdot\vec{\delta}} \tag{3.13c}$$

is a geometric factor that depends on the crystal structure. The state $|1k\rangle$ represents one magnon with energy $E_{1k}$ and momentum $\hbar\vec{k}$. It is the Bloch spin wave [1]. If one attempts to find the state of two magnons by applying the operator $\exp(i\vec{k} \cdot \vec{r}_i)S_i^-$ twice to the ground state, the result is not an eigenstate of the Hamiltonian. This can be interpreted as due to the interaction between two states of the type in Eq. (3.11). Hence, in order to understand the excitations of the system when more than one spin deviation exist, it is necessary to use a different formalism. The most convenient method to study spin waves is the one developed by Holstein and Primakoff (HP) [2], presented in the next subsection. Some authors state that this method is valid only at low temperatures. Actually, as we will show later, by introducing magnon interactions, the HP formalism is valid at quite high temperatures and explains quantitatively several experimental data.

### 3.1.2   Magnon Operators

In order to study the collective quantum excitations of the ferromagnetic spin system it is desirable to use a technique that can treat a system with many quanta and also is easy to generalize to more complicated Hamiltonians. The so-called second quantization used in many-body problems constitutes the technique we want. The basic idea is to look for canonical transformations that cast the Hamiltonian in the form

$$H = E_0 + \sum_k \hbar\omega_k a_k^\dagger a_k + H_{int},\tag{3.14}$$

i.e., the sum of a ground state energy, a term that represents a collection of independent harmonic oscillator type of excitations, and a term that represents the
interactions among these excitations. Provided that the relaxation times of the
excitations due to their interactions are much longer than the periods of oscillations,
the concept of excitation is meaningful and the form (3.14) is perfectly good. The
statistics of the excitations is crucial in determining the properties of the spin wave
operators. Magnons behave as bosons, therefore, the operators we want must obey
the boson commutation relations

$$[a_k, a_{k'}^\dagger] = \delta_{kk'}, \quad [a_k, a_{k'}] = 0, \quad [a_k^\dagger, a_{k'}^\dagger] = 0.\tag{3.15}$$

The starting point to find the magnon operators is the first HP transformation that
introduces the spin-deviation operators. The operator for the number of spin
deviations at site $i$ is

$$n_i = S - S_i^z.\tag{3.16}$$

Clearly the eigenstates of this operator are the states $\left|S_i^z\right\rangle$. The spin-deviation
creation operator $a_i^\dagger$ is the operator that creates a quantum of spin deviation, i.e.,
which reduces $S_i^z$ by one unit. Likewise, the spin-deviation annihilation operator $a_i$
increases $S_i^z$ by one unit. Denoting by $\left|n_i\right\rangle \equiv \left|S_i^z\right\rangle$ the eigenstates of $n_i = a_i^\dagger a_i$, we
require them to have the harmonic oscillator type of properties

$$a_i^\dagger\left|n_i\right\rangle = (n_i + 1)^{1/2}\left|n_i + 1\right\rangle,\tag{3.17a}$$

$$a_i\left|n_i\right\rangle = (n_i)^{1/2}\left|n_i - 1\right\rangle,\tag{3.17b}$$

and also that they satisfy the Bose commutation relations

$$[a_i, a_j^\dagger] = \delta_{ij}, \quad [a_i, a_j] = 0, \quad [a_i^\dagger, a_j^\dagger] = 0.\tag{3.18}$$

Now we rewrite Eq. (3.5) with the help of (3.16)

$$S_i^-\left|n_i\right\rangle = (2S)^{1/2}(n_i + 1)^{1/2}\left(1 - \frac{n_i}{2S}\right)^{1/2}\left|n_i + 1\right\rangle,\tag{3.19a}$$

$$S_i^+\left|n_i\right\rangle = (2S)^{1/2}\left(1 - \frac{n_i - 1}{2S}\right)^{1/2}(n_i)^{1/2}\left|n_i - 1\right\rangle.\tag{3.19b}$$

Note that the $n_i$ inside the parentheses in Eqs. (3.19a and 3.19b) are the
eigenvalues. However, as they appear in front of the eigenstates $\left|n_i\right\rangle$, they can be

replaced by the operators $n_i$. Comparison of Eqs. (3.19a and 3.19b) with (3.17a and 3.17b) leads to the relations between the spin and the spin-deviation operators.

$$S_i^+ = (2S)^{1/2} \left( 1 - \frac{a_i^\dagger a_i}{2S} \right)^{1/2} a_i, \tag{3.20a}$$

$$S_i^- = (2S)^{1/2} a_i^\dagger \left( 1 - \frac{a_i^\dagger a_i}{2S} \right)^{1/2}, \tag{3.20b}$$

$$S_i^z = S - a_i^\dagger a_i = S - n_i. \tag{3.20c}$$

It is not difficult to show that the transformations (3.20a–3.20c) satisfy the commutation relations in Eqs. (3.4 and 3.18). For the transformations (3.20a–3.20c) to be useful, we expand the operators in the square roots as follows:

$$S_i^+ = (2S)^{1/2} (a_i - \frac{1}{4S} a_i^\dagger a_i a_i + \ldots .), \tag{3.21a}$$

$$S_i^- = (2S)^{1/2} (a_i^\dagger - \frac{1}{4S} a_i^\dagger a_i^\dagger a_i + \ldots .). \tag{3.21b}$$

Clearly, we want to approximate the expansions by the lowest order terms. This is valid only if $\langle n_i \rangle / 2S \ll 1$, i.e., if the expectation value of the number of deviations in spin $\vec{S}_i$ is much smaller than the maximum possible value, which is $2S$. We have shown that one magnon in a system of $N$ spins corresponds to a local spin deviation of $\langle n_i \rangle = 1/N$. Thus, the criterion for the validity of the expansion in Eqs. (3.21a and 3.21b) is that the number of magnons is such that $n_k \ll 2SN$. This condition is comfortably satisfied at low temperatures, $T \ll T_c$. However, as we will show later, by taking into account the second-order terms in Eqs. (3.21a and 3.21b), we can extend the validity of the spin wave theory to higher temperatures.

The first term in (3.21a and 3.21b) will give origin to the quadratic part of the Hamiltonian. The others will result in terms with three, four, or more operators, which represent the interactions between the harmonic oscillator type states. The interactions will be treated later in this chapter. Here we restrict ourselves to the terms that give the quadratic part of the Hamiltonian, namely

$$S_i^+ = (2S)^{1/2} a_i, \tag{3.22a}$$

$$S_i^- = (2S)^{1/2} a_i^\dagger, \tag{3.22b}$$

$$S_i^z = S - a_i^\dagger a_i. \tag{3.22c}$$

Using the transformations (3.22a–3.22c) in Eq. (3.8), we obtain for the quadratic part of the Hamiltonian

$$H^{(2)} = \sum_{i,j=i+\delta} A_{ij} a_i^\dagger a_j, \tag{3.23}$$

where

$$A_{ij} = (g\mu_B H_z + 2zJS)\delta_{ij} - 2JS. \tag{3.24}$$

Equation (3.23) is still not in the diagonal form of Eq. (3.14). It is necessary to use a transformation from the local field operators $a_i^\dagger$ and $a_i$ to normal mode collective creation and annihilation operators $a_k^\dagger$ and $a_k$. Here we use a transformation that is more general than the one of Holstein–Primakoff

$$a_i = \sum_k \phi_k^i a_k, \tag{3.25a}$$

$$a_k^\dagger = \sum_k \phi_k^{i*} a_k^\dagger, \tag{3.25b}$$

which substituted in the Hamiltonian (3.23) leads to the form

$$H^{(2)} = \sum_{i,j} \sum_{k,k'} A_{ij} \phi_k^{i*} \phi_{k'}^j \, a_k^\dagger a_{k'}. \tag{3.26}$$

This expression acquires the desired diagonal form

$$H^{(2)} = \sum_k E_k a_k^\dagger a_k, \tag{3.27}$$

with the operators satisfying the boson commutation relations in Eqs. (3.15), if the eigenfunctions satisfy the eigenvalue equation

$$E_k \phi_k^i = \sum_j A_{ij} \phi_k^j. \tag{3.28}$$

From Eqs. (3.27 and 3.28) it can be shown that the eigenfunctions form a complete set and satisfy the orthonormality conditions

$$\sum_i \phi_k^i \phi_{k'}^{i*} = \delta_{kk'}, \tag{3.29a}$$

$$\sum_i \phi_k^i \phi_k^{j*} = \delta_{ij}. \tag{3.29b}$$

The more general transformations in Eqs. (3.25a and 3.25b) should be used to diagonalize the quadratic Hamiltonian when either $H_z$ or any other parameter varies in space. In the case of translational symmetry, $\phi_k^i = \exp(i\vec{k} \cdot \vec{r})$ is a solution of

(3.28) so that Eqs. (3.25a and 3.25b) reduce to the form used by HP, namely, a Fourier transformation

$$a_i = N^{-1/2} \sum_k e^{i\vec{k}\cdot\vec{r}_i} a_k, \qquad a_i^\dagger = N^{-1/2} \sum_k e^{-i\vec{k}\cdot\vec{r}_i} a_k^\dagger. \tag{3.30}$$

The inverse transformation is

$$a_k = N^{-1/2} \sum_k e^{-i\vec{k}\cdot\vec{r}_i} a_i, \qquad a_i^\dagger = N^{-1/2} \sum_k e^{i\vec{k}\cdot\vec{r}_i} a_i^\dagger. \tag{3.31}$$

The completeness and orthogonality relations of the eigenfunctions become

$$\sum_i e^{i(\vec{k}-\vec{k}')\cdot\vec{r}_i} = N\delta_{kk'}, \qquad \sum_i e^{i\vec{k}\cdot(\vec{r}_i-\vec{r}_j)} = N\delta_{ij}. \tag{3.32}$$

The energy eigenvalues can be obtained directly from Eqs. (3.28 and 3.30). Instead of using this procedure, we follow other textbooks and use the transformation (3.30) to replace the local operators in the Hamiltonian (3.23) to obtain

$$H^{(2)} = \sum_{i,j} A_{ij} a_i^\dagger a_j = \sum_{i,j} \sum_{k,k'} \frac{1}{N} A_{ij} e^{-i\vec{k}\cdot\vec{r}_i} a_k^\dagger e^{-i\vec{k}'\cdot\vec{r}_j} a_{k'}$$

$$= \sum_{k,k'} \sum_{i,j=i+\delta} \frac{1}{N} A_{ij} e^{-i(\vec{k}-\vec{k}')\cdot\vec{r}_i} e^{i\vec{k}'\cdot(\vec{r}_i-\vec{r}_j)} a_k^\dagger a_{k'}, \tag{3.33}$$

which, with the use of the orthonormality relation in Eq. (3.32) gives

$$\sum_\delta A_{i,i+\delta} \sum_{k,k'} e^{-i\vec{k}'\cdot\vec{\delta}} a_k^\dagger a_{k'} \delta_{kk'} = \sum_k \sum_\delta \left( A_{i,i+\delta} e^{-i\vec{k}\cdot\vec{\delta}} \right) a_k^\dagger a_k = \sum_k A_k\, a_k^\dagger a_k.$$

Thus, the quadratic Hamiltonian in Eq. (3.33) has the expected form

$$H^{(2)} = \sum_k E_k a_k^\dagger a_k, \tag{3.34}$$

where, with the use of (3.24), the energy eigenvalue follows

$$E_k = A_k = g\mu_B H_z + 2zJS(1 - \gamma_k), \tag{3.35}$$

where $\gamma_k$ is the structure factor given by

$$\gamma_k = \frac{1}{z} \sum_\delta e^{i\vec{k}\cdot\vec{\delta}}, \tag{3.36}$$

and $z$ is the number of nearest neighbors. In the case of a simple cubic crystal with lattice parameter $a$ and Cartesian coordinates along $<100>$ axes, this becomes

$$\gamma_k = \frac{1}{3}\left(\cos k_x a + \cos k_y a + \cos k_z a\right). \tag{3.37}$$

The eigenstates of the Hamiltonian (3.34) are the magnon states. The state with one magnon with wave vector $\vec{k}$ generated by the application of the creation operator $a_k^\dagger$ to the vacuum state is

$$|1k\rangle = a_k^\dagger |0\rangle. \tag{3.38}$$

Using Eqs. (3.31, 3.22a and 3.22b) one can show that the state in (3.38) is the same as the Bloch state in Eq. (3.11). The energy in Eq. (3.35) is the same as the one obtained for the Bloch spin wave. The advantage of using the HP formalism is that other contributions to the energy can be introduced systematically, as we shall see in the next section. Equation (3.35) represents the energy of one magnon, thus, it can be written as $E_k = \hbar\omega_k$, so that the magnon frequency is

$$\omega_k = \gamma H_z + \frac{2zJS}{\hbar}(1 - \gamma_k), \tag{3.39}$$

where $\gamma = g\mu_B/\hbar$ is the gyromagnetic ratio. Figure 3.2 shows the magnon dispersion curve for the wave vector along a $<111>$ axis. At the center of the Brillouin zone ($k = 0$), $\gamma_k = 1$ so that the frequency is $\omega_0 = \gamma H_z$. At the zone boundary edge $k_x a = k_y a = k_z a = \pi/2$, $\gamma_k = 0$, and the frequency is dominated by the exchange energy $\omega_{ZB} \approx 2zJS/\hbar$.

**Fig. 3.2** Magnon dispersion curve in the first Brillouin zone, for a simple ferromagnet with wave vector along a $<111>$ axis of a simple cubic crystal

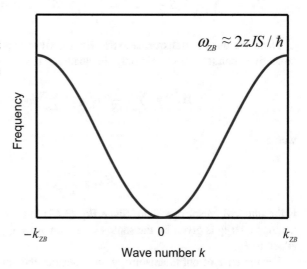

For magnons with wave vector near the BZ center, $ka \ll 1$, $\gamma_k \approx 1 - (ka)^2/z$, and the dispersion relation (3.39) can be written approximately as

$$\omega_k \approx \gamma\left(H_z + Dk^2\right), \tag{3.40}$$

where $D = 2JSa^2/\gamma\hbar$ is the exchange parameter, consistently with the result (2.24) obtained with the semiclassical treatment of long-wavelength spin waves.

As we shall study in the next section, the Holstein–Primakoff formalism can be applied in a systematic manner for ferromagnets with other interaction energies that are important in several magnonic phenomena.

## 3.2   Magnons with Anisotropy and Dipolar and Interactions

Consider initially the effect of the magnetocrystalline anisotropy on the magnon energy that is simpler to treat than the dipolar interaction. We take first the uniaxial anisotropy that has energy density given by Eq. (1.54). Considering only the first-order contribution, the uniaxial anisotropy Hamiltonian becomes

$$H_A^u = -\sum_i \frac{K_{u1}}{S^2}\left(S_i^z\right)^2, \tag{3.41}$$

where we have considered that the static field is along the axis of the anisotropy. In order to express this Hamiltonian in terms of the operators $a_k$ and $a_k^\dagger$, we first substitute the transformation in Eq. (3.22c) into (3.41)

$$H_A^u = -\sum_i \frac{K_{u1}}{S^2}\left(S - a_i^\dagger a_i\right)^2 = \sum_i \frac{K_{u1}}{S^2}\left(S^2 - 2Sa_i^\dagger a_i + a_i^\dagger a_i a_i^\dagger a_i\right). \tag{3.42}$$

Next, we use the transformation in Eq. (3.30) to express (3.42) in terms of the collective operators and retain only the quadratic term

$$H_A^{u(2)} = \sum_k \frac{2K_{u1}}{S} a_k^\dagger a_k = \sum_k g\mu_B H_{Au} a_k^\dagger a_k, \tag{3.43}$$

where

$$H_{Au} = \frac{2K_{u1}}{g\mu_B S} \tag{3.44}$$

is the uniaxial anisotropy field. Since Eq. (3.43) has the same form of (3.34), the magnon energy is given by the same expression as (3.35), with the anisotropy field added to $H_z$.

For the case of cubic anisotropy, we consider the energy density in Eq. (1.55) with only the first-order term. Since the components of the spin operators do not commute, the Hamiltonian is cast in the form

$$H_A^c = \sum_i \frac{K_{c1}}{2S^4} \left( S_i^{x2} S_i^{y2} + S_i^{y2} S_i^{z2} + S_i^{z2} S_i^{x2} + S_i^{y2} S_i^{x2} + S_i^{z2} S_i^{y2} + S_i^{x2} S_i^{z2} \right). \quad (3.45)$$

This is treated similarly to the uniaxial anisotropy. The components of the spin operators are replaced by the operators $a_i$ and $a_i^\dagger$ with Eqs. (3.22a–3.22c), and these are substituted by the magnon operators $a_k$ and $a_k^\dagger$ with the transformation (3.30). Assuming that the internal field is in the direction of either a $<100>$ or a $<111>$ axis, the only term quadratic in the magnon operators has the same form as the Hamiltonian in Eq. (3.43), with anisotropy field, that is, depending on the direction of the applied field, either $H_{A100} = 2K_{c1}/M$ or $H_{A111} = -4K_{c1}/3M$.

The treatment of the dipolar interaction is quite more involved, for two reasons. First, the interaction between the spins is of long range. Second, the Hamiltonian has other types of terms quadratic in magnon operators, so that an additional transformation is required to diagonalize the Hamiltonian. The Hamiltonian for the dipole–dipole interaction between the spins is

$$H_{dip} = \frac{1}{2} (g\mu_B)^2 \sum_{ij} \left[ \frac{\vec{S}_i \cdot \vec{S}_j}{r_{ij}^3} - \frac{3(\vec{S}_i \cdot \vec{r}_{ij})(\vec{S}_j \cdot \vec{r}_{ij})}{r_{ij}^5} \right], \quad (3.46)$$

where $\vec{r}_{ij} = \vec{r}_i - \vec{r}_j$. Using the transformations in (3.22a–3.22c) to replace the spin operators by spin deviation operators and keeping only terms with two operators we obtain

$$H_{dip}^{(2)} = \frac{1}{2} (g\mu_B)^2 S \sum_{ij} \frac{1}{r_{ij}^3} \left[ -2 \left( 1 - \frac{3z_{ij}^2}{r_{ij}^2} \right) a_i^\dagger a_i + \left( 1 - \frac{3}{2} \frac{r_{ij}^+ r_{ij}^-}{r_{ij}^2} \right) a_i^\dagger a_j \right.$$
$$\left. + \left( 1 - \frac{3}{2} \frac{r_{ij}^+ r_{ij}^-}{r_{ij}^2} \right) a_i a_j^\dagger - \frac{3}{2} \frac{(r_{ij}^-)^2}{r_{ij}^2} a_i a_j - \frac{3}{2} \frac{(r_{ij}^+)^2}{r_{ij}^2} a_i^\dagger a_j^\dagger \right] \quad (3.47)$$

The next steps consist in using transformations (3.30) to replace the operators $a_i, a_i^\dagger$ by $a_k, a_k^\dagger$, and use the orthonormality relation (3.32) that results from the sum over the lattice sites $i$ to eliminate one of the sums over wave vectors. The quadratic dipolar Hamiltonian then becomes

$$H_{dip}^{(2)} = \frac{1}{2} (g\mu_B)^2 S \sum_k \sum_r \frac{1}{r^3} \left[ -2 \left( 1 - \frac{3z^2}{r^2} \right) a_k^\dagger a_k + 2e^{i\vec{k}\cdot\vec{r}} \left( 1 - \frac{3}{2} \frac{r^+ r^-}{r^2} \right) a_k^\dagger a_k \right.$$
$$\left. - \frac{3}{2} e^{i\vec{k}\cdot\vec{r}} \frac{(r^-)^2}{r^2} a_k a_{-k} - \frac{3}{2} e^{-i\vec{k}\cdot\vec{r}} \frac{(r^+)^2}{r^2} a_k^\dagger a_{-k}^\dagger \right], \quad (3.48)$$

where $\vec{r} = \vec{r}_i - \vec{r}_j$. The sum over $\vec{r}$ now is broken in two parts, one inside a sphere with radius much larger than the lattice parameter, and another outside that is approximated by a volume integral $(N/V) \int d\vec{r}$. This is convenient because for long wavelength spin waves, the dipolar sum within the sphere is zero for a cubic

**Fig. 3.3** Coordinates used to express the magnon wave vector

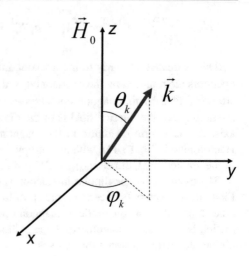

crystal. For nonzero wave vectors, the integrals of the functions of $\vec{r}$ in Eq. (3.48) are [3]

$$D^{\alpha\beta}(\vec{k}) = \int d\vec{r} \frac{1}{r^3} e^{i\vec{k}\cdot\vec{r}} \left(3\frac{r^\alpha r^\beta}{r^2} - \delta_{\alpha\beta}\right) = \frac{4\pi}{3} - 4\pi \frac{k^\alpha k^\beta}{k^2}, \qquad (3.49)$$

where $\alpha, \beta, \gamma$ stand for the coordinates $x, y, z$ and also for $x \pm iy$. With this result, we obtain for the quadratic dipolar Hamiltonian

$$H_{dip}^{(2)} = \frac{1}{2} g\mu_B \sum_k [4\pi M \sin^2\theta_k a_k^\dagger a_k + (2\pi M \sin^2\theta_k e^{-i2\varphi_k} a_k a_{-k} + H.c.)], \qquad (3.50)$$

where $M = g\mu_B SN/V$ is the static magnetization, $\theta_k$ and $\varphi_k$ are the polar and azimuthal angles of the wave vector shown in Fig. 3.3. Note that in Eq. (3.50) we have not considered the contribution from the sample surfaces because we assume that the wavelength is much smaller that the sample dimensions and that the demagnetizing effect in the $z$-direction is already taken into account in $H_z$. With Eqs. (3.35, 3.43, and 3.50), the quadratic Hamiltonian becomes

$$H^{(2)} = \sum_k A_k a_k^\dagger a_k + \frac{1}{2} B_k a_k a_{-k} + \frac{1}{2} B_k^* a_k^\dagger a_{-k}^\dagger, \qquad (3.51a)$$

$$A_k = g\mu_B H_z + 2zJS(1 - \gamma_k) + g\mu_B 2\pi M \sin^2\theta_k, \qquad (3.51b)$$

$$B_k = g\mu_B 2\pi M \sin^2\theta_k e^{-i2\varphi_k}, \qquad (3.51c)$$

where $H_z = H_0 - N_z 4\pi M + H_A$ includes the effects of the external field, demagnetization and effective anisotropy.

Due to the presence of the terms with $B_k$, the Hamiltonian (3.50) is no longer in the diagonal form. This means that the dipolar interaction not only changes the energy of the magnon but also changes its character. We must then look for new normal mode operators to cast the quadratic Hamiltonian in the diagonal form. In order to find the appropriate transformation, we use the equations of motion for the operators $a_k$ and $a_k^\dagger$ with the Heisenberg equation $i\hbar(dA/dt) = [A, H]$. Using the Hamiltonian (3.51a–3.51c) and the commutation relations (3.15) we have

$$i\hbar \frac{da_k}{dt} = A_k a_k + B_k^* a_{-k}^\dagger, \tag{3.52a}$$

$$i\hbar \frac{da_{-k}^\dagger}{dt} = A_k a_{-k}^\dagger + B_k a_k. \tag{3.52b}$$

Note that if $B_k = 0$ we would have $a_k(t) = a_k(0) \exp(-A_k t/\hbar)$, so that $a_k$ and $a_{-k}^\dagger$ would be normal mode operators. Equations (3.52a and 3.52b) suggest that we look for new operators $c_{-k}^\dagger$ and $c_k$ which diagonalize the Hamiltonian (3.51a–3.51c). Thus we make

$$a_k = u_k c_k - v_k c_{-k}^\dagger, \tag{3.53a}$$

$$a_{-k}^\dagger = u_{-k} c_{-k}^\dagger - v_{-k}^* c_k, \tag{3.53b}$$

which is a Bogoliubov transformation. Since the eigenstates of the new Hamiltonian must be bosons, we require that the new operators satisfy the commutation relations

$$[c_k, c_{k'}^\dagger] = \delta_{kk'}, \qquad [c_k, c_{k'}] = 0, \qquad [c_k^\dagger, c_{k'}^\dagger] = 0. \tag{3.54}$$

Transformation of the type (3.53a and 3.53b) between operators that have the same commutation relations are called canonical. Upon substitution of Eqs. (3.53a and 3.53b) into (3.51a–3.51c) and imposing that the off-diagonal terms with the new operators vanish, we can determine the coefficients $u_k$, $u_{-k}$, $v_k$, $v_{-k}$ and the new energy. Instead of doing this, we prefer at this point to indicate a more general procedure to use the Bogoliubov transformation to diagonalize quadratic Hamiltonians [4] that will be useful later. Most quadratic Hamiltonians involving one or more operators can be written as

$$H = \sum_{k>0} H_k, \tag{3.55}$$

where $H_k$ is a product of matrices of the type

$$H_k = (X)^\dagger [H](X), \tag{3.56}$$

where $(X)$ is a column matrix with elements taken from the operators that appear in H, and [H] is a square matrix formed by the coefficients of H. In the present case we have

$$(X) = \begin{pmatrix} a_k \\ a_{-k}^{\dagger} \end{pmatrix}, \quad [H] = \begin{bmatrix} A_k & B_k^* \\ B_k & A_k \end{bmatrix}. \tag{3.57}$$

With Eq. (3.57), the matrix product on the right-hand side of Eq. (3.56) gives

$$H_k = (X)^{\dagger}[H](X) = A_k a_k^{\dagger} a_k + A_k a_{-k}^{\dagger} a_{-k} + B_k a_k a_{-k} + B_k^* a_k^{\dagger} a_{-k}^{\dagger}. \tag{3.58}$$

The commutation relations of the operators involved can also be written in matrix form

$$[X, X^{\dagger}] = X(X^*)^{\mathrm{T}} - (X^* X^T)^{T} = [g], \tag{3.59}$$

where $X(X^*)^{\mathrm{T}} = X(X)^{\dagger}$ is a direct product of the matrices. In our case

$$[g] = \begin{pmatrix} a_k \\ a_{-k}^{\dagger} \end{pmatrix} \begin{pmatrix} a_k^{\dagger} & a_{-k} \end{pmatrix} - \left\{ \begin{pmatrix} a_k^{\dagger} \\ a_{-k} \end{pmatrix} \begin{pmatrix} a_k & a_{-k}^{\dagger} \end{pmatrix} \right\},$$

which gives

$$[g] = \begin{bmatrix} a_k a_k^{\dagger} - a_k^{\dagger} a_k & a_k a_{-k} - a_{-k} a_k \\ a_{-k}^{\dagger} a_k^{\dagger} - a_k^{\dagger} a_{-k}^{\dagger} & a_{-k}^{\dagger} a_k - a_{-k} a_{-k}^{\dagger} \end{bmatrix} = \begin{bmatrix} 1 & 0 \\ 0 & -1 \end{bmatrix}. \tag{3.60}$$

The next step consists in expressing the new operators that diagonalize the Hamiltonian (3.56) in matrix form

$$(Y) = \begin{pmatrix} c_k \\ c_{-k}^{\dagger} \end{pmatrix}, \quad [Y, Y^{\dagger}] = [g'], \tag{3.61}$$

which should be such that (3.58) becomes

$$H_k = (Y)^{\dagger}[\Omega_H](Y), \quad [\Omega_H] = \hbar \begin{bmatrix} \omega_1 & 0 \\ 0 & \omega_2 \end{bmatrix} \tag{3.62}$$

where $[\Omega_H]$ is a diagonal matrix that contains the desired normal mode frequencies. The transformation of the Hamiltonian to a diagonal form is made possible by a canonical transformation of the operators

$$(X) = [S](Y). \tag{3.63}$$

In order to find [S] and $[\Omega_H]$ we substitute (3.63) into (3.56) and use (3.61)

$$H_k = (X)^\dagger[H](X) = (Y)^\dagger[S]^\dagger[H][S](Y) = (Y)^\dagger[\Omega_H](Y),$$

which leads to

$$[\Omega_H] = [S]^\dagger[H][S]. \tag{3.64}$$

Note that usually the transformation (3.63) is not unitary and therefore (3.64) does not reduce to the usual matrix diagonalization problem. However, (3.64) can be solved more easily by means of a relation between $[S]^\dagger$ and $[S]^{-1}$ derived as follows

$$[X, X^\dagger] = [SY, Y^\dagger S^\dagger] = [SY][Y^\dagger S^\dagger] - [S^* Y^* Y^T S^T]^T = [g].$$

Since $[S]$ is a c-number matrix, this can be written as

$$[S]\{YY^\dagger - [Y^* Y^T]^T\}[S]^\dagger = [S][g'][S]^\dagger.$$

Therefore,

$$[S][g'][S]^\dagger = [g], \quad \text{and} \quad [S]^{-1} = [g'][S]^\dagger[g]^{-1}. \tag{3.65}$$

Using (3.65) in (3.64) we have

$$[H][S] = [g]^{-1}[S][g'][\Omega_H]. \tag{3.66}$$

This is a set of eigenvalue equations that can be solved by standard techniques to give $[S]$ and $[\Omega_H]$.

In the present case, the use of this general procedure does not save any work. However, for more complicated Hamiltonians, such as in antiferromagnetic magnons, it is very useful and has been largely used. We use it here for the sake of training. From the previous discussion, we see that for the Hamiltonian (3.51a–3.51c) the relevant matrices are

$$(X) = \begin{pmatrix} a_k \\ a_{-k}^\dagger \end{pmatrix}, \quad [H] = \begin{bmatrix} A_k & B_k^* \\ B_k & A_k \end{bmatrix}, \quad (Y) = \begin{pmatrix} c_k \\ c_{-k}^\dagger \end{pmatrix}, \tag{3.67}$$

and

$$[g] = [g'] = [g]^{-1} = \begin{bmatrix} 1 & 0 \\ 0 & -1 \end{bmatrix}. \tag{3.68}$$

Using

$$[\Omega_H] = \hbar \begin{bmatrix} \omega_1 & 0 \\ 0 & \omega_2 \end{bmatrix} \quad \text{and} \quad [S] = \begin{bmatrix} S_{11} & S_{12} \\ S_{21} & S_{22} \end{bmatrix} \tag{3.69}$$

in the eigenvalue Eq. (3.66), we obtain the following set

$$(A_k + \hbar\omega_1)S_{11} + B_k^* S_{21} = 0, \tag{3.70a}$$

$$B_k S_{11} + (A_k - \hbar\omega_1)S_{21} = 0. \tag{3.70b}$$

They give for the energy

$$\hbar\omega_k = (A_k^2 - |B_k|^2)^{1/2}. \tag{3.71}$$

Using Eq. (3.65) we also find

$$S_{11} = S_{22}^* \equiv u_k = u_k^*, \qquad S_{12} = S_{21}^* \equiv -v_k, \tag{3.72a}$$

and

$$\det[S] = u_k^2 - |v_k|^2 = 1. \tag{3.72b}$$

These equations, together with (3.70a and 3.70b), give the coefficients of the canonical transformation

$$u_k = \left(\frac{A_k + \hbar\omega_k}{2\hbar\omega_k}\right)^{1/2}, \qquad v_k = \left(\frac{A_k - \hbar\omega_k}{2\hbar\omega_k}\right)^{1/2} e^{i2\varphi_k}. \tag{3.73}$$

Equations (3.71 and 3.73) completely characterize the Hamiltonian

$$H^{(2)} = \sum_k \hbar\omega_k \, c_k^\dagger c_k \tag{3.74}$$

and the transformation between the operators

$$\begin{pmatrix} a_k \\ a_{-k}^\dagger \end{pmatrix} = \begin{bmatrix} u_k & -v_k \\ -v_k^* & u_k \end{bmatrix} \begin{pmatrix} c_k \\ c_{-k}^\dagger \end{pmatrix}, \qquad \begin{pmatrix} c_k \\ c_{-k}^\dagger \end{pmatrix} = \begin{bmatrix} u_k & v_k^* \\ v_k & u_k \end{bmatrix} \begin{pmatrix} a_k \\ a_{-k}^\dagger \end{pmatrix}. \tag{3.75}$$

Notice that if the dipolar interaction is neglected, $A_k = \hbar\omega_k$, $v_k = 0$, and then $c_k = a_k$, as expected.

Using for the parameters of the total Hamiltonian the expressions in Eqs. (3.51a–3.51c), Eq. (3.71) gives for the magnon energy

$$\hbar\omega_k = \hbar\gamma \left[H_z + \frac{2zJS}{\hbar\gamma}(1 - \gamma_k)\right]^{1/2} \left[H_z + \frac{2zJS}{\hbar\gamma}(1 - \gamma_k) + 4\pi M \sin^2\theta_k\right]^{1/2}. \tag{3.76}$$

For magnons with small wave vector, $ka \ll 1$, $\gamma_k \approx 1 - (ka)^2/z$, and the dispersion relation (3.76) agrees with the one obtained with the classical spin wave theory given by Eq. (2.36).

## 3.3 Properties of Magnons: Coherent Magnon States

### 3.3.1 Eigenstates of the Hamiltonian

The Hamiltonian (3.74) represents a collection of independent harmonic oscillators. Since $n_k = c_k^\dagger c_k$ is the number operator, the eigenstates of (3.74) are harmonic-oscillator type states, which can be represented by

$$|n_1, n_2, \ldots n_k \ldots n_N\rangle, \tag{3.77}$$

where $n_k$ is the number of magnons with wave vector $\vec{k}$ and energy $\hbar\omega_k$. The wave vector $\vec{k}$ assumes $N$ discrete values determined by the sample boundary conditions. These states form a complete orthogonal set. Magnons are bosons and therefore $n_k$ are integer numbers that can vary from zero to infinite. $c_k^\dagger$ and $c_k$ are, respectively, creation and annihilation magnon operators, which have the following properties

$$c_k^\dagger |n_1, n_2, \ldots n_k \ldots n_N\rangle = (n_k + 1)^{1/2} |n_1, n_2, \ldots n_k + 1 \ldots n_N\rangle, \tag{3.78a}$$

$$c_k |n_1, n_2, \ldots n_k \ldots n_N\rangle = (n_k)^{1/2} |n_1, n_2, \ldots n_k - 1 \ldots n_N\rangle, \tag{3.78b}$$

and, therefore, the number operator satisfies the eigenvalue relation

$$c_k^\dagger c_k |n_1, n_2, \ldots n_k \ldots n_N\rangle = n_k |n_1, n_2, \ldots n_k \ldots n_N\rangle. \tag{3.79}$$

The state with zero magnon at all wave vectors is the vacuum state, represented by $|0_1, 0_2, \ldots 0_k \ldots 0_N\rangle$. For simplicity, the state with $n_k$ magnons with wave vector $\vec{k}$ and zero magnon with any other wave vector is represented by $|n_k\rangle$. Using the relation (3.78a) we see that, starting from the vacuum state, $|n_k\rangle$ can be generated by the successive application of the creation operator so that the normalized state is

$$|n_k\rangle = \frac{1}{(n_k!)^{1/2}} (c_k^\dagger)^{n_k} |0\rangle. \tag{3.80}$$

Using the expression (3.79) in the Hamiltonian (3.74) we have

$$H^{(2)} |n_k\rangle = n_k \hbar\omega_k |n_k\rangle. \tag{3.81}$$

Thus, the states with well-defined number of magnons with wave vector $\vec{k}$ and energy $\hbar\omega_k$ are the eigenstates of the Hamiltonian (3.74). Using the relations (3.78a) and (3.78b), one can show that the states (3.77) are normalized and orthogonal to states with different $\vec{k}$, that is, $\langle n_k | n_{k'}\rangle = \delta_{k,k'}$. The vacuum state is defined by the condition $c_k |0\rangle = 0$. In order to understand other properties of the magnon states, let us relate the operators $c_k^\dagger$ and $c_k$ with the original spin variables. With Eqs. (3.3, 3.22a, 3.22b and 3.30) we have

$$S_i^x = \frac{1}{2}(S_i^+ + S_i^-) = \left(\frac{S}{2N}\right)^{1/2} \sum_k (a_k e^{i\vec{k}\cdot\vec{r}_i} + H.c.),$$

which, with the transformation (3.75) gives

$$S_i^x = \left(\frac{S}{2N}\right)^{1/2} \sum_k [(u_k - v_k^*)e^{i\vec{k}\cdot\vec{r}_i} c_k + H.c.], \qquad (3.82a)$$

and similarly, for the $y$ component

$$S_i^y = -i\left(\frac{S}{2N}\right)^{1/2} \sum_k [(u_k + v_k^*)e^{i\vec{k}\cdot\vec{r}_i} c_k - H.c.]. \qquad (3.82b)$$

Using the relation between the magnetization and the spin, $\vec{M} = (N/V)g\mu_B\vec{S}$, one can find the total magnon linear momentum given by the volume integration of Eq. (2.85), $\vec{p}_m = \int \vec{g}_m d\vec{r}$. With Eqs. (3.82a and 3.82b) one can show that the magnon momentum operator is

$$\vec{p}_m = \sum_k \hbar \vec{k} c_k^\dagger c_k. \qquad (3.83)$$

Hence, the magnon number state $|n_k\rangle$ is also an eigenstate of the linear momentum operator, and has momentum $n_k\hbar\vec{k}$, which is the result anticipated in Chap. 2.

Finally, let us calculate the expectation values of the spin components in the eigenstates. Using Eqs. (3.78a and 3.78b) and the orthogonality relation $\langle n_k|n_{k'}\rangle = \delta_{k,k'}$, we see that $\langle n_k|c_k|n_k\rangle = \langle n_k|c_k^\dagger|n_k\rangle = 0$. Thus, the expectation values of the transverse components of the spin operators in Eqs. (3.82) are

$$\langle n_k|S_i^x(t)|n_k\rangle = \langle n_k|S_i^y(t)|n_k\rangle = 0, \qquad (3.84)$$

which means that the states with well-defined number of magnons have no spin component transverse to the applied field and thus have uncertain phase. This result reveals that the eigenstates of the magnon Hamiltonian do not correspond to the classical picture of the spin wave, in which the spins precess about the direction of the field. As we shall show in Sect. 3.3.3, the quantum states that correspond to the macroscopic spin waves are the coherent magnon states.

## 3.3.2   Thermal Magnons

Magnons are excited by the lattice thermal vibrations of a ferromagnetic material. As mentioned in Chap. 2, the insight of Bloch regarding the lowest lying spin excitation led him to predict that at low temperatures the magnetization would behave differently than expected with the Weiss molecular field theory. Since the excitation of one magnon corresponds to a spin deviation of one unit, the thermal excitation of magnons has an effect on the temperature dependence of the magnetization. The calculation of the effect starts with Eq. (3.22c) relating the change of the spin in the equilibrium direction with the number of spin deviations, $S_i^z = S - a_i^\dagger a_i$. Using the Fourier transform in (3.30) and considering for simplicity $u_k = 1$, $v_k = 0$, we have

$$\langle S^z \rangle = S - \frac{1}{N} \sum_k \langle n_k \rangle. \tag{3.85}$$

Since magnons are boson quasiparticles, their population in thermal equilibrium is given by the Bose–Einstein distribution

$$\langle n_k \rangle = \frac{1}{e^{\hbar \omega_k / k_B T} - 1}, \tag{3.86}$$

where $k_B$ is Boltzmann constant and $T$ the temperature. In this book we will denote the thermal average either by $\langle n_k \rangle$ or by $\bar{n}_k$. In order to evaluate the sum over wave vectors in Eq. (3.85) we consider periodic boundary conditions in which the components of $\vec{k}$ have values

$$k_x = \pm n_x \frac{2\pi}{L_x}, \quad k_y = \pm n_y \frac{2\pi}{L_y}, \quad k_z = \pm n_z \frac{2\pi}{L_z}, \tag{3.87}$$

where $n_x$, $n_y$, $n_z$ are zero or integers and $L_x$, $L_y$, $L_z$ are the sample dimensions. Since the allowed $\vec{k}$ values are closely spaced, we assume that the wave vector varies continuously and replace the summation by an integration

$$\frac{1}{N} \sum_k \rightarrow \frac{\Omega}{(2\pi)^3} \int d\vec{k}, \tag{3.88}$$

where $N$ is the number of allowed $\vec{k}$ values in the first Brillouin zone and $\Omega$ is the volume of the unit cell. Note that $N$ is also the number of spins in the sample, because in the ferromagnet there is one spin per unit cell. Assuming a spherical Brillouin zone, i.e., a magnon energy that does depend on the direction of $\vec{k}$, and considering that the change in magnetization is proportional to the number of spin deviations, we obtain with Eqs. (3.85–3.88)

$$\frac{M(0) - M(T)}{M(0)} = \frac{S - \langle S^z \rangle}{S} = \frac{\Omega}{S(2\pi)^3} \int_0^{2\pi} d\varphi \int_0^\pi \sin\theta \, d\theta \int_0^{k_m} \frac{k^2 dk}{e^{\hbar \omega_k / k_B T} - 1}$$

which gives

$$\frac{M(0) - M(T)}{M(0)} = \frac{a^3}{S2\pi^2} \int_0^{k_m} \frac{k^2 dk}{e^{\hbar\omega_k/k_B T} - 1}, \tag{3.89}$$

where $a = \Omega^{1/3}$ is the lattice parameter and $k_m$ is the radius of the spherical Brillouin zone. At low temperatures only magnons with low energies are thermally excited so that we can use two approximations to evaluate (3.89). The first consists in considering the quadratic dispersion relation valid for $ka \ll 1$, $\omega_k = \gamma D k^2$, neglecting dipolar and Zeeman contributions. The second is to set the upper limit of the integral to infinite, because the population of magnons with high wave numbers is negligible, since it decreases exponentially with increasing energy. Introducing the normalized energy $x = \hbar\omega_k/k_B T$ in Eq. (3.89) we have

$$\frac{M(0) - M(T)}{M(0)} = \frac{a^3}{S4\pi^2} \left(\frac{k_B T}{\gamma \hbar D}\right)^{3/2} \int_0^\infty \frac{x^{1/2} dx}{e^x - 1}. \tag{3.90}$$

The dimensionless integral in (3.90) is equal to $(\sqrt{\pi}/2)\zeta(3/2)$, where $\zeta$ is the Riemann zeta function. Therefore

$$\frac{M(0) - M(T)}{M(0)} = \zeta(3/2) \frac{a^3}{S} \left(\frac{k_B}{4\pi\gamma\hbar D}\right)^{3/2} T^{3/2}. \tag{3.91}$$

This is the celebrated Bloch $T^{3/2}$ law [1] that was first confirmed experimentally in low temperatures magnetization measurements of EuO, one of the very few known ferromagnetic insulators [5].

### 3.3.3   Coherent Magnon States

As presented in Sect. 3.3.1, the eigenstates $|n_k\rangle$ of the Hamiltonian (3.74) have well-defined number of magnons and uncertain phase. These states form a complete orthonormal set, which can be used as a basis for the expansion of any state of spin excitation. They are used in nearly all quantum treatments of thermodynamic properties, relaxation mechanisms, and other phenomena involving magnons. However, as shown by expression (3.84), they have zero expectation value for the small-signal transverse magnetization operators and thus do not have a macroscopic wave function.

In order to establish a correspondence between classical and quantum spin waves one should use the concept of coherent magnon states [6, 7], defined in analogy to the coherent photon states introduced by Glauber [8]. A coherent magnon state is the eigenket of the circularly polarized magnetization operator $m^+ = m_x + im_y$. It can be written as the direct product of single-mode coherent states, defined as the eigenstates of the annihilation operator

$$c_k|\alpha_k\rangle = \alpha_k|\alpha_k\rangle, \tag{3.92}$$

where the eigenvalue $\alpha_k$ is a complex number. Although the coherent states are not eigenstates of the unperturbed Hamiltonian and as such do not have a well-defined number of magnons, they have nonzero expectation values for the magnetization $m^+$ with a well-defined phase. The first important property of the coherent states is that they can be expanded in terms of the eigenstates of the magnon Hamiltonian [8]

$$|\alpha_k\rangle = e^{-|\alpha_k|^2/2} \sum_{n_k=0}^{\infty} \frac{(\alpha_k)^{n_k}}{(n_k!)^{1/2}} |n_k\rangle. \tag{3.93}$$

The expectation value of the magnon number operator $c_k^\dagger c_k$ is obtained directly from (3.92) and its Hermitian conjugate

$$\left\langle \alpha_k \middle| c_k^\dagger c_k \middle| \alpha_k \right\rangle = |\alpha_k|^2. \tag{3.94}$$

The probability of finding $n_k$ magnons in the coherent state $|\alpha_k\rangle$, calculated with (3.93), is given by

$$\rho_{\mathrm{coh}}(n_k) = |\langle n_k|\alpha_k\rangle|^2 = (|\alpha_k|^{2n_k}/n_k!)e^{-|\alpha_k|^2}. \tag{3.95}$$

This function is a Poisson distribution that exhibits a peak at the expectation value of the occupation number operator $\langle n_k \rangle = |\alpha_k|^2$ in the coherent state, illustrated in Fig. 3.4 for $\langle n_k \rangle = 50$. It can be shown that coherent states are not orthogonal to one another, but they form a complete set, so that they constitute a basis for the expansion of an arbitrary state.

**Fig. 3.4**  Distribution of magnons in a system in thermal equilibrium and in a coherent state with $\langle n_k \rangle = 50$

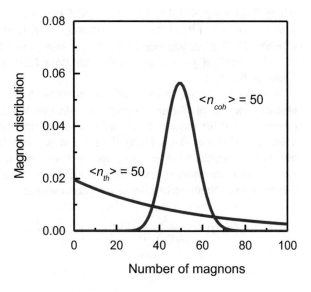

The distribution (3.95) is very different from the one for magnons in thermal equilibrium, which cannot be described by pure quantum states. Instead, they are described by a mixture in which the average number of magnons $\langle n_k \rangle$ with energy $\hbar\omega_k$ is given by the Bose–Einstein distribution (3.86). In this case, the probability of finding $n_k$ magnons with energy $\hbar\omega_k$ in the mixture describing the thermal equilibrium can be shown to be [8]

$$\rho_{th}(n_k) = \frac{\langle n_k \rangle^{n_k}}{[1 + \langle n_k \rangle]^{n_k+1}}. \qquad (3.96)$$

This function is a Gaussian distribution, very different from the Poisson distribution (3.95), as shown in Fig. 3.4.

Another important property of a coherent state is that it can be generated by the application of a displacement operator to the vacuum [8]

$$|\alpha_k\rangle = D(\alpha_k)|0\rangle, \qquad (3.97a)$$

where

$$D(\alpha_k) = \exp\left(\alpha_k c_k^\dagger - \alpha_k^* c_k\right). \qquad (3.97b)$$

In order to study the coherence properties of a magnon system, it is convenient to use the density matrix operator $\rho$ and its representation as a statistical mixture of coherent states

$$\rho = \int P(\alpha_k)|\alpha_k\rangle\langle\alpha_k|d^2\alpha_k, \qquad (3.98)$$

where $P(\alpha_k)$ is a probability density, called $P$ representation, satisfying the normalization condition $\int P(\alpha_k)d^2\alpha_k = 1$ and $d^2\alpha_k = d(\mathrm{Re}\ \alpha_k)d(\mathrm{Im}\ \alpha_k)$. As shown by Glauber [8], if $\rho$ corresponds to a coherent state, $P(\alpha_k)$ is a Dirac $\delta$-function. On the other hand, if $\rho$ represents a thermal Bose–Einstein distribution, $P(\alpha_k)$ is a Gaussian function.

In order to see the correspondence between the quantum treatment and the classical picture of the spin wave, we calculate the expectation values of the components of the magnetization operators for a single coherent state with eigenvalue $\alpha_k = |\alpha_k| \exp(i\phi_k)$. Using the relation between the magnetization and the transverse components of the spin operators, expressing (3.82a and 3.82b) in the Heisenberg representation, with $c_k(t) = c_k \exp(-i\omega_k t)$, with the definition (3.92) one can show that the magnetization in a coherent state is

$$\langle m_x(\vec{r}, t)\rangle = \frac{M}{(NS/2)^{1/2}}|\alpha_k|(u_k - v_k^*)\cos\left(\vec{k}\cdot\vec{r} - \omega_k t + \phi_k\right), \qquad (3.99a)$$

$$\langle m_y(\vec{r},t)\rangle = \frac{M}{(NS/2)^{1/2}}|\alpha_k|(u_k + v_k^*)\sin(\vec{k}.\vec{r} - \omega_k t + \phi_k). \qquad (3.99b)$$

Thus, the transverse components of the magnetization in (3.99a and 3.99b), together with $\hat{z}M_z$, correspond to the classical view of a spin wave, namely, the magnetization precesses around the equilibrium direction with a phase that varies along the direction of propagation and with an ellipticity given by

$$\frac{m_y^{max}}{m_x^{max}} = \left|\frac{u_k + v_k^*}{u_k - v_k^*}\right| = \frac{A_k + |B_k|}{\hbar\omega_k}. \qquad (3.100)$$

Using the expressions for the coefficients in (3.51a–3.51c), one can see that for $ka \ll 1$ this result is in complete agreement with the ellipticity in Eq. (2.38) obtained with the classical treatment. As will be shown in Chap. 6, magnons excited by microwave magnetic fields are described by quantum coherent states.

## 3.4   Magnon Interactions and Relaxation Mechanisms

### 3.4.1   Magnon Interactions

The magnon states we studied in the previous sections do not actually have infinite lifetimes, as implied in Eq. (3.99a and 3.99b). The absence of a decay factor in the spin wave amplitude is due to the fact that the quadratic Hamiltonian (3.74) characterizes a system of independent collective modes. Actually, the magnon modes interact among themselves and with other elementary excitations in the crystal, such as phonons, electrons, and plasmons. Typically, a more realistic Hamiltonian could be written as

$$
\begin{aligned}
H = &\sum_k \hbar\omega_k\, c_k^\dagger c_k \\
&+ \sum_{k_1,k_2,k_3} [V_{m-m}^{(3)}(k_1,k_2,k_3)\, c_{k_1}^\dagger c_{k_2} c_{k_3} \Delta(\vec{k}_1 - \vec{k}_2 - \vec{k}_3) + H.c.] \\
&+ \sum_{k_1,k_2,k_3,k_4} [V_{m-m}^{(4)}(k_1,k_2,k_3,k_4)\, c_{k_1}^\dagger c_{k_2}^\dagger c_{k_3} c_{k_4} \Delta(\vec{k}_1 + \vec{k}_2 - \vec{k}_3 - \vec{k}_4) \\
&+ \sum_{k_1,k_2,k_3} [V_{m-p}^{(3)}(k_1,k_2,k_3)\, c_{k_1} c_{k_2}^\dagger b_{k_3}^\dagger \Delta(\vec{k}_1 - \vec{k}_2 - \vec{k}_3) + H.c.] \\
&+ \ldots\ldots
\end{aligned}
\qquad (3.101)
$$

where the second term represents 3-magnon processes, the third term represents 4-magnon processes, and the last one is a 2-magnon-1-phonon process. The magnon interactions are responsible for the scattering processes that account for the magnon energy renormalization and relaxation, or damping, and consequently for its finite lifetime. In addition, they provide the means for the nonlinear dynamic interactions among magnons and other excitations, leading to very interesting phenomena. Among them are the parametric excitation of magnons, spin wave chaos, and Bose–Einstein condensation of magnons, that will be studied in Chaps. 6 and 7.

The term of the 3-magnon interaction in Eq. (3.101) arises from the dipolar energy. It is obtained from the Hamiltonian (3.46), using the transformations (3.20a–3.20c) and the expansions (3.21a and 3.21b) in the terms $S_i^z S_j^+$ and $S_i^z S_j^-$. It is not difficult to see that these terms give

$$H_{dip}^{(3)} = (g\mu_B)^2 \frac{(2S)^{1/2}}{4} \sum_{ij} \frac{3}{r_{ij}^5} [r_{ij}^- z_{ij}(a_i a_j^\dagger a_j + a_i^\dagger a_i^\dagger a_i/4 + a_i^\dagger a_i a_i + a_j^\dagger a_j^\dagger a_j/4)$$

$$\times r_{ij}^+ z_{ij}(a_i^\dagger a_j^\dagger a_j + a_i^\dagger a_i a_i/4 + a_i^\dagger a_i a_i^\dagger + a_j^\dagger a_j a_j/4)].$$

The next steps consist in using the transformations (3.30) to replace the operators $a_i, a_i^\dagger$ by $a_k, a_k^\dagger$, and use the orthonormality relation (3.32) that results from the sum over the lattice sites $i$ to eliminate two sums over wave vectors. Two approximations are used to simplify the final expression. We consider that the terms with factor ¼ are much smaller than the others and also $u_k \approx 1$, $v_k \approx 0$ so that the third HP transformation into the magnon operators is disregarded. The 3-magnon dipolar Hamiltonian then becomes

$$H_{dip}^{(3)} \approx \frac{\sqrt{2S}(g\mu_B)^2}{4\sqrt{N}} \sum_{k_1, k_2, k_3} \Delta(\vec{k}_1 - \vec{k}_2 - \vec{k}_3)$$

$$\times \sum_r \frac{3}{r^5} [r^- z\, c_{k_1}^\dagger c_{k_2} c_{k_3} (e^{-i\vec{k}_2 \cdot \vec{r}} + e^{-i\vec{k}_3 \cdot \vec{r}}) + H.c.], \qquad (3.102)$$

where $\vec{r} = \vec{r}_{ij}$ and $\Delta(\vec{k}_1 - \vec{k}_2 - \vec{k}_2)$ is the Kronecker delta that arises from the summation over the lattice sites $i$ and represents momentum conservation in the scattering process. With the same procedure used to obtain the quadratic dipolar Hamiltonian, we break the sum over $\vec{r}$ into a sum inside a sphere and an integration outside. Considering long-wavelength magnons in a cubic crystal, neglecting the effect of the surfaces, and using Eq. (3.49), it can be shown that Eq. (3.102) gives [See Keffer]

$$H_{dip}^{(3)} \approx -g\mu_B \frac{\pi M}{\sqrt{2SN}} \sum_{k_1, k_2, k_3} \Delta(\vec{k}_1 - \vec{k}_2 - \vec{k}_3)$$

$$\times [(\sin 2\theta_{k_2} e^{-i\varphi_{k_2}} + \sin 2\theta_{k_3} e^{-i\varphi_{k_3}}) c_{k_1}^\dagger c_{k_2} c_{k_3} + H.c.], \qquad (3.103)$$

where we have used $M = g\mu_B SN/V$. Notice that this has a magnitude that is smaller than that in the quadratic dipolar Hamiltonian (3.50) by a factor $1/\sqrt{N}$.

The 4-magnon interaction arises mainly from the dipolar and exchange interactions between the spins. The contribution from the magnetocrystalline anisotropy is usually not important. In order to calculate the 4-magnon dipolar Hamiltonian we follow the same steps used for the 3-magnon case. Using the transformations (3.20a–3.20c) and the expansions (3.21a and 3.21b) in the Hamiltonian (3.46), and keeping the terms with four spin-deviation operators in $S_i^z S_j^z$ and $S_i^+ S_j^-$, we obtain

$$H_{dip}^{(4)} = \frac{1}{2}(g\mu_B)^2 \sum_{ij} \left( -\frac{1}{4r_{ij}^3} + \frac{3r_{ij}^- r_{ij}^+}{8r_{ij}^5} \right) (a_i a_j^\dagger a_j^\dagger a_j + a_i^\dagger a_i a_i a_j^\dagger + a_i^\dagger a_j^\dagger a_j a_j + a_i^\dagger a_i^\dagger a_i a_j)$$

$$+ \left( \frac{1}{r_{ij}^3} - \frac{3 z_{ij}^2}{r_{ij}^5} \right) a_i^\dagger a_i a_j^\dagger a_j.$$

Following the same steps as before, the 4-magnon dipolar Hamiltonian becomes

$$H_{dip}^{(4)} \approx g\mu_B \frac{\pi M}{2SN} \sum_{\substack{k_1, k_2, \\ k_3, k_4}} \Delta(\vec{k})[4\cos^2\theta_{k_3-k_1} - \sin^2\theta_{k_1} - \sin^2\theta_{k_3}]c_{k_1}^\dagger c_{k_2}^\dagger c_{k_3} c_{k_4},$$

$$(3.104)$$

where $\Delta(\vec{k}) = \Delta(\vec{k}_1 + \vec{k}_2 - \vec{k}_3 - \vec{k}_4)$. For $k_1 = 0$ or $k_3 = 0$, the corresponding $\sin^2\theta_k$ should be replaced by $N_x + N_y$. The contribution of the exchange energy for the 4-magnon interaction is obtained from Eq. (3.7) with the transformations (3.20a–3.20c) and the expansions (3.21a and 3.21b). Keeping the terms with four spin-deviation operators in $S_i^+ S_j^-$, $S_i^- S_j^+$ and $S_i^z S_j^z$, we obtain

$$H_{exc}^{(4)} = -\frac{J}{2} \sum_{i,j=i+\delta} \left[ -\frac{1}{2}(a_i a_j^\dagger a_j^\dagger a_j + a_i^\dagger a_i a_i a_j^\dagger + a_i^\dagger a_j^\dagger a_j a_j + a_i^\dagger a_i^\dagger a_i a_j) + a_i^\dagger a_i a_j^\dagger a_j \right].$$

As in the previous derivations, using the transformations (3.30) to replace the operators $a_i, a_i^\dagger$ by $a_k, a_k^\dagger$, using the orthonormality relation (3.32) that results from the sum over the lattice sites $i$, introducing the structure factor (3.36), and considering $u_k \approx 1$, $v_k \approx 0$ the 4-magnon exchange Hamiltonian becomes

$$H_{exc}^{(4)} = \frac{zJ}{4N} \sum_{\substack{k_1, k_2, \\ k_3, k_4}} \Delta(\vec{k})[\gamma_{k1} + \gamma_{k2} + \gamma_{k3} + \gamma_{k4} - 4\gamma_{k4-k1}] c_{k_1}^\dagger c_{k_2}^\dagger c_{k_3} c_{k_4}, \quad (3.105)$$

where $\gamma_k$ is the structure factor defined in Eq. (3.36) and we have omitted the vector symbol in the subscripts to simplify the notation. The magnon interaction Hamiltonians in (3.103–3.105) will be used in several occasions in this book.

### 3.4.2  Magnon Energy Renormalization

One important consequence of the magnon interactions when the number of magnons increases is a change in the magnon energy. This happens when the temperature increases and becomes a sizeable fraction of the Curie temperature. The magnon energy renormalization can be calculated by writing the 4-magnon Hamiltonian approximately in the quadratic form that contains the energy correction. Let us do this only with the exchange Hamiltonian since it has a magnitude much larger than the dipolar one. Consider the Hamiltonian (3.105) for only two modes, the magnon of interest with wave vector $\vec{k}$ and another magnon $\vec{k}'$. Since $\vec{k}$ and $\vec{k}'$ can be any of the four wave vectors in (3.105), the Hamiltonian becomes

$$\mathrm{H}_{exc}^{(4)} = \frac{2zJ}{N} \sum_{k,k'} [\gamma_k + \gamma_{k'} - \gamma(0) - \gamma_{k-k'}] c_k^\dagger c_k c_{k'}^\dagger c_{k'}.$$

Using the random-phase approximation, $c_{k'}^\dagger c_{k'} \rightarrow \left\langle c_{k'}^\dagger c_{k'} \right\rangle = \bar{n}_{k'}$, we can write this Hamiltonian as

$$\mathrm{H}_{exc}^{(4)} = \sum_k \hbar \Delta \omega_k c_k^\dagger c_k, \tag{3.106a}$$

where

$$\hbar \Delta \omega_k = -\frac{2zJ}{N} \sum_{k'} (1 + \gamma_{k-k'} - \gamma_k - \gamma_{k'}) \bar{n}_{k'} \tag{3.106b}$$

is the energy renormalization of the $k$-magnon due to the 4-magnon interaction. This result shows that, as the temperature increases, the $k$-magnon energy decreases because the thermal number of magnons $k'$ increases. This magnon energy renormalization has been observed experimentally in many materials by magnetic resonance, and by light and neutron scattering. It also has an indirect effect on the temperature dependence of the magnetization, since the lower energy causes an increase in the number of thermal magnons and thus a decrease in the magnetization. In Sect. 3.7.3, we will present a calculation of the energy renormalization for YIG.

### 3.4.3  Magnon Relaxation

The interactions (3.101) represent processes that tend to share the energy of a certain magnon mode with other magnons and with phonons. This means that, if by some external driving, the occupation number of a magnon mode is changed from its thermal equilibrium value, the interactions act as to reestablish the thermalization. This is the same as to say that the lifetime of the magnon is finite, or that magnons undergo relaxation, or damping. It is not difficult to see that if the magnons excited are coherent, the relaxation makes the expectation value of the spins to precess in a spiraling motion toward the thermal equilibrium mean direction.

Figure 3.5 illustrates the 3- and 4-magnon relaxation processes described by the Hamiltonian (3.101). Let us consider initially the 3-magnon confluence process of Fig. 3.5a. The Hamiltonian for this process is

$$H^{3m} = \sum_{k_1, k_2, k_3} (V_{123} c_{k_1} c_{k_2} c_{k_3}^\dagger + V_{123}^* c_{k_1}^\dagger c_{k_2}^\dagger c_{k_3}) \Delta(\vec{k}_1 - \vec{k}_2 - \vec{k}_3), \qquad (3.107)$$

where $V_{123}$ is originated by the dipolar interaction (3.103). In this process, a magnon with a wave vector $\vec{k}_1$ interacts with a thermal magnon $\vec{k}_2$ to generate a third magnon $\vec{k}_3$, conserving momentum and energy. The probability that one magnon in the $n_k$ mode is destroyed and one magnon $\vec{k}_3$ is created can be calculated with first-order perturbation theory. The matrix element of (3.107) that corresponds to this process is, according to (3.78a and 3.78b)

$$\langle n_{k_1} - 1, n_{k_2} - 1, n_{k_3} + 1 | H^{3m} | n_{k_1}, n_{k_2}, n_{k_3} \rangle = [n_{k_1} n_{k_2} (n_{k_3} + 1)]^{1/2} (V_{123} + V_{213}),$$

where we have considered that $\vec{k}$ can be either $\vec{k}_1$ or $\vec{k}_2$. Thus, the probability per unit time for the number of magnons $n_k$ to decrease by one unit, given by the Fermi Golden rule, is

$$W_{n_k \to n_k - 1} = \frac{2\pi}{\hbar^2} \sum_{k_2} (V_{123} + V_{213})^2 [n_k n_2 (n_3 + 1)] \delta(\omega_k + \omega_2 - \omega_3), \qquad (3.108)$$

where the sum runs only over $\vec{k}_2$ because of the momentum conservation relation $\vec{k}_3 = \vec{k} + \vec{k}_2$. In the subscripts of the occupation number and of the frequency we have dropped $k$ to simplify the notation. The reverse process by which the number of magnons increases by one unit is calculated in a similar manner, so that the time rate of change of the magnon number is given by

$$\frac{dn_k}{dt} = W_{n_k \to n_k + 1} - W_{n_k \to n_k - 1}. \qquad (3.109)$$

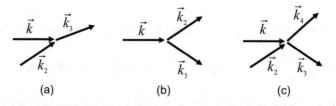

(a)                         (b)                         (c)

**Fig. 3.5** Illustration of magnon relaxation processes. (**a**) 3-magnon confluence process. (**b**) 3-magnon splitting process. (**c**) 4-magnon process

We consider that modes $\vec{k}_2$ and $\vec{k}_3$ stay in thermal equilibrium during the relaxation of mode $\vec{k}$. Then, with Eqs. (3.108 and 3.109), we find the decay rate for the number of magnons in mode $\vec{k}$ due to the 3-magnon confluence process

$$\frac{dn_k}{dt} = \frac{2\pi}{\hbar^2} \sum_{k_2} (V_{123} + V_{213})^2 [(n_k + 1)(\bar{n}_2 + 1)\bar{n}_3 - n_k \bar{n}_2 (\bar{n}_3 + 1)] \delta(\omega_k + \omega_2 - \omega_3).$$

(3.110)

In thermal equilibrium $dn_k/dt = 0$, so that we have the following relation for the thermal numbers

$$(\bar{n}_k + 1)(\bar{n}_2 + 1)\bar{n}_3 - \bar{n}_k \bar{n}_2 (\bar{n}_3 + 1) = 0.$$  (3.111)

This result can also be demonstrated with the use of the Bose–Einstein distribution (3.86) for the occupation numbers and the relation for energy conservation $\omega_3 = \omega_k + \omega_2$. By subtracting Eq. (3.111) from (3.110) we obtain

$$\frac{dn_k}{dt} = -(n_k - \bar{n}_k) \frac{2\pi}{\hbar^2} \sum_{k_2} (V_{123} + V_{213})^2 [\bar{n}_2(\bar{n}_3 + 1) - (\bar{n}_2 + 1)\bar{n}_3] \delta(\omega_k + \omega_2 - \omega_3).$$

(3.112)

This expression can be written in the form

$$\frac{dn_k}{dt} = -\eta_k (n_k - \bar{n}_k),$$  (3.113)

showing that the number of $k$-magnons decays exponentially in time toward thermal equilibrium with a relaxation rate, or damping rate, $\eta_k$. The relaxation time, or lifetime, is the inverse of the relaxation rate, $T_k = 1/\eta_k$. From Eq. (3.86) it is easy to show that $(\bar{n} + 1) = \bar{n} \exp(\hbar\omega/k_B T)$. Using this relation in Eq. (3.112), one can show that the relaxation rate of the $k$-magnon due to the 3-magnon *confluence* process becomes

$$\eta_{k-3mc} = \frac{2\pi}{\hbar^2} (e^{\hbar\omega_k/k_B T} - 1) \sum_{k_2} (V_{123} + V_{213})^2 e^{\hbar\omega_2/k_B T} \bar{n}_2 \, \bar{n}_3 \, \delta(\omega_k + \omega_2 - \omega_3).$$

(3.114)

A similar calculation gives the relaxation rate for the 3-magnon *splitting* process

$$\eta_{k-3ms} = \frac{\pi}{\hbar^2} (e^{\hbar\omega_k/k_B T} - 1) \sum_{k_2} (V_{123} + V_{132})^2 \, \bar{n}_2 \, \bar{n}_3 \, \delta(\omega_k - \omega_2 - \omega_3). \quad (3.115)$$

As will be shown in Sect. 3.7.2, in YIG at room temperature, the 3-magnon processes are responsible for the relaxation of long-wavelength spin waves with wave numbers up to about $5 \times 10^5 \text{ cm}^{-1}$. For spin waves with larger wave numbers, the relaxation is dominated by 4-magnon process originating in the exchange interaction. Considering the 4-magnon Hamiltonian (3.105), a calculation similar to the one just presented gives for the 4-magnon process, illustrated in Fig. 3.5c, the relaxation rate

$$
\eta_{k-4m} = \frac{2\pi}{\hbar^2} \left[ e^{\hbar\omega_k/k_B T} - 1 \right] \sum_{k_2, k_3, k_4} (V_{1234} + V_{2134})^2 e^{\hbar\omega_2/k_B T} \bar{n}_2 \bar{n}_3 \bar{n}_4 \, \Delta(k) \, \delta(\omega),
$$

(3.116)

where $\Delta(k) = \Delta(\vec{k} + \vec{k}_2 - \vec{k}_3 - \vec{k}_4)$, $\delta(\omega) = \delta(\omega_k + \omega_2 - \omega_3 - \omega_4)$, and

$$
V_{1234} = \frac{zJ}{4N} (\gamma_{k1} + \gamma_{k2} + \gamma_{k3} + \gamma_{k4} - 4\gamma_{k4-k1}).
$$

(3.117)

In Sect. 3.7.2, we will show that the 3- and 4-magnon scattering processes explain the wave number dependence of the magnon relaxation rate in crystalline yttrium iron garnet. The "residual" damping at $k = 0$ is mainly due to 2-magnon scattering induced by surface roughness and, in bulk samples, radiation damping. In crystals with impurities there are other relaxation channels, especially with impurities of rare earth elements that have large spin–orbit coupling [9]. In metallic ferromagnets the spin wave damping tends to be larger than in insulators, produced mainly by eddy current effects and magnon-conduction electron scattering [10, 11].

## 3.5 Magnon Hybrid Excitations with Phonons and Photons

### 3.5.1 Quantization of Elastic Waves: Phonons

In this section, we present a quantum theory for the magnetoelastic waves studied in Chap. 2. The quantization of spin waves was presented earlier. Here we present the quantization of elastic waves, following closely the book of Kittel, *Quantum Theory of Solids*. Let us consider that the ferromagnetic crystal is a continuous solid, elastically isotropic, with average mass density $\rho$. We also assume that it is a cubic crystal so that, within the linear approximation, the relation between the stress tensor and the strain tensor involves only two different elastic constants, $c_{12}$ and $c_{44}$. The elastic deformations of the solid are expressed in terms of the displacement vector $\vec{u} = \vec{r} - \vec{r}'$, where $\vec{r}$ is the initial position of an atom or of a volume element, and $\vec{r}'$ is the position after deformation. The contributions of the elastic system to the Hamiltonian arise from the kinetic and potential energies. Introducing the momentum density conjugate to the displacement, $\rho \partial u_i/\partial t$, in the linear approximation the elastic Hamiltonian can be written as

$$
H_e = \int d^3 r \left( \frac{\rho}{2} \frac{\partial u_i}{\partial t} \frac{\partial u_i}{\partial t} + \frac{\alpha}{2} \frac{\partial u_i}{\partial x_i} \frac{\partial u_j}{\partial x_j} + \frac{\beta}{2} \frac{\partial u_i}{\partial x_j} \frac{\partial u_i}{\partial x_j} \right),
$$

(3.118)

where the elastic constants are written as $\alpha = c_{12} + c_{44}$, $\beta = c_{44}$, for a Cartesian coordinate system chosen with axes lying along the <100> crystallographic directions. In order to obtain the collective excitation operators for the elastic system, we use the canonical transformation

$$u_i(\vec{r}, t) = \left(\frac{\hbar}{V}\right)^{1/2} \sum_{k,\mu} \varepsilon_{i\mu}(\vec{k}) Q_k^\mu(t) e^{i\vec{k}\cdot\vec{r}}, \qquad (3.119a)$$

$$\rho \dot{u}_i(\vec{r}, t) = \left(\frac{\hbar}{V}\right)^{1/2} \sum_{k,\mu} \varepsilon_{i\mu}(\vec{k}) P_k^\mu(t) e^{-i\vec{k}\cdot\vec{r}}, \qquad (3.119b)$$

where $\varepsilon_{i\mu} = \hat{x}_i \cdot \hat{\varepsilon}(\vec{k}, \mu)$ and $\hat{\varepsilon}(\vec{k}, \mu)$ are unitary polarization vectors. We will denote by $\mu = 1$, 2 the two polarizations transverse to the wave vector $\vec{k}$, and $\mu = 3$ the longitudinal one. Notice that from Hermiticity it follows that $Q_k^i = Q_{-k}^{i\ \dagger}$ and $P_k^i = P_{-k}^{i\ \dagger}$. The quantization of the elastic vibrations is made through the commutation relations involving $u_i(\vec{r})$ and is conjugate momentum density $\rho \partial \vec{u}/\partial t$. The only non-commuting pair is such that

$$[u_i(\vec{r}), \rho \dot{u}_j(\vec{r}')] = i\hbar \delta_{ij} \delta(\vec{r} - \vec{r}'), \qquad (3.120a)$$

which leads to

$$[Q_k^\mu, P_{k'}^\nu] = i\hbar \delta_{kk'} \delta_{\mu\nu}. \qquad (3.120b)$$

In order to diagonalize the elastic Hamiltonian, it is necessary to introduce the canonical transformation

$$Q_k^\mu = \left[\frac{\hbar}{2\rho\omega_{p\mu}(k)}\right]^{1/2} (b_{\mu-k}^\dagger + b_{\mu k}), \qquad (3.121a)$$

$$P_k^\mu = i\left[\frac{\rho\hbar\omega_{p\mu}(k)}{2}\right]^{1/2} (b_{\mu k}^\dagger - b_{\mu-k}), \qquad (3.121b)$$

where

$$\omega_{p\mu}(k) = k[(\beta + \alpha\delta_{\mu 3})/\rho]^{1/2} \qquad (3.122)$$

is the phonon frequency, as in Eqs. (2.49 and 2.50) obtained with the classical treatment. With transformations (3.119a and 3.119b) and (3.121a and 3.121b), the Hamiltonian (3.118) becomes

$$H_e = \sum_{k,\mu} \hbar\omega_{p\mu}(k)(b_{\mu k}^\dagger b_{\mu k} + 1/2). \qquad (3.123)$$

The new operators satisfy the boson commutation relations

$$[b_{\mu k}, b_{\nu k'}] = 0, \quad [b_{\mu k}, b_{\nu k'}^{\dagger}] = \delta_{\mu\nu}\delta_{kk'}, \tag{3.124}$$

and are interpreted as creation and annihilation operators of lattice vibrations, whose quanta are called *phonons*. In terms of these operators, the displacement and the momentum density operators are

$$u_i = \left(\frac{\hbar}{2\rho V}\right)^{1/2} \sum_{k,\mu} \varepsilon_{i\mu}(\vec{k})\omega_{p\mu}^{-1/2}(b_{\mu k}^{\dagger} e^{-i\vec{k}\cdot\vec{r}} + b_{\mu k} e^{i\vec{k}\cdot\vec{r}}), \tag{3.125a}$$

$$\rho \dot{u}_i = \left(\frac{\rho\hbar}{2V}\right)^{1/2} \sum_{k,\mu} i\varepsilon_{i\mu}(\vec{k})\omega_{p\mu}^{1/2}(b_{\mu k}^{\dagger} e^{i\vec{k}\cdot\vec{r}} - b_{\mu k} e^{-i\vec{k}\cdot\vec{r}}). \tag{3.125b}$$

Note that $\rho \dot{u}_i$ is the canonical momentum density associated with the elastic displacement. One can also introduce a linear momentum density carried by the elastic waves

$$g_p^i = \frac{\rho}{2}\left(\frac{\partial^2 \vec{u}}{\partial x_i \partial t} \cdot \vec{u} - \frac{\partial \vec{u}}{\partial t} \cdot \frac{\partial \vec{u}}{\partial x_i}\right). \tag{3.126}$$

Using the transformations to phonon operators given by Eqs. (3.119a, 3.119b, 3.121a, and 3.121b), integration of Eq. (3.126) in the volume gives the total phonon linear momentum

$$\vec{P}_p = \sum_{k,\mu} \hbar b_{\mu k}^{\dagger} b_{\mu k}, \tag{3.127}$$

where $\hbar \vec{k}$ is the momentum of one phonon. As shown in [12], phonons may also carry angular momentum. The angular momentum of an elastic solid is the sum of two components, an orbital angular momentum corresponding to the macroscopic rotation, and a spin angular momentum corresponding to small-radius circular shear displacements. For a rigid solid, only the elastic spin angular momentum exists, which is given by [12]

$$\vec{S}_p = \int d^3 r \rho \vec{u} \times \frac{\partial \vec{u}}{\partial t}. \tag{3.128}$$

Using the transformations (3.119a and 3.119b) and (3.121a and 3.121b) in Eq. (3.128), one can write the elastic spin angular momentum in terms of the two transverse phonon operators [12]

$$\vec{S}_p = i\hbar \sum_k \frac{\vec{k}}{k}(b_{2k}^{\dagger} b_{1k} - b_{1k}^{\dagger} b_{2k}), \tag{3.129}$$

In order to write Eq. (3.129) in a diagonal form, we introduce creation operators for transverse circularly polarized phonons, denoted by (+) and (−)

$$b^{\dagger}_{k(+)} = 2^{-1/2}(b^{\dagger}_{1k} + ib^{\dagger}_{2k}), \tag{3.130a}$$

$$b^{\dagger}_{k(-)} = 2^{-1/2}(b^{\dagger}_{1k} - ib^{\dagger}_{2k}). \tag{3.130b}$$

The expressions for annihilation operators for circularly polarized phonons are given by the Hermitian conjugates of Eqs. (3.130a and 3.130b). With the circular polarization operators, one can show that the elastic Hamiltonian and the commutation relations have the same form as Eqs. (3.123 and 3.124), respectively, while the spin angular momentum given by Eq. (3.129) becomes

$$\vec{S}_p = \hbar \sum_k \frac{\vec{k}}{k}(b^{\dagger}_{k(+)}b_{k(+)} - b^{\dagger}_{k(-)}b_{k(-)}). \tag{3.131}$$

This result shows that a circularly polarized (+) or (−) phonon carries an angular momentum parallel or antiparallel to its wave vector that can be interpreted as the spin of the phonon [12]. As expected, a linearly polarized phonon carries no angular momentum since it is a superposition of (+) and (−) phonons.

### 3.5.2  Interacting Magnons and Phonons

As we saw in Sect. 2.3, spin waves couple with elastic waves by means of the magnetoelastic interaction. This interaction can be expressed by a phenomenological energy which is a function of the magnetization $\vec{M}$ and the elastic displacement $\vec{u}$. For simplicity, we will treat here only waves propagating in the direction of the magnetic field that have no contribution from the dipolar interaction. More general treatments can be found in [13, 14]. Considering only first-order terms in the small-signal magnetization components, for a cubic crystal, with the static field applied along one of the <100> directions, the magnetoelastic Hamiltonian corresponding the energy density in Eq. (2.52) is

$$H_{me} = \frac{b_2}{M} \int d^3r \left( m_y \frac{\partial u_y}{\partial z} + m_x \frac{\partial u_x}{\partial z} \right). \tag{3.132}$$

Using the transformations (3.22a and 3.22b) and (3.125a and 3.125b), this Hamiltonian can be written in terms of the magnon and phonon operators

$$H_{me} = i\left( \frac{b_2^2 \gamma \hbar^2}{4\rho} \right)^{1/2} \sum_k [k\omega_t^{-1/2} c_k (b^{\dagger}_{1k} - ib^{\dagger}_{2k} + b_{1-k} - ib_{2-k}) - H.c.], \tag{3.133}$$

where $\omega_t$ is the shear phonon frequency and $\rho$ is the mass density. As we saw in Sect. 2.3, longitudinal phonons do not couple with magnons propagating along the magnetic field. Using the transformations to circularly polarized phonons given by Eqs. (1.131), the total Hamiltonian for the magnon–phonon system becomes

$$H_t = H_m + H_e + H_{me}$$

$$= \sum_k \hbar\omega_k\, c_k^\dagger c_k + \sum_{k,\mu} \hbar\omega_t\, b_{k\mu}^\dagger b_{\mu k}$$

$$+ \sum_k i\,\hbar(\sigma_k/2)[c_k^\dagger(b_{k(+)} + b_{-k(-)}^\dagger) - c_k(b_{k(+)}^\dagger + b_{-k(-)})] \qquad (3.134a)$$

where

$$\sigma_k = b_2\left(\frac{2\gamma k}{\rho v_t M}\right)^{1/2} \qquad (3.134b)$$

is a parameter that expresses the strength of the magnon-phonon interaction, and $v_t$ is the velocity of the transverse phonon, given by $v_t = (c_{44}/\rho)^{1/2}$ for a wave propagating along a <100> axis in a cubic crystal.

### 3.5.3   Eigenstates of the Magnon–phonon System

In this section, we study some properties of the normal mode collective excitations of a magnetoelastic crystal under a static uniform magnetic field. We consider magnetoelastic waves propagating along the z-direction, so that spin waves are coupled only to shear elastic waves. Initially, we calculate the eigenfrequencies using the equations of motion for the magnon and phonon operators in the Heisenberg representation. With the Heisenberg equation

$$\frac{dA}{dt} = \frac{\partial A}{\partial t} + \frac{1}{i\hbar}[A, H_t], \qquad (3.135)$$

we obtain with the total Hamiltonian (3.134a)

$$\frac{dc_k^\dagger}{dt} = i\omega_k c_k^\dagger + \frac{\sigma_k}{2} b_{k(+)}^\dagger + \frac{\sigma_k}{2} b_{-k(-)}, \qquad (3.136a)$$

$$\frac{db_{k(+)}^\dagger}{dt} = i\omega_t b_{k(+)}^\dagger - \frac{\sigma_k}{2} c_k^\dagger, \qquad (3.136b)$$

$$\frac{db_{k(-)}}{dt} = -i\omega_t b_{k(-)} + \frac{\sigma_k}{2} c_{-k}^\dagger. \qquad (3.136c)$$

In the stationary state, all operators have a $\exp(i\omega t)$ variation, so that the magnon–phonon frequencies are given by the roots of the equation

$$\begin{vmatrix} i(\omega - \omega_k) & -\sigma_k/2 & -\sigma_k/2 \\ \sigma_k/2 & i(\omega - \omega_t) & 0 \\ -\sigma_k/2 & 0 & i(\omega + \omega_t) \end{vmatrix} = 0. \qquad (3.137)$$

Expansion of the determinant gives for the magnetoelastic dispersion relation

$$\left(\omega^2 - \omega_t^2\right)\left(\omega - \omega_k\right) - \frac{1}{2}\omega_t\sigma_k^2 = 0, \qquad (3.138)$$

which agrees with the result of the semiclassical treatment (2.68). If there is no magnetoelastic coupling, $\sigma_k = 0$, so that the three roots of Eq. (3.138) are $\omega_k$ and $\pm\omega_t$, the two signs corresponding to (+) and (−) circularly polarized phonons. As studied in Sect. 2.3, the dispersion relation has three branches because the magnetoelastic excitation involves one magnon and two transverse phonon modes. Now it is clear that the negative linear dispersion corresponds to the (−) circularly polarized phonon that has negligible coupling with magnons. The two positive branches correspond to the hybridized magnon-(+) circularly polarized phonon, that exhibit a splitting at the wave number at which the magnon and phonon curves intercept, as shown in Fig. 2.7. The fact that magnons couple with (+) circularly polarized phonons has an interesting consequence. In an experiment of magnon–phonon conversion in a nonuniform static field, as illustrated in Fig. 2.11, the phonons resulting from the conversion are circularly polarized, and hence they have spin. This has been demonstrated experimentally in a YIG film strip where magnons are excited by microwave radiation and the phonons are detected with Brillouin light scattering techniques [15]. The light scattered by phonons is circularly polarized, differently from the linearly polarized light scattered by thermal phonons as in Fig. 2.19.

To study the modes of the coupled magnon–phonon system we can neglect the negative phonon operators in Eq. (3.134a and 3.134b). Dropping the (+) index in the phonon operators left, we can write the Hamiltonian as

$$H_t = \sum_k [\hbar\omega_k c_k^\dagger c_k + \hbar\omega_t b_k^\dagger b_k + i\frac{1}{2}\hbar\sigma_k(c_k^\dagger b_k - b_k^\dagger c_k)]. \qquad (3.139)$$

This Hamiltonian can be diagonalized by a canonical transformation to new operators obtained by linear combinations of magnon and phonon operators,

$$d_{ak} = u_k b_k - iv_k c_k, \qquad (3.140a)$$

$$d_{bk} = u_k c_k - iv_k b_k, \qquad (3.140b)$$

where

$$u_k = \left(\frac{\omega_s + \omega_\delta}{2\omega_s}\right)^{1/2}, \quad \nu_k = \left(\frac{\omega_s - \omega_\delta}{2\omega_s}\right)^{1/2}, \quad u_k^2 + \nu_k^2 = 1, \quad (3.141a)$$

and

$$\omega_\delta = (\omega_t - \omega_k)/2, \quad \omega_s = (\omega_\delta^2 + \sigma_k^2/4)^{1/2}. \quad (3.141b)$$

The transformation (3.140a and 3.140b) is such that the new operators satisfy the boson commutation relations

$$[d_{\nu k}, d_{\lambda k'}^\dagger] = \delta_{kk'}\delta_{\nu\lambda}, \quad [d_{\nu k}^\dagger, d_{\lambda k}^\dagger] = [d_{\nu k}, d_{\lambda k}] = 0, \quad (3.142)$$

and the Hamiltonian has the diagonal form

$$H = \sum_k [\hbar\omega_a(k)d_{ak}^\dagger d_{ak} + \hbar\omega_b(k)d_{bk}^\dagger d_{bk}], \quad (3.143a)$$

where

$$\omega_a(k) = (\omega_t + \omega_k)/2 + \omega_s, \quad (3.143b)$$

$$\omega_b(k) = (\omega_t + \omega_k)/2 - \omega_s, \quad (3.143c)$$

which are the same normal mode frequencies obtained in Sect. 2.3 and shown in Fig. 2.7. Equations (3.143a–3.143c) lead to the interpretation that $d_{ak}^\dagger$, $d_{ak}$, $d_{bk}^\dagger$, and $d_{bk}$ are the creation and annihilation operators of quanta of collective magnetoelastic excitations, or magnon–phonon hybrid excitations, with energies $\hbar\omega_a(k)$ and $\hbar\omega_b(k)$.

Note that far from the crossover, i.e., in the region where the difference between the magnon and phonon frequencies is much larger than the splitting of the two branches, $|\omega_t - \omega_k| \gg \sigma_k$, we have the following limits:

$$\begin{aligned} &\omega_t > \omega_k \quad \omega_a \to \omega_t \quad \text{and} \quad \omega_b \to \omega_k \\ &(\nu_k \to 0) \quad d_{ak} \to b_k \qquad\qquad d_{bk} \to c_k \end{aligned} \quad (3.144a)$$

$$\begin{aligned} &\omega_k > \omega_t \quad \omega_b \to \omega_k \quad \text{and} \quad \omega_a \to \omega_t \\ &(u_k \to 0) \quad d_{ak} \to -ic_k \qquad\quad d_{bk} \to -ib_k. \end{aligned} \quad (3.144b)$$

The stationary states of the Hamiltonian (3.143a) may be obtained by applying integral powers of the creation operators to the vacuum state. The single-mode states can be written in normalized form as

$$|n_{ak}\rangle = [(d_{ak}^\dagger)^{n_k}/(n_{ak}!)^{1/2}]|0\rangle, \quad (3.145a)$$

$$|n_{bk}\rangle = [(d_{bk}^\dagger)^{n_k}/(n_{bk}!)^{1/2}]|0\rangle. \quad (3.145b)$$

The mean occupation numbers of magnons and phonons in these states are given by

$$\langle n_{ak} | c_k^\dagger c_k | n_{ak} \rangle = \langle n_{bk} | b_k^\dagger b_k | n_{bk} \rangle = v_k^2 n_k, \tag{3.146a}$$

$$\langle n_{ak} | b_k^\dagger b_k | n_{ak} \rangle = \langle n_{bk} | c_k^\dagger c_k | n_{bk} \rangle = u_k^2 n_k, \tag{3.146b}$$

which are in agreement with the limits (3.144a and 3.144b). Note also that, since $u_k^2 + v_k^2 = 1$, the mean number of magnons plus the mean number of phonons in any state is the total number of the magnetoelastic quanta in that state.

The stationary states (3.145a and 3.145b) can also be expanded in terms of pure magnon and pure phonon eigenstates. As discussed in Sect. 3.3, these states have well-defined number of quanta and uncertain phase. Coherent magnetoelastic waves should have well-defined phase and involve a large and uncertain number of magnons and phonons. In order to establish a correspondence between classical and quantum magnetoelastic waves one must use the magnetoelastic coherent states, defined as the eigenstates of the annihilation operators

$$d_{ak} | \alpha_{ak} \rangle = \alpha_{ak} | \alpha_{ak} \rangle, \qquad d_{bk} | \alpha_{bk} \rangle = \alpha_{bk} | \alpha_{bk} \rangle. \tag{3.147}$$

These can be expanded in terms of the eigenstates of the Hamiltonian

$$|\alpha_{ak}\rangle = e^{-|\alpha_{ak}|^2/2} \sum_{n_{ak}} (\alpha_{ak})^{n_{ak}} / (n_{ak}!)^{1/2} |n_{ak}\rangle, \tag{3.148a}$$

$$|\alpha_{bk}\rangle = e^{-|\alpha_{bk}|^2/2} \sum_{n_{bk}} (\alpha_{bk})^{n_{bk}} / (n_{bk}!)^{1/2} |n_{bk}\rangle, \tag{3.148b}$$

and they have magnetization and elastic displacement components with well-defined phase, as expected for a classical wave.

### 3.5.4   Magnon–photon Hybrid Excitation

The coupled spin–electromagnetic wave, the magnetic polariton, studied in Sect. 2.5 using semiclassical equations, can also be treated with a quantum approach such as the one just presented. As in the classical approach, we consider that the interaction between magnons and photons results from the direct coupling between the magnetic moments of the spins $\vec{S}_i$ at sites $i$ of the ferromagnet with the magnetic field $\vec{h}$ of the radiation [16]. For simplicity, we shall consider waves propagating in the $z$-direction, so that the dipolar energy of the spin waves vanishes. Thus, the interaction Hamiltonian can be written as

$$H_{m-p} = -\gamma \hbar \sum_i \vec{h} \cdot \vec{S}_i = -\gamma \hbar \sum_i \left( h_x S_i^x + h_y S_i^y \right). \tag{3.149}$$

Using Eqs. (3.82a and 3.82b), we express the spin components in terms of the magnon operators as

$$S_i^x = \left(\frac{S}{2N}\right)^{1/2} \sum_k (e^{i\vec{k}\cdot\vec{r}_i} c_k + e^{-i\vec{k}\cdot\vec{r}_i} c_k^\dagger), \tag{3.150a}$$

$$S_i^y = -i\left(\frac{S}{2N}\right)^{1/2} \sum_k (e^{i\vec{k}\cdot\vec{r}_i} c_k - e^{-i\vec{k}\cdot\vec{r}_i} c_k^\dagger), \tag{3.150b}$$

where we have set $u_k = 1$ and $v_k = 0$. The quantization of the magnetic field $\vec{h}$ with frequency $\omega$ is made by expressing its components in terms of the photon operators, as follows [16]

$$h_x = -i\sum_q \left(\frac{2\pi\hbar\omega}{V}\right)^{1/2} (e^{i\vec{k}\cdot\vec{r}_i} a_{qx} - e^{-i\vec{k}\cdot\vec{r}_i} a_{qx}^\dagger), \tag{3.151a}$$

$$h_y = i\sum_q \left(\frac{2\pi\hbar\omega}{V}\right)^{1/2} (e^{i\vec{k}\cdot\vec{r}_i} a_{qy} - e^{-i\vec{k}\cdot\vec{r}_i} a_{qy}^\dagger), \tag{3.151b}$$

where $a_{q\lambda}^\dagger$ and $a_{q\lambda}$ are the creation and annihilation operators for photons with wave number $q$ and polarization $\lambda$, and $V$ is the volume. Similar to what we did for phonons, we introduce the creation operators for circularly polarized photons

$$a_{q(+)}^\dagger = 2^{-1/2}(a_{qx}^\dagger + i a_{qy}^\dagger), \tag{3.152a}$$

$$a_{q(-)}^\dagger = 2^{-1/2}(a_{qx}^\dagger - i a_{qy}^\dagger), \tag{3.152b}$$

and the annihilation operators are given by the Hermitian conjugates of (3.152a and 3.152b). Substituting Eqs. (3.150a–3.152b) in (3.149) and using the orthogonality relation (3.32), we obtain the magnon–photon interaction Hamiltonian

$$H_{m-p} = i\hbar \sum_k \delta_k [c_k(a_{-k(-)} - a_{k(+)}^\dagger) + c_k^\dagger(a_{k(+)} - a_{-k(-)}^\dagger)], \tag{3.153}$$

where $\delta_k = (\omega_M \omega/2)^{1/2}$, and we have used $\omega_M = \gamma 4\pi M$ and $M = \gamma\hbar SN/V$. Thus, adding to Eq. (3.153) the free Hamiltonians for magnons and photons, we have for the total Hamiltonian of the magnon–photon system

$$H_t = \sum_k \hbar\omega_k c_k^\dagger c_k + \sum_k \hbar\omega_p (a_{k(+)}^\dagger a_{k(+)} + 1/2) + \hbar\omega_p (a_{k(-)}^\dagger a_{k(-)} + 1/2)$$

$$+ \sum_k i\hbar\delta_k [c_k(a_{-k(-)} - a_{k(+)}^\dagger) + c_k^\dagger(a_{k(+)} - a_{-k(-)}^\dagger)], \tag{3.154}$$

which is similar, but not equal, to the Hamiltonian for the magnon–phonon system. We can calculate the eigenfrequencies by writing the Heisenberg equations of motion for the magnon and photon operators and following the same procedure that led to Eqs. (3.136a–3.136c) and (3.137). One can show (Problem 3.8) that the magnon–photon dispersion relation is given by

$$\left(\omega^2 - \omega_p^2\right)(\omega - \omega_k) - 2\omega\delta_k^2 = 0. \tag{3.155}$$

It can be shown that with $\omega_k = \omega_0$, this equation is the same as Eq. (2.94), obtained with the classical treatment of the magnetic polariton. Similarly to the magnon–phonon system, if there is no magnon–photon coupling, $\delta_k = 0$, and the three roots of (3.155) are $\omega_k$ and $\pm\omega_p$, the two signs corresponding to (+) and (−) circularly polarized photons. As in Sect. 2.5, the dispersion relation has three branches because the excitation involves one magnon and two photon modes. Clearly, the negative linear dispersion corresponds to the (−) circularly polarized photon that has negligible coupling with magnons. The two positive branches correspond to the hybridized magnon-(+) circularly polarized photon, that exhibit a splitting at the wave number at which the magnon and photon curves intercept, as shown in Fig. 2.8.

## 3.6   Light Scattering by Magnons

### 3.6.1   Magneto-Optical Interaction

In Sect. 2.6, we presented the experimental aspects of the technique of inelastic scattering of light by elementary excitations in solids. Figure 3.6 illustrates the two simplest processes involving magnons. Photons of a laser beam with frequency $\omega_L$

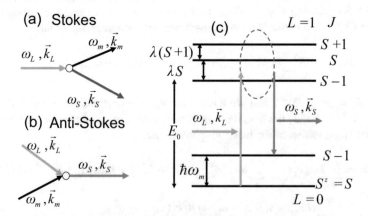

**Fig. 3.6** (a) and (b) Representation of the Stokes and anti-Stokes inelastic light scattering by magnons. (c) Illustration of the transitions involved in the Stokes process

and wave vector $\vec{k}_L$ interact in a magnetic material with magnons of frequency $\omega_m$ and wave vector $\vec{k}_m$. In the Stokes process, shown in Fig. 3.6a, one magnon is created and one laser photon is destroyed, so that by conservation of energy and momentum the scattered photons have frequency $\omega_S = \omega_L - \omega_m$ and wave vector $\vec{k}_S = \vec{k}_L - \vec{k}_m$. On the other hand, in the anti-Stokes process, illustrated in Fig. 3.6b, one magnon is destroyed, so that the scattered photons have frequency $\omega_S = \omega_L + \omega_m$ and wave vector $\vec{k}_S = \vec{k}_L + \vec{k}_m$.

The light scattering processes of Fig. 3.6 requires an interaction mechanism involving at least two photons and one magnon. The origin of this mechanism lies in the indirect coupling between the electric field of the photons and the atomic spins via the spin–orbit interaction [17, 18]. The mechanism can be understood with the help of the energy level diagram of Fig. 3.6c, where we assume, for simplicity, that the crystal has magnetic ions with $L = 0$ ground state. The ground state is split by the magnetic interactions (Zeeman, exchange, etc.), with energy separation corresponding to the magnon energy $\hbar\omega_m$. Since visible photons have energy much larger than magnons, the electric field of the radiation is coupled to the electrons of the magnetic ions through the electric–dipole interaction

$$\mathrm{H}_{ED} = -e\sum_i \vec{E}_i \cdot \vec{r}_i, \tag{3.156}$$

where $\vec{E}_i$ is the electric field of the radiation at site $i$. In an excited state with $L \neq 0$, the electron orbital is coupled to the spins through the spin–orbit interaction

$$\mathrm{H}_{LS} = \sum_i \lambda \vec{L}_i \cdot \vec{S}_i, \tag{3.157}$$

where $\vec{L}_i$ is the electronic orbital angular momentum and $\lambda$ is the spin–orbit interaction parameter. The total Hamiltonian for the radiation-spin system is the sum of the magnetic, photon, electric dipole, and spin–orbit contributions

$$\mathrm{H} = \mathrm{H}_{mag} + \mathrm{H}_{photon} + \mathrm{H}_{ED} + \mathrm{H}_{LS}. \tag{3.158}$$

Using perturbation theory one can obtain an effective Hamiltonian for the electric field–spin interaction. For the Stokes process, illustrated in Fig. 3.6a, this is calculated assuming that initially the magnetic system is in the ground state, $S_i^z = S$, and that the radiation system has $n_L$ laser photons and $n_S$ scattered photons. Thus, the initial state can be represented by

$$|I\rangle = |S_i^z = S, n_L, n_S\rangle. \tag{3.159}$$

The final state has one spin deviation, $S_i^z = S - 1$, $n_L - 1$ laser photons, and $n_S + 1$ scattered photons. With third-order perturbation theory, one can find the transition probability for the system to go from the initial to the final state via the possible intermediate states in which there is a mixing between the spin and electronic orbital wave functions, as illustrated in Fig. 3.6c. The result for the Stokes process can be written in the form of an effective Hamiltonian [19]

$$H^{Stokes} = K \sum_i (E_L^z E_S^+ - E_L^+ E_S^z) S_i^-,  \tag{3.160}$$

where $E_{L,S}^\pm = E_{L,S}^x \pm i E_{L,S}^y$ and

$$K = \frac{e^2 \lambda}{2^{3/2}} \langle S,0|r^z|P,0\rangle \langle P,-1|r^-|S,0\rangle \left[ \frac{1}{(E_0 - \hbar\omega_L)^2} - \frac{1}{(E_0 + \hbar\omega_S)^2} \right],  \tag{3.161}$$

where $E_0$ is the energy separation between the ground state and the first excited electronic state, $r^- = x - iy$, and $|L, L^z\rangle$ denotes the orbital state. Analogously, one can obtain for the anti-Stokes process

$$H^{AS} = -K \sum_i (E_L^z E_S^- - E_L^- E_S^z) S_i^+.  \tag{3.162}$$

The sum of Eqs. (3.160 and 3.162) gives the total Hamiltonian for the interaction between the radiation and the spin system in first order [19]

$$H^{(1)}_{rad-S} = i2K \sum_i [(E_L^y E_S^z - E_L^z E_S^y) S_i^x + (E_L^z E_S^x - E_L^x E_S^z) S_i^y].  \tag{3.163}$$

which can also be written as

$$H^{(1)}_{rad-S} = i2K \sum_i (\vec{E}_L \times \vec{E}_S) \cdot \vec{S}_i.  \tag{3.164}$$

Notice that $\vec{S}_i$ is a pseudo-vector, and the vector product of two vectors is also a pseudo-vector. Thus, the product in Eq. (3.164) is the only scalar that can be formed with $\vec{E}_L$, $\vec{E}_S$, and $\vec{S}_i$ which is invariant under time reversal, since this operation changes the signs of both $i$ (the imaginary unit) and $\vec{S}_i$. Equation (3.164) shows that for the magnetic system the radiation acts as an effective magnetic field

$$H^{eff}_{rad} = \frac{i2K}{g\mu_B} \vec{E}_L \times \vec{E}_S.  \tag{3.165}$$

On the other hand, from the point of view of the radiation, the spins act as to modify the dielectric constant of the medium and therefore influence its propagation. For a medium magnetized macroscopically in a given direction, the interaction (3.164) gives origin to the Faraday rotation of light propagating along the magnetization [20, 21].

The Hamiltonian (3.164) represents only the first-order term of an expansion of the interaction between the radiation and the spin system. Actually, one can expect to find in the interaction energy terms with all powers of the electric field and of the spin, with coefficients that tend to decrease as the power increases. The coefficient in (3.164) is the linear, or first-order, magneto-optical constant. The next term of importance in the expansion should involve two-electric field operators and two spin operators. It is not difficult to see that the only scalar product invariant under rotation and time reversal that can be formed with two vectors and two pseudo-vectors has the form

$$H^{(2)}_{rad-S} = G \sum_i (\vec{E}_L \times \vec{S}_i) \cdot (\vec{E}_S \times \vec{S}_i), \tag{3.166}$$

where $G$ is an appropriate quadratic magneto-optical constant, which is related to the birefringence properties of the material [22].

The expressions (3.164 and 3.166) for the linear and quadratic magneto-optical interactions are valid only for the particular case of rotational symmetry of the magnetic ions. A general description of the interaction can be made by means of the electronic polarizability tensor defined by

$$p_i^\beta = M_i^{\alpha\beta} E_L^\alpha, \tag{3.167}$$

where $p_i^\beta$ is the $\beta$-component of the electric dipole moment created by the $\alpha$-component of the incident electric field in the atom at site $i$. $M_i^{\alpha\beta}$ are the elements of the polarizability tensor, which depend on physical quantities of the material that interact with the radiation, such as strain, spin, etc. We can, therefore, expand $M_i^{\alpha\beta}$ in powers of the spin components [23]

$$M_i^{\alpha\beta}(S) = M_i^{\alpha\beta}(0) + K^{\alpha\beta\gamma} S_i^\gamma + \sum_j G_{ij}^{\alpha\beta\gamma\delta} S_i^\gamma S_j^\delta \ldots, \tag{3.168}$$

where $M_i^{\alpha\beta}(0)$ is the spin-independent part, $K^{\alpha\beta\gamma}$ is associated with the linear magneto-optical interaction, $G_{ij}^{\alpha\beta\gamma\delta}$ with the quadratic, and so on. The magneto-optical interaction Hamiltonian can then be constructed from (3.167). It represents the energy change in the medium due to the interaction between the scattered field and the electric dipole moment created by the laser field, and takes the form

$$H_{rad-S} = \sum_i \vec{p}_i \cdot \vec{E}_S = \sum_{i,\beta} p_i^\beta E_S^\beta, \tag{3.169}$$

which, with (3.167), gives

$$H_{rad-S} = \sum_{i,\alpha,\beta} E_L^\alpha M_i^{\alpha\beta} E_S^\beta. \tag{3.170}$$

Note that the polarizability $M_i^{\alpha\beta}$ has a spin-dependent form, with elements characterized by the crystal symmetry [23]. For example, for cubic crystals, the only nonvanishing components for the linear magneto-optical effect are

$$K^{\alpha\beta\gamma} = \varepsilon_{\alpha\beta\gamma}K, \tag{3.171}$$

where $\varepsilon_{\alpha\beta\gamma}$ is the Levi-Civita antisymmetric tensor.

### 3.6.2   One-Magnon Light Scattering in Ferromagnets

The classical picture of inelastic light scattering is illustrated in Fig. 3.7 for the case of the anti-Stokes scattering. Due to the magneto-optical effect, the presence of a spin wave with frequency $\omega_m$ and wave number $k_m$ modulates the dielectric constant of the medium in space and time, creating a diffraction grating that oscillates in time. The incident light with wave number $k_L$ at an angle $\theta$ with the phase planes produces a diffracted light with the same angle according to the Bragg law of diffraction, $k_m = 2k_L \sin \theta$. This corresponds to the condition of momentum conservation, $\vec{k}_S = \vec{k}_L + \vec{k}_m$. Since the electric–dipole moment of the medium is also modulated in time, the diffraction grating reradiates a field with frequency $\omega_L + \omega_m$. Thus, in this classical picture, the frequency and wave vector of the scattered light are related to the frequency and wave vector of the incident light by the same conservation equations obtained with the quantum picture.

The quantum description of the light scattering by magnons can be formulated by quantizing both the spin and the electric field operators in the radiation–spin interaction Hamiltonian (3.170). For simplicity, let us consider only the Stokes scattering process with the first-order magneto-optical constant, governed by the Hamiltonian (3.160). The spin operator is transformed into magnon operators as in Eqs. (3.82a and 3.82b). Here we neglect the effect of the dipolar interaction and use

**Fig. 3.7** Classical picture of the anti-Stokes inelastic scattering of light by spin waves

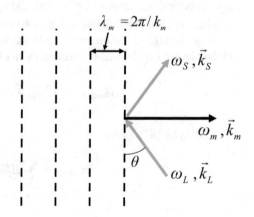

$$S_i^- = (2S/N)^{1/2} \sum_{k_m} c_{k_m}^\dagger e^{-i(\vec{k}_m \cdot \vec{r}_i - \omega_m t)}. \tag{3.172}$$

The electric field operator is

$$\vec{E}_\lambda(\vec{r}, t) = i\hat{\varepsilon}_\lambda (2\pi\hbar/V\eta)^{1/2} \sum_k \omega^{1/2} a_k e^{i(\vec{k}\cdot\vec{r} - \omega t)} + H.c., \tag{3.173}$$

where $a_k$ is the annihilation operator for photons with wave vector $\vec{k}$, $\hat{\varepsilon}_\lambda$ is the polarization vector and $\eta$ is the index of refraction at the light wavelength. By substituting Eqs. (3.172 and 3.173) into (3.160), keeping only the creation operator for the scattered field, and noting that the summation over lattice sites in (3.160) leads to a delta function for the wave vectors, we obtain for the Stokes process

$$H^{Stokes} = \frac{2\pi\hbar(2SN\omega_L\omega_S)^{1/2}}{\eta_S\eta_L V} K \sum_{k_L, k_S, k_m} (\varepsilon_L^z \varepsilon_S^+ - \varepsilon_L^+ \varepsilon_S^z) a_{k_L} a_{k_S}^\dagger c_{k_m}^\dagger \Delta(\vec{k}_L - \vec{k}_S - \vec{k}_m), \tag{3.174}$$

where $\varepsilon_\lambda^\alpha = \hat{\varepsilon}_\lambda \cdot \hat{\alpha}$ denotes the components of the photon polarization. The Hamiltonian (3.174) represents a process in which one laser photon is destroyed, while one magnon and one scattered photon are created. The selection rules governing the scattering are given in (3.174). For example, if the incident light is linearly polarized along the z-direction, the scattered light is *circularly* polarized in the xy plane, as observed experimentally [24]. As mentioned in Sect. 2.6.2, this polarization provides a manner for distinguishing the scattering from thermal magnons or phonons, since in the latter case the scattering is predominantly linearly polarized.

The Hamiltonian (3.174) can be used to derive experimentally measurable quantities. One quantity of interest is the *differential scattering cross section*, defined as the rate of change in the number of photons in the incident light as a result of the scattering into the solid angle $d\Omega$, expressed by

$$\frac{d\sigma}{d\Omega} = \frac{W}{N_L}, \tag{3.175}$$

where $N_L = n_L c/V\eta_L$ is the number of incident laser photons per unit area per unit time and $W d\Omega$ is the number of photons scattered into a cone with solid angle $d\Omega$ per unit time. For the one-magnon Stokes scattering, described by the Hamiltonian (3.174), the probability of creating one scattered photon into the solid angle $d\Omega$ due to $n_L$ incident laser photons is given by the Fermi Golden Rule

$$W d\Omega = \frac{2\pi}{\hbar^2} \sum_{k_S(d\Omega)} |\langle n_L - 1, n_S + 1, n_m + 1|H^{Stokes}|n_L, n_S, n_m\rangle|^2 \delta(\omega_L - \omega_S - \omega_m),$$

which gives

$$W\,d\Omega = \frac{(2\pi)^3}{\eta_L^2\eta_S^2 V^2}(2SN)\,\omega_L\omega_S K^2 \sum_{k_S(d\Omega)}\left|\varepsilon_L^z\varepsilon_S^+ - \varepsilon_L^+\varepsilon_S^z\right|^2 n_L(n_S+1)$$

$$\times\,(n_m+1)\delta(\omega_L - \omega_S - \omega_m) \tag{3.176}$$

where $n_m$ is the occupation number of magnons with frequency $\omega_m$ and wave vector $\vec{k}_m$. Now we replace the summation over wave vector by an integral using

$$\sum_{k_S(d\Omega)} \to \frac{V}{(2\pi)^3}\int_{d\Omega} d^3k_S = \frac{V}{(2\pi)^3}\,d\Omega\int k_S^2 dk_S, \tag{3.177}$$

and use a transformation in the argument of the delta function that gives

$$\int k_S^2 dk_S\delta(\omega_L - \omega_S - \omega_m) = \int k_S^2 dk_S\frac{\eta_S}{c}\delta\left(k_S - \frac{\omega_L - \omega_m}{c/\eta_S}\right) = \frac{\eta_S^3}{c^3}\omega_S^2. \tag{3.178}$$

Assuming that $n_S = 0$ because the scattered photons have sufficiently high energy so that their thermal number is negligible, we obtain

$$W\,d\Omega = d\Omega\frac{2SN\,\eta_S\omega_L\omega_S^3}{\eta_L^2 c^3 V}K^2\left|\varepsilon_L^z\varepsilon_S^+ - \varepsilon_L^+\varepsilon_S^z\right|^2 n_L(n_m+1), \tag{3.179}$$

which substituted in (3.175) gives the differential scattering cross section for one-magnon Stokes process

$$\left(\frac{d\sigma}{d\Omega}\right)^{Stokes} = \frac{2SN\,\eta_S\omega_L\omega_S^3}{c^4}K^2\left|\varepsilon_L^z\varepsilon_S^+ - \varepsilon_L^+\varepsilon_S^z\right|^2(n_m+1). \tag{3.180}$$

The scattering cross section for the anti-Stokes process can be derived in a similar manner. The result is

$$\left(\frac{d\sigma}{d\Omega}\right)^{AS} = \frac{2SN\,\eta_S\omega_L\omega_S^3}{c^4}K^2\left|\varepsilon_L^z\varepsilon_S^- - \varepsilon_L^-\varepsilon_S^z\right|^2 n_m. \tag{3.181}$$

Thus, the ratio between the magnitudes of the Stokes and anti-Stokes scattering cross sections is $\sigma^S/\sigma^{AS} = (n_m+1)/n_m$. For scattering by thermal magnons, we find with Eq. (3.86)

$$\frac{\sigma^S}{\sigma^{AS}} = e^{\hbar\omega_m/k_B T}. \tag{3.182}$$

In order to calculate this ratio, we consider magnons with frequency of a few gigahertz, corresponding to an energy $\hbar\omega_m \sim 10^{-17}$ erg, and use for room temperature $k_B T \approx 4.2 \times 10^{-14}$ erg. Thus $\hbar\omega_m/k_B T \sim 10^{-4}$, so that the ratio (3.182) is approximately one, which means that the Stokes and anti-Stokes processes have approximately the same cross section. However, this is not what is observed experimentally in YIG [24]. As can be seen in Fig. 2.18, the measured anti-Stokes BLS peak is nearly three times larger than the Stokes peak. This discrepancy is due to the combined effect of the linear and quadratic magneto-optical effects. As shown in Sandercock and Wettling [24], the contributions of the two effects subtract in the Stokes process and add up in the anti-Stokes, so that the ratio of the intensities is actually

$$\frac{\sigma^S}{\sigma^{AS}} \approx \frac{[(G_{11} - G_{12}) - K]^2}{[(G_{11} - G_{12}) + K]^2}, \tag{3.183}$$

where $G_{11}$, $G_{12}$ are two elements of the quadratic magneto-optical tensor for a cubic crystal. It turns out that in yttrium iron garnet, the magneto-optical constants $G$ and $K$ have comparable magnitudes, so that the Stokes peak is considerably smaller than the anti-Stokes peak.

## 3.7    Magnons in Yttrium Iron Garnet

Since its discovery in 1956 by Bertaut and Forrat [25], the ferrimagnetic insulator yttrium iron garnet ($Y_3Fe_5O_{12}$–YIG) has played an outstanding role in magnetism [26]. Due to its very low magnetic and acoustic losses and long spin-lattice relaxation time, YIG has been for several decades the prototype material for the study of the basic physics of a variety of magnonic phenomena [27], such as spin wave resonances and relaxation, propagation and processing of coherent magnon packets, microwave spin wave instabilities, solitons, nonlinear dynamics and chaotic behavior, and Bose–Einstein condensation of magnons. YIG also has attracted technological attention for its possible use in many devices, such as parametric microwave amplifiers, variable delay lines, tunable microwave filters, magnetic bubble memories, and magneto-optical devices, some of which led to commercial products. More recently, YIG has gained renewed attention as a key material for insulator-based magnon spintronics [28]. Due to its importance in magnonics, this section is entirely devoted to the properties of magnons in YIG.

YIG has a complex cubic crystal structure. Its conventional bcc unit cell, shown in Fig. 3.8, contains eight formula units of $Y_3 \, Fe^{3+}{}_2 \, Fe^{3+}{}_3 \, O^{2-}{}_{12}$, with the magnetic ferric $Fe^{3+}$ ions occupying two inequivalent positions with respect to their $O^{2-}$ ligands, tetrahedral ($d$) and octahedral ($a$). In each of the two primitive cells, there are twenty magnetic ions, the majority (12) in sites $d$, and the minority (8) in sites $a$. The three nearest-neighbor exchange interactions between the magnetic ions are negative, favoring an antiferromagnetic alignment. But the $a$–$d$ interaction dominates, so that in the ordered state the 12 $d$-spins are parallel to each other, as are the 8 $a$-spins, and the two sets are aligned in opposite directions, forming a *ferrimagnetic* arrangement.

**Fig. 3.8** Conventional bcc unit cell of YIG, with the majority $Fe^{+3}$ ions in tetrahedral sites in green and the minority $Fe^{+3}$ in octahedral sites in blue. Black spheres represent yttrium and red spheres oxygen [29]. Reproduced with kind permission from [29]

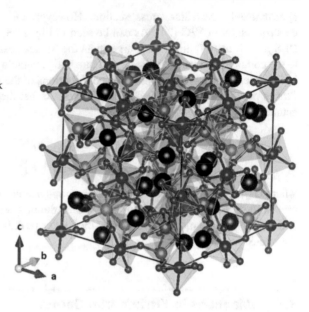

Since the $Fe^{3+}$ ions have $S = 5/2$ and zero orbital angular momentum, the spin–orbit coupling is negligible, hence the magnetic moments have very small interaction with the lattice. This is one of the reasons for the exceptional magnetic properties of YIG, despite its complex structure. The magnetization of YIG can be calculated considering that the unit cell with lattice parameter $a$ has net magnetic moment $2 \times 4 \times (5/2)g\mu_B$, so that $M = 20g\mu_B/a^3$. Using the value $a = 1.24$ nm at $T = 0$, we obtain $M = 194$ G. At T $= 300$ K the magnetization reduces to $M = 140$ G. The Curie temperature is $T_c = 559$ K. YIG does not exist in natural minerals. Single crystal bulk YIG samples are grown from the melt by several methods, while films are grown by liquid-phase epitaxy, pulsed laser and sputter deposition [See Wu and Hoffmann].

### 3.7.1    Magnon Dispersion Relations in YIG

So far, we have studied spin waves only in ferromagnets having one spin per unit cell. In this case, for a specific wave vector direction, there is only one magnon dispersion curve. With 20 spins per primitive cell, YIG is expected to have 20 magnon dispersion curves. One of them corresponds to a mode that has all 20 spins precessing in phase, so it is called acoustic mode, in analogy to the elastic vibration modes. The other 19 modes have higher energies and are called optical modes. The acoustic mode corresponds to the spin waves studied in the previous sections, since the spins precessing in phase behave like one spin as in a ferromagnet. This mode can be probed by microwave techniques and inelastic light scattering only for wave numbers $k < 5 \times 10^5 \text{cm}^{-1}$.

**Fig. 3.9** Schematic diagram
of a triple-axis neutron
spectrometer. The letters
represent: C—Collimators;
n—neutron beam; M—
Monochromator; S—Sample;
A—Analyzer; D—Detector

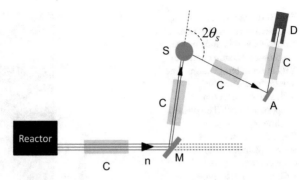

In order to probe acoustic modes with higher $k$ and to investigate optical modes, the standard experimental technique is inelastic neutron scattering. The uniqueness of this technique arises from the fact that thermal neutrons at room temperature have a wavelength $\lambda \approx 0.16$ nm, corresponding to a wave number $k \approx 4 \times 10^8 \text{cm}^{-1}$, that covers the whole Brillouin zone. The corresponding energy $(\hbar k)^2/2m$ is about 30 meV, comparable to those of the excitations in solids. Figure 3.9 shows a schematic diagram of a *tripple-axis spectrometer* used in neutron scattering experiments. An energetic neutron beam, produced in a nuclear reactor, passes through a collimator and a monochromator, and is directed towards the sample with energy $E_{inc}$ and momentum $\vec{p}_{inc}$. Since neutrons have no electric charge, they easily penetrate into the sample. Neutrons have spin, so they interact with the spins in magnetic samples, and are inelastically scattered by magnons. The energy $E_{sc}$ of the scattered neutrons are measured by the analyzer and the momentum $\vec{p}_{sc}$ is defined by the scattering angle, that can be varied mechanically. The conservation laws determine the energy and momentum of the magnons created in the scattering process

$$\hbar\omega_m = E_{inc} - E_{sc}, \quad \hbar\vec{k}_m = \vec{p}_{inc} - \vec{p}_{sc}. \tag{3.184}$$

By varying the energy of the incident neutron beam and scanning the scattering angle, one can measure the dispersion relations of excitations in solids. The first calculation of the magnon dispersion relations for YIG was made in the 1960s, using three nearest-neighbor exchange parameters inferred from magnetic measurements [30]. The first neutron scattering measurements, made two decades later [31], were in reasonable agreement with the calculations. Three decades later, data obtained in a modern neutron scattering facility [29], allowed the determination of more precise exchange parameters than the ones previously used for YIG. Figure 3.10 shows the twenty magnon dispersion curves for the wave vector varying along high symmetry directions in the reciprocal lattice. The $\Gamma$ point denotes the center of the Brillouin zone, $k = 0$, while the N and H points represent zone boundaries. The shaded curves represent the data and the solid lines are calculated with the parameters that provide

**Fig. 3.10** Neutron scattering measurements and fits with numerical simulations of the magnon dispersion curves in YIG at room temperature. The horizontal dashed line indicates $k_B T$ at room temperature [29]. The dashed red line represents Eq. (3.186), used as an approximate acoustic magnon dispersion in a spherical Brillouin zone. Reproduced with kind permission from [29]

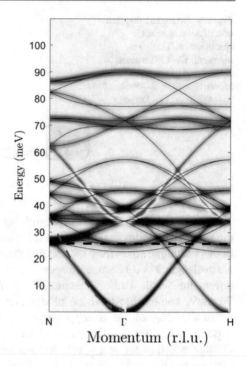

the best fit [29]. The fitting was made considering the exchange interaction between six nearest neighbors. The parameters corresponding to the first three nearest-neighbor interactions are $J_{ad} = -3.4$ meV, $J_{dd} = -0.26$ meV, and $J_{aa} = -0.27$ meV.

The dispersion curve for the acoustic magnon in Fig. 3.10 resembles the curve obtained for a ferromagnet, given by Eqs. (3.35 and 3.37), and shown in Fig. 3.2. Thus, near the center of the Brillouin zone, the magnon energy can be described by the parabolic approximation $\hbar\omega_k = g\mu_B D k^2$. Considering the definition in Eq. (2.19), $D = 2J_{eff}Sa^2/g\mu_B$, it can be shown [30] that the effective exchange interaction parameter $J_{eff}$ is related to the three nearest-neighbor parameters by

$$J_{eff} = \frac{1}{16}(8J_{aa} + 3J_{dd} - 5J_{ad}). \tag{3.185}$$

Using the values $D = 5.4 \times 10^{-9}$ Oe cm$^2$, $S = 5/2$, $a = 1.237 \times 10^{-7}$ cm, and $g\mu_B = 1.85 \times 10^{-20}$ erg G$^{-1}$, we obtain from (2.19) $J_{eff} = 0.82$ meV. Using in Eq. (3.185) the parameters of Princep et al. [29] we find $J_{eff} = 0.87$ meV, which is in quite good agreement considering that in Princep et al. [29] the fit was made using the parameters for the first six nearest neighbors.

The parabolic approximation for the dispersion relation works very well for calculating magnetic quantities at low temperatures, because only acoustic magnons with small wave numbers are thermally excited. However, for

$T = 300$ K, $k_B T = 25.8$ meV, so that one has to consider thermal magnons with large wave vectors. Calculations using the dispersion relations as in Fig. 3.10 are complicated and have to be done with sophisticated numerical methods [32]. In the following subsections, we will simplify the calculations assuming a spherical Brillouin zone and considering for the acoustic magnon branch the approximate dispersion relation [33]

$$\omega_k = \omega_{ZB}\left(1 - \cos\frac{\pi k}{2k_m}\right), \tag{3.186}$$

where $\omega_{ZB}$ is the zone boundary frequency and $k_m$ is the radius of the spherical Brillouin zone. The dashed red curve in Fig. 3.10 is plotted for $\omega_{ZB}/2\pi = 8$ THz, corresponding to an energy of 37.4 meV, which is intermediate between the values at the zone boundary points N and H. In regard to $k_m$, using Eq. (3.88), one can see that the radius of a sphere in $k$-space that has $N$ allowed wave vectors, that equals the number of spins in a volume $V = Na^3$, is given by $4\pi k_m^3/3 = (2\pi/a)^3$. Actually, since the spherical Brillouin zone is only an approximation used to simplify the calculations, $\omega_{ZB}$ and $k_m$ can be considered adjustable parameters about the values indicated, so as to provide good fits of theory to experimental data.

### 3.7.2  Magnon Relaxation in YIG

The mechanisms of magnon relaxation in YIG have been extensively studied for five decades and are well understood. In single crystals with no impurities, the spin-lattice coupling is very weak so that the relaxation takes place in two steps. In the first, the magnon mode excited by some external driving transfers its energy to other magnetic modes, in a time span that varies from tens of picoseconds to hundreds of nanoseconds, depending on the wave vector of the mode, the temperature, and the intensity of the applied magnetic field. In other words, the energy is redistributed in the magnetic system. In the second step the magnetic energy relaxes to the lattice, with a characteristic time of a few microseconds at room temperature, or longer at lower temperatures. In experiments, what is usually measured is the relaxation rate of the first step, because it is the one that determines the energy decay of the excited mode.

For magnons with frequencies in the gigahertz range and wave numbers up to about $5 \times 10^5$ cm$^{-1}$, the dominant relaxation mechanism is the 3-magnon confluence process illustrated in Fig. 3.5a. The excited mode with frequency $\omega_k$ and wave vector $\vec{k}$ interacts with a thermal magnon $\vec{k}_2$ by means of the 3-magnon dipolar interaction, resulting in a third magnon with wave vector $\vec{k}_3 = \vec{k}_2 + \vec{k}$ and frequency $\omega_3 = \omega_k + \omega_2$. With suitable approximations, the relaxation rate for this process, calculated with Eq. (3.114), is proportional to the wave number and to the temperature, and can be written as [34, 35]

$$\eta_{k-3m} = A(M, \omega_k, \theta_k)\, kT, \qquad (3.187)$$

where $A$ is a parameter that depends on the magnetization, the magnon frequency, and the polar angle of the wave vector.

An important experimental method to measure the relaxation rate of magnons with frequencies in the gigahertz range and wave numbers up to about $5 \times 10^5\,\mathrm{cm}^{-1}$ is the parallel pumping technique, that will be studied in detail in Chap. 6. It consists of applying a microwave magnetic field parallel to the static field to excite magnons with half the pumping frequency. This is a parametric process characterized by a threshold microwave pumping power. Magnons propagating perpendicularly to the field ($\theta_k = \pi/2$) are excited only when the microwave power exceeds a threshold value that is proportional to the magnon relaxation rate. Thus, by measuring the threshold as a function of the static magnetic field intensity, one can measure the relaxation rate as a function of the magnon wave number. Figure 3.11 shows the data obtained with microwave pumping of frequency 11.4 GHz in a sample of single-crystal YIG sphere with diameter 0.5 mm and highly polished surface. The measured magnon relaxation rate increases linearly proportional to the wave number $k$, according to Eq. (3.187) for the 3-magnon confluence process [36].

The data in Fig. 3.11 shows that in addition to the 3-magnon confluence process, the magnon relaxation has a contribution independent of $k$, given by the value at $k = 0$, $\eta_0 = 1/\tau_0 = 2.4 \times 10^6\,\mathrm{s}^{-1}$. This corresponds to a full linewidth of the magnon lineshape $\Delta H_0 = \eta_0/\gamma$ of only 0.14 Oe, the smallest of all known ferromagnetic materials. In small samples, the origin of the residual linewidth is a two-magnon process induced by pit scattering at the sample surface.

The fabrication of a spherical sample consists of tumbling a small piece of bulk YIG inside a cylindrical surface with a grinding compound. This is done in various steps, each new one with a grinding compound of finer pitch until the surface is

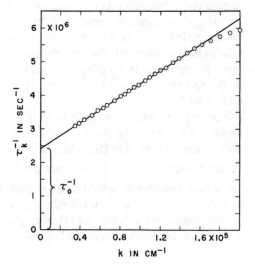

**Fig. 3.11** Relaxation rate versus wave number for spin waves in single-crystal YIG sphere. Magnons with frequency 5.7 GHz are driven by parallel pumping with a microwave field of frequency 11.4 GHz [36]. Reprinted with permission from T. Kasuya and R. C. Le Craw, Phys. Rev. Lett. **6**, 223, (1961). Copyright (1961) by the American Physical Society

optically polished. This process leaves pits on the surface, such that the magnetization precession associated with a magnon with wave vector $\vec{k}$ produces a demagnetizing field that has a Fourier transform in a range of wave vectors that depends on $\vec{k}$ and on the shape and size of the pit. This provides a magnetic dipolar mechanism for coupling two magnons without momentum conservation, that can be represented by the Hamiltonian

$$H_{dip}^{(2)} = \sum_{k,k'} V_{kk'}^{(2)} (c_k c_{k'}^\dagger + c_{k'}^\dagger c_k), \qquad (3.188)$$

where the interaction strength $V_{kk'}^{(2)}$ depends on the shape and size of the pits and on the wave vectors of the magnons involved [34]. The two-magnon pit-scattering mechanism is responsible for the FMR linewidth of small YIG spheres. Using a transition probability calculation one can show that the decay rate of the number of $k = 0$ magnons due to this process is

$$\frac{dn_0}{dt} = -\eta_0(n_0 - \bar{n}_0), \qquad (3.189)$$

where the relaxation rate is

$$\eta_0 = \frac{2\pi}{\hbar^2} \sum_k \left| V_{0k}^{(2)} \right|^2 \delta(\omega_0 - \omega_k). \qquad (3.190)$$

Calculation of (3.190) assuming a hemispherical shape for the pits provides good agreement with experimental data in YIG [34].

Another source of broadening of the FMR line in larger samples is radiation damping. From classical electromagnetic theory, one knows that the total power radiated by a magnetic dipole moment $\vec{\mu}$ precessing at a frequency $\omega_0$ in free space is

$$P_{rad} = \frac{2}{3} \frac{1}{c^3} \omega_0^4 \mu_\perp^2, \qquad (3.191)$$

where $\mu_\perp$ is the component of the magnetic moment in the plane perpendicular to the axis of precession and $c$ is the speed of light. This radiated power represents a loss of energy in the magnetic system, thus it contributes to the magnetic damping. The magnetic energy of the uniform precession mode in a sphere with magnetization $M$ and volume $V$, given by Eq. (2.73), can be written as

$$U_m = \frac{1}{2\gamma MV} \omega_0 \mu_\perp^2, \qquad (3.192)$$

where $\gamma$ is the gyromagnetic ratio and $\omega_0 = \gamma H$ is the frequency of the uniform mode. The relaxation rate for the radiation damping process is $\eta_{rad} = P_{rad}/U_m$, and the corresponding linewidth is $\Delta H_{rad} = \eta_{rad}/\gamma$, so that with Eqs. (3.191 and 3.192) we obtain

$$\Delta H_{rad} = \frac{2\pi M d^3}{9c^3}\,\omega_0^3.\tag{3.193}$$

For a YIG sphere with diameter $d$ in mm, using $M = 140$ G, Eq. (3.193) gives for a frequency of 10 GHz, a linewidth of $\Delta H_{rad} \approx 0.9\,d^3$Oe. This shows that in FMR experiments with YIG in the X-band microwave frequency range, the radiation-induced linewidth is larger than the contributions from other mechanisms, even in spheres with diameter as small as 1 mm. The radiation damping is clearly demonstrated in an arrangement that employs two horn microwave antennas, one for driving the FMR and the other for detecting the radiation from the sample, as shown in Fig. 3.12 [37]. A highly polished YIG sphere is held at the tip of a plastic rod placed between the poles of an electromagnet, such that the static field is perpendicular to the magnetic field of the driving microwave. As the static field is swept, the spin wave resonances in the sphere are excited and the precessing magnetization emits radiation that is picked up by the other antenna. Figure 3.12 shows the radiation spectra of two spheres with diameter 0.5 mm and 3 mm, excited by microwave radiation of frequency 8.9 GHz. The spectrum of the smaller sphere is weak because the radiating magnetic moment is small, and it exhibits only the peak of the uniform mode, with a linewidth smaller than 1 Oe. However, the spectra of the 3-mm sphere exhibit peaks of several spin wave resonances, also called magneto-static modes. The peak corresponding to the uniform precession mode, labeled (110), has a much larger linewidth of 15 Oe, providing a clear demonstration of the radiation-induced damping [37].

While the relaxation processes of magnons with $k < 5 \times 10^5\,\text{cm}^{-1}$ in YIG have been investigated experimentally and theoretically in detail, few studies have been

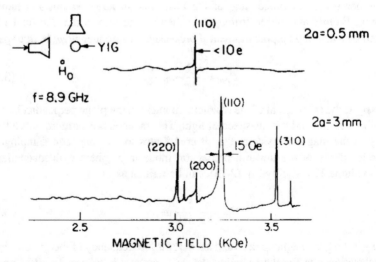

**Fig. 3.12** Radiation spectra of YIG spheres with diameter 0.5 and 3 mm excited by microwave radiation with frequency 8.9 GHz. The uniform mode, labeled (110), is shifted in field in the larger sample due to propagation effects [37]. Reproduced with kind permission from Elsevier Inc.

made for larger wave numbers. However, experiments with higher microwave frequencies indicate that four-magnon processes are important for $k \sim 10^6 \, \text{cm}^{-1}$ [38] and should dominate the relaxation for larger wave numbers. Next, we calculate the four-magnon relaxation rate for YIG using Eq. (3.116). We assume spherical energy surfaces in $k$-space with maximum radius $k_m$ and dispersion relation as in Eq. (3.186), $\omega_k = \omega_{ZB}(1 - \gamma_k)$. The 4-magnon interaction is considered to be dominated by exchange, as in Eq. (3.117), with the geometrical structure factor given by $\gamma_k = \cos(\pi k/2k_m)$. Define $\vec{k}_S = \vec{k} + \vec{k}_2 = \vec{k}_3 + \vec{k}_4$, so that the energy delta function in Eq. (3.116) can be transformed into

$$\delta(\omega) = \frac{2k_m k_4 \delta(u - u_0)}{\omega_{ZB} \pi k_S k_3 \sin(\pi k_4/2k_m)}, \qquad (3.194)$$

where $u = \cos\theta_3$, $\theta_3$ being the angle between $\vec{k}_3$ and $\vec{k}_S$, and $u_0$ is the value of $u$ for which energy is conserved $\omega_k + \omega_2 = \omega_3 + \omega_4$. The calculation proceeds as follows: the sum over $k_4$ is eliminated by momentum conservation and the other sums are converted into integrals over the Brillouin zone with $\sum \to N a^3/(2\pi)^3 \int d^3k$. Then the integrals over the azimuthal angles $\varphi_3$ and $\varphi_4$ are freely evaluated giving $(2\pi)^2$ so that the relaxation rate becomes

$$\eta_{k-4m} = (e^{\hbar\omega_k/k_B T} - 1) \frac{\omega_{ZB} a^6}{64 S^2 \pi^4} \int k_2^2 dk_2 \int_{-\pi}^{\pi} \sin\theta_2 d\theta_2 \int k_3^2 dk_3$$

$$\times \int_{-1}^{1} du \, \delta(u - u_0) |C_{1234}|^2 \bar{n}_{k2} \bar{n}_{k3} \bar{n}_{k4} \frac{k_m k_4}{k_S k_3 \sin(\pi k_4/2k_m)}, \qquad (3.195a)$$

where

$$C_{1234} = (\gamma_{k1} + \gamma_{k2} + \gamma_{k3} + \gamma_{k4} - 4\gamma_{k4-k2}). \qquad (3.195b)$$

The integrals in $k_2$, $k_3$, and $\theta_2$ can be evaluated numerically by coarse sums over a spherical Brillouin zone for a fixed temperature $T$ and varying wave number $k$, or for fixed $k$ and varying $T$. The sums are carried out by dividing the integration range in a large number of points, letting $k_2$, $k_3$, and $\theta_2$ assume all possible values and considering for the sums only those values for which $|u_0| \leq 1$, where $u_0$ is calculated from $k_4^2 = k_S^2 + k_3^2 - 2k_S k_3 u_0$.

The calculation gives a relaxation rate that increases monotonically with $k$, but drops near the zone boundary. This drop can be eliminated by considering umklapp processes, in which the momentum conservation assumes the form $\vec{k} + \vec{k}_2 = \vec{k}_3 + \vec{k}_4 + \vec{G}$, where $\vec{G}$ is a vector of the reciprocal lattice. The difficulty in defining $\vec{G}$ in a spherical Brillouin zone is resolved by assuming it either parallel or perpendicular to $\vec{k}$ and having amplitude $2k_m$. The dashed curve in Fig. 3.13 represents the calculated 4-magnon relaxation rate as a function of the wave number in YIG, for $T = 300$ K, considering $\omega_{ZB}/2\pi = 7$ THz and $k_m a = 2.8$ [33], to account

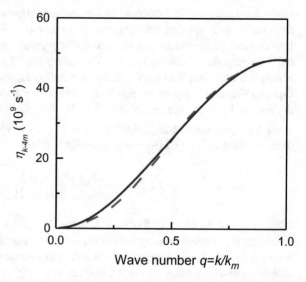

**Fig. 3.13** The dashed curve represents the calculated 4-magnon relaxation rate in YIG at $T = 300$ K as a function of the normalized wave number. The solid curve is a polynomial fit

for the energy renormalization not considered in [39]. It is noticeable that the relaxation rate for $k \sim 10^7$ cm$^{-1}$ is on the order of $\eta_k \sim 10^{11}$s$^{-1}$, which corresponds to a magnon lifetime of $\tau_k \sim 10^{-11}$ s. The solid line in Fig. 3.13 represents a simple polynomial fit to the calculated 4-magnon relaxation rate, given by

$$\eta_{k-4m} = (15.2q^2 - 9.8q^3) \times 10^{10} \text{ s}^{-1}. \tag{3.196}$$

where $q = k/k_m$. This expression can be used in analytical calculations involving the magnon relaxation rate. Calculation of the relaxation rate for a fixed wave number in the middle of the Brillouin zone gives an approximate $T^2$ temperature dependence.

To conclude this subsection, we note that considering the mechanisms presented here, the total magnon relaxation rate in YIG at 300 K can be represented approximately by the expression

$$\eta_k = \eta_0 + (1.5q + 15.2q^2 - 9.8\,q^3) \times 10^{10} \text{ s}^{-1}, \tag{3.197}$$

where the coefficient of the 3-magnon term was obtained from calculations and experimental data for YIG. Considering the temperature dependencies of the 3- and 4-magnon relaxation processes, one can write for the total relaxation rate

$$\eta_k(T) = \eta_0(T) + \left[ 1.5\,q\left(\frac{T}{300}\right) + (15.2\,q^2 - 9.8\,q^3)\left(\frac{T}{300}\right)^2 \right] \times 10^{10} \text{ s}^{-1}.$$

$$\tag{3.198}$$

This expression will be used in this book to calculate quantities in YIG that depend on the magnon relaxation rate. Note, however, that the value of the relaxation rate $\eta_0$ at the zone center is sample dependent. In YIG films it can have an additional contribution from two-magnon scattering at the surfaces and interfaces. Also, the presence of impurities creates an additional temperature-dependent contribution [9, 40].

### 3.7.3 Thermal Properties of Magnons in YIG

As we saw in Sect. 3.3, the decrease of the magnetization with increasing temperature is caused by the excitation of thermal magnons. At low temperatures, the temperature dependence of the magnetization is described quite well by a calculation using the quadratic magnon dispersion. The same is true for other thermodynamic quantities, such as the magnetic specific heat and thermal conductivity. In YIG, this calculation is expected to give good results only for temperatures up to 100 K.

One difficulty for testing the spin wave calculations for some thermodynamic quantities in YIG is that, for temperatures above about 20 K, its thermal properties are dominated by the behavior of phonons, so that the contributions from magnons are very difficult to measure separately. As a result, the thermal properties of magnons are well characterized and explained by theory only at low temperatures, where magnons with small wave numbers are involved. This section is devoted to the calculation of the temperature dependencies of the magnetization, magnetic specific heat and thermal conductivity in YIG. We will present the calculations for low temperatures that can be found in other textbooks, but will also show results for elevated temperatures.

Initially, we present a calculation of the energy renormalization, because the knowledge of the temperature dependence of the magnon energy is essential for calculating the thermodynamic quantities at higher temperatures. Replacing the sum over wave vectors by an integral in $k$-space in Eq. (3.106b), using $2zJ = \hbar\omega_{ZB}/S$, and considering a spherical Brillouin zone with radius $k_m$, the change in energy of the $k$-magnon becomes

$$\hbar\Delta\omega_k = \frac{\hbar\omega_{ZB}(k_m a)^3}{2\pi^2 S} \int\limits_0^1 q^2 \, dq \frac{(1 + \gamma_{k-q} - \gamma_k - \gamma_q)}{e^x - 1}, \tag{3.199}$$

where $x = \hbar\omega_k/k_B T$ is the normalized energy and $q = k/k_m$ is the normalized wave number, that varies from 0 to 1. In order to obtain Eq. (3.199), the dependence of the integrand on the angle between the wave vectors was neglected and the integration on the polar and azimuthal angles was freely evaluated. For each temperature the integral in (3.199) was evaluated numerically by a discrete sum over a spherical Brillouin zone, considering the unrenormalized frequencies given by the dispersion relation in Eq. (3.186) and the corresponding geometric factor, and with the approximation $\gamma_{k-q} = \gamma_k\gamma_q$. The numbers used for

YIG are $S = 5/2$, $\omega_{ZB}/2\pi = 8.0\,\text{THz}$ (at $T = 0$ K), and $k_m a = 2.8$, which are the values that give a good fit of theory to the magnetization data, as will be shown later.

In the first cycle of the evaluation of $\Delta\omega_k$, the frequencies and the Bose factors are calculated for each point $k$ in the Brillouin zone without renormalization. In the following cycles, the new Bose factors are calculated with the magnon frequencies $\omega_k(T) = \omega_k(0) + \Delta\omega_k$, using $\Delta\omega_k$ from the previous cycle. The process is repeated until the change in frequency at all points is smaller than 0.1%. Figure 3.14 shows the calculated dispersions for three temperatures. From 0 to 300 K, the zone boundary magnon frequency reduces by about 12%, and from 300 to 400 K it reduces by 8%. This variation is very similar to the one calculated for YIG with the full magnon dispersion, using three nearest-neighbors exchange interactions, and the exact Brillouin zone [32].

In the calculation of the temperature dependence of the magnetization, we first consider the approximation for low temperatures. As shown in Sect. 3.3.2, using the quadratic dispersion $\omega_k \propto k^2$ and assuming thermal excitation of magnons with small wave numbers, the magnetization decreases with increasing temperature according to Bloch's law, $\Delta M(T) = -a_0 T^{3/2}$. Using for YIG $a = 1.24 \times 10^{-7}\,\text{cm}^{-1}$ and $D = 5.4 \times 10^{-9}\,\text{Oe}\,\text{cm}^2$, we obtain with Eq. (3.91) the dashed red curve in Fig. 3.15. The calculated variation agrees with the experimental data of Anderson [41] only at temperatures below 50 K. Considering the magnon interactions, Dyson has shown that the change in magnetization can be expressed in a power series of temperature in the form $\Delta M(T) = -a_0 T^{3/2} - a_1 T^{5/2} - a_2 T^{7/2} - a_3 T^4 + O(T^{2})$, where all coefficients are positive [42]. This power series improves the agreement between theory and experimental data only to about 100 K. For higher temperatures, calculations with the quadratic dispersion relation and integration to infinity underestimate the variation of the magnetization.

**Fig. 3.14** Magnon dispersion relations in YIG at various temperatures, calculated with energy renormalization due to 4-magnon interaction

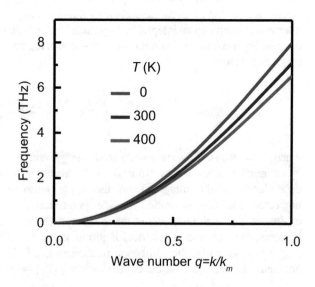

**Fig. 3.15** Variation of the spontaneous magnetization of YIG with temperature. Circles represent the data of Anderson [41]. Dashed red curve is calculated with the Bloch's low temperature expression Eq. (3.91). Wine curve is calculated with Eq. (3.200) without magnon energy renormalization. Solid blue curve is calculated with Eq. (3.200) with magnon energy renormalization

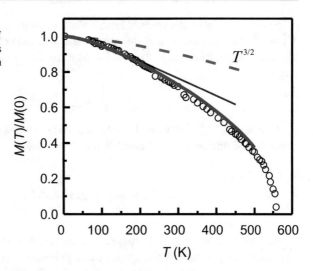

Two steps are taken to improve the spin wave calculation of the thermal properties, namely, use the actual finite upper limit in the integration over the Brillouin zone, and use Eq. (3.186) for the magnon dispersion over the whole Brillouin zone. From Eq. (3.89), we write the temperature dependence of the magnetization as

$$\frac{M(0) - M(T)}{M(0)} = \frac{(k_m a)^3}{S 2\pi^2} \int\limits_0^1 dq\, q^2\, \frac{1}{e^x - 1}. \tag{3.200}$$

The integration in (3.200) is evaluated by a coarse sum, initially with the unrenormalized magnon energies. Figure 3.15 shows the solid wine curve obtained with $\omega_{ZB}/2\pi = 8.0$ THz and $k_m a = 2.8$, which is the value that gives best agreement with the experimental data at low temperatures. The result is much better than the one obtained with the quadratic dispersion and integration to infinity, but the agreement is good only up to about 200 K.

For higher temperatures, it is essential to consider the magnon energy renormalization. The calculation done with the renormalization as in Eq. (3.199), represented by the solid blue curve in Fig. 3.15, is in very good agreement with the experimental data up to 500 K, which is 90% of the Curie temperature in YIG. This result demonstrates that spin wave theory is not restricted to low temperatures. It can be used to calculate magnetic thermodynamic properties at quite high temperatures, as long as the magnon interactions are taken into account. It also shows that calculations considering only the thermal excitation of acoustic magnons in YIG give good agreement with experimental data at room temperature.

Next we calculate the specific heat, or heat capacity per unit volume, defined by $C_V = (\partial U/\partial T)/V$, where $U$ is the energy and $V$ is the volume of a thermodynamic system. The contribution to the specific heat from the magnetic system is given by

$$C_m = \frac{1}{V}\frac{\partial}{\partial T}\sum_k \bar{n}_k \hbar \omega_k. \tag{3.201}$$

To evaluate (3.201), we replace the sum over wave vectors by an integral in the usual way. Considering a spherical Brillouin zone we have

$$C_m = \frac{k_B}{2\pi^2}\int\limits_0^{k_m} dk\, k^2 \frac{e^x x^2}{(e^x - 1)^2}. \tag{3.202}$$

At low temperatures only magnons with small wave numbers are thermally excited, so that we can use the approximate quadratic dispersion relation $\omega_k = \gamma D k^2$. One can then express $k$ in terms of $x$ so that Eq. (3.202) becomes

$$C_m = \frac{k_B^{5/2} T^{3/2}}{4\pi^2 (\gamma D \hbar)^{3/2}}\int\limits_0^{x_m} dx\, \frac{e^x x^{5/2}}{(e^x - 1)^2}. \tag{3.203}$$

Since the number of thermal magnons with wave number $k_m$ is negligible at low temperatures, the upper limit in Eq. (3.203) can be set to infinity and the integral solved analytically to give

$$C_m = \frac{15}{4}\zeta(5/2)\frac{k_B^{5/2} T^{3/2}}{(4\pi\gamma\hbar D)^{3/2}}, \tag{3.204}$$

where $\zeta(5/2)$ is Riemann zeta function. Equation (3.204) with the $T^{3/2}$ dependence is a well-known result for the magnetic specific heat at low temperatures. Calculation of (3.204) with the parameters for YIG and $\zeta(5/2) \approx 1.34$ gives the dashed red curve in Fig. 3.16. At $T = 300$ K the value is $1.4 \times 10^4$ J/(K m$^2$). For elevated temperatures the calculation is done numerically with the dispersion relation (3.186). Using Eq. (3.202) with the normalized wave number we have

$$C_m = \frac{k_B k_m^3}{2\pi^2}\int\limits_0^1 dq\, q^2 \frac{e^x x^2}{(e^x - 1)^2}. \tag{3.205}$$

The solid blue curve in Fig. 3.16 represents the result of the numerical evaluation of (3.205) for YIG, with $\omega_{ZB}/2\pi = 8$ THz and $k_m a = 2.8$ cm$^{-1}$, which are the same values used to calculate the magnetization. The magnon energy renormalization was introduced in the calculation, but its effect is quite small for temperatures below 350 K.

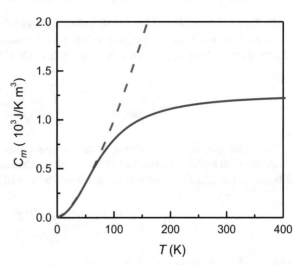

**Fig. 3.16** Variation with temperature of magnetic specific heat in YIG. Dashed red curve is calculated with the low-temperature expression Eq. (3.204). Solid blue curve is calculated with Eq. (3.205) with magnon energy renormalization [33]. Reprinted with permission from S. M. Rezende et al., Phys. Rev. B **89**, 134406 (2014). Copyright (2014) by the American Physical Society

As seen in Fig. 3.16, while Eq. (3.204) predicts a continuous increase in $C_m$ with increasing temperature in the form $T^{3/2}$, the calculation with the appropriate treatment of the magnon system shows that $C_m$ tends to saturate at higher $T$. This is due to the fact that the number of magnon modes in the Brillouin zone is finite and above certain temperature they are almost all thermally excited, while in the calculation for low temperatures there is no limit to the number of excited magnons. The calculation with Eq. (3.205) gives for YIG at $T = 300$ K $C_m \approx 1.2 \times 10^3$ J/(K m$^3$), which is one order of magnitude smaller than the value predicted by Eq. (3.204).

To close this section, we present calculations of the magnon thermal conductivity. Consider that a temperature gradient $\nabla T$ is applied in a ferromagnet with volume $V$ creating a nonequilibrium magnon distribution $n_k(\vec{r}, T)$. The flow of magnons due to $\nabla T$ generates a heat–current density given by $\vec{J}_Q = V^{-1}\sum_k \delta n_k \hbar \omega_k \vec{v}_k$, where $\delta n_k = n_k - \bar{n}_k$ is the magnon number in excess of equilibrium and $\vec{v}_k$ is the $k$-magnon group velocity. Using the Boltzmann transport equation [39] one can write a first-order expression for the excess magnon number in the steady-state and in the relaxation approximation, $\delta n_k = -\tau_k \vec{v}_k \cdot \nabla \bar{n}_k$, where $\tau_k$ is the $k$-magnon relaxation time. Assuming spherical magnon energy surfaces one obtains a heat–current density in the form $\vec{J}_Q = -K_m \nabla T$, where $K_m$ is the magnon thermal conductivity given by

$$K_m = \frac{k_B}{6\pi^2} \int dk\, k^2 \tau_k \nu_k^2 \frac{e^x x^2}{(e^x - 1)^2}. \tag{3.206}$$

In the usual treatment of Eq. (3.206), one employs a quadratic dispersion relation $\omega_k = \gamma D k^2$ which implies a magnon velocity $v_k = \partial \omega_k / \partial k = 2\gamma D k$. Assuming that the magnon lifetime $\tau_m$ is independent of $k$ and $T$, we obtain

$$K_m = \frac{\tau_m k_B^{7/2} T^{5/2}}{3\pi^2 (\gamma D)^{1/2} \hbar^{5/2}} \int_0^{x_m} dx \frac{e^x x^{7/2}}{(e^x - 1)^2}. \qquad (3.207)$$

As in the previous calculations, at low temperatures the occupation number of magnons with higher energies can be neglected and one can set the upper limit of the integral in Eq. (3.207) to infinity so that it can be solved analytically to give

$$K_m = \frac{\Gamma(9/2)\zeta(7/2)}{3\pi^2} \frac{\tau_m k_B^{7/2} T^{5/2}}{(\gamma D)^{1/2} \hbar^{5/2}}, \qquad (3.208)$$

which is the well-known result of Yellon and Berger [43]. Using for YIG the same parameters given before and $\tau_m = 10^{-10}$ s, with $\zeta(7/2) \approx 3.38$ one obtains from Eq. (3.208) at $T = 300$ K the value $K_m = 1.97 \times 10^3$ W/(K m).

The calculation for elevated temperatures is done with the magnon dispersion relation in Eq. (3.186), that gives a group velocity $v_k = \omega_{ZB}(\pi/2k_m) \sin(\pi k/2k_m)$. We also consider that the magnon lifetime $\tau_k = 1/\eta_k$ varies with wave number and temperature. From Eq. (3.207) we have

$$K_m = \frac{\omega_{ZB}^2 \tau_0 k_m k_B}{24} \int_0^1 dq\, q^2 \sin^2\left(\frac{\pi q}{2}\right) \frac{e^x x^2}{\eta_q (e^x - 1)^2}, \qquad (3.209)$$

where the magnon lifetime has been written as $\tau_k = \tau_0/\eta_q$, where $\tau_0$ is the lifetime of $k \approx 0$ magnons and $\eta_q = \tau_0 \eta_k$ is a dimensionless relaxation rate. The evaluation of Eq. (3.209) for YIG, with $\omega_{ZB}/2\pi = 8$ THz and $k_m a = 2.8$ cm$^{-1}$, using the relaxation rate in Eq. (3.198), with $\tau_0 = 5 \times 10^{-7}$ s, and considering the magnon energy renormalization, is represented by the curve in Fig. 3.17. Note that the initial increase of $K_m$ with temperature follows the $T^{5/2}$ power law of Eq. (3.208). However, as the temperature increases further two facts contribute for $K_m$ to peak at $T = 50$ K and drop fast at higher $T$: (1) The number of magnons involved in the heat transport saturates at higher temperatures, as a result of the finite size of the Brillouin zone, represented by the upper limit of the integral in Eq. (3.209); (2) The lifetime of magnons with large wave numbers decreases with increasing temperature as $1/T^2$, contributing to a fast decrease in $K_m$. Note that the value of the magnon thermal conductivity for YIG at $T = 300$ K is $K_m \approx 5$ W/(m K), which is about two orders of magnitude smaller than the value obtained with Eq. (3.208).

To close this section, we note that the calculations of the temperature dependencies of the magnetization, magnetic specific heat, and magnon thermal conductivity demonstrate that spin wave theory is not restricted to low temperatures.

**Fig. 3.17** Variation with temperature of the magnon thermal conductivity in YIG calculated with Eq. (3.209) with magnon energy renormalization [33]. Reprinted with permission from S. M. Rezende et al., Phys. Rev. B **89**, 134406 (2014). Copyright (2014) by the American Physical Society

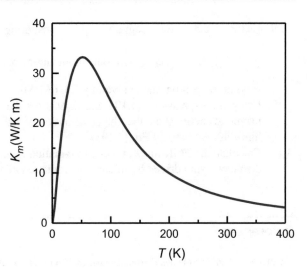

It can be used to calculate magnetic thermodynamic properties at quite high temperatures, as long as the magnon interactions are properly taken into account.

## Problems

3.1. Show that the spin state in Eq. (3.11) is an eigenstate of the Hamiltonian (3.8).

3.2. Using Eqs. (3.17a, 3.17b and 3.19a, 3.19b) for the spin-deviation operators, demonstrate Eq. (3.20a–3.20c) that establish their relation with the spin operators.

3.3. Using Eqs. (3.31 and 3.22a–3.22c) show that the spin wave state in (3.38) is the same as the Bloch state in Eq. (3.11).

3.4. Using the relations (3.78a and 3.78b) show that the eigenstates of the magnon Hamiltonian (3.74) are normalized and orthogonal to states with different $\vec{k}$.

3.5. With Eqs. (2.85 and 3.82a and 3.82b), and the relation between the magnetization and the spin, show that the linear momentum operator is given by

$$\vec{P}_m = \sum_k \hbar \vec{k}\, c_k^\dagger c_k.$$

3.6. Using the Bose–Einstein distribution for magnons in Eq. (3.86), demonstrate the relation in (3.111) for the thermal magnon numbers considering energy conservation $\omega_{k\,+\,k'} = \omega_k + \omega_{k'}$.

3.7. Show that the transformations in (3.140) diagonalize the Hamiltonian (3.139) for the coupled magnon–phonon system.

3.8. Using the Heisenberg equations of motion for the magnon and photon operators with the Hamiltonian (3.154), calculate the dispersion relation for the magnon–photon hybrid excitation. Show that Eq. (3.155) is the same as Eq. (2.94) for the magnetic polariton (with $\omega_k = \omega_0$).

3.9. Using the two-magnon pit scattering Hamiltonian $H_{dip}^{(2)} =$ $\sum_{k,k'} V_{kk'}^{(2)} (c_k c_{k'}^\dagger + c_{k'}^\dagger c_k)$, without momentum conservation, show that the magnon relaxation rate is given by Eq. (3.190).

3.10. Using the expressions (3.191 and 3.192) for the radiated power and stored magnetic energy show that the contribution of radiation damping to the FMR linewidth is given by Eq. (3.193).

3.11. Calculate the FMR linewidth due to radiation damping for a YIG sphere with diameter 1 mm driven by microwave radiation of frequency 30 GHz.

## References

1. Bloch, F.: Zur Theorie des Ferromagnetismus. Z. Physik. **61**, 206 (1930)
2. Holstein, T., Primakoff, H.: Field dependence of the intrinsic domain magnetization of a ferromagnet. Phys. Rev. **58**, 1098 (1940)
3. Cohen, M.H., Keffer, F.: Dipolar sums in the primitive cubic lattices. Phys. Rev. **99**, 1128 (1955)
4. White, R.M., Sparks, M., Ortenburger, I.: Diagonalization of the antiferromagnetic magnon-phonon interaction. Phys. Rev. **139**, A450 (1965)
5. Low, G.G.: Application of spin wave theory to 3 magnetic salts. Proc. Phys. Soc. Lond. **82**, 992 (1963)
6. Rezende, S.M., Zagury, N.: Coherent magnon states. Phys. Lett. A. **29**, 47 (1969)
7. Zagury, N., Rezende, S.M.: Theory of macroscopic excitations of magnons. Phys. Rev. B. **4**, 201 (1971)
8. Glauber, R.J.: Coherent and incoherent states of the radiation field. Phys. Rev. **131**, 2766 (1963)
9. Sparks, M.: Effect of impurities on the microwave properties of yttrium iron garnet. J. Appl. Phys. **38**, 1031 (1967)
10. Korenman, V., Prange, R.E.: Anomalous damping of spin waves in magnetic metals. Phys. Rev. B. **6**, 2769 (1972)
11. Mills, D.L.: Ferromagnetic resonance relaxation in ultrathin metal films: the role of the conduction electrons. Phys. Rev. B. **68**, 014419 (2003)
12. Garanin, D.A., Chudnovsky, E.M.: Angular momentum in spin-phonon processes. Phys. Rev. B. **92**, 024421 (2015)
13. Rückriegel, A., Kopietz, P., Bozhko, D.A., Serga, A.A., Hillebrands, B.: Magnetoelastic modes and lifetime of magnons in thin yttrium iron garnet films. Phys. Rev. B. **89**, 184413 (2014)
14. Guerreiro, S.C., Rezende, S.M.: Magnon-phonon interconversion in a dynamically reconfigurable magnetic material. Phys. Rev. B. **92**, 214437 (2015)
15. Holanda, J., Maior, D.S., Azevedo, A., Rezende, S.M.: Detecting the phonon spin in magnon–phonon conversion experiments. Nat. Phys. **14**, 500 (2018)
16. Chaudhuri, S., Keffer, F.: Classical and quantum theories of radiation damping in ferromagnetic and antiferromagnetic resonance. J. Phys. Chem. Solids. **45**, 47 (1984)
17. Elliot, R.J., Loudon, R.: The possible observation of electronic Raman transitions in crystals. Phys. Lett. **3**, 189 (1963)
18. Shen, Y.R., Bloembergen, N.: Interaction between light waves and spin waves. Phys. Rev. **143**, 372 (1966)
19. Fleury, P.A., Loudon, R.: Scattering of light by one- and two-magnon excitations. Phys. Rev. **166**, 514 (1968)
20. Pershan, P.S.: Magneto-optical effects. J. Appl. Phys. **38**, 1482 (1967)

21. Le Gall, H., Jamet, J.P.: Theory of the elastic and inelastic scattering of light by magnetic crystals. I. First-order processes. Phys. Stat. Solidi (b). **46**, 467 (1971)
22. Le Gall, H., Vien, T.K., Desormiére, B.: Theory of the elastic and inelastic scattering of light by magnetic crystals. II. Second-order processes. Phys. Stat. Solidi (b). **47**, 591 (1971)
23. Moriya, T.: Theory of light scattering by magnetic crystals. J. Phys. Soc. Jpn. **23**, 490 (1967)
24. Sandercock, J.R., Wettling, W.: Light scattering from thermal acoustic magnons in yttrium iron garnet. Solid State Comm. **13**, 1729 (1973)
25. Bertaut, F., Forrat, F.: Structure des ferrites ferrimagnetiques des terres rares. Compt. Rend. **242**, 382 (1956)
26. Cherepanov, V., Kolokolov, I., L'vov, V.: The saga of YIG: spectra, thermodynamics, interaction and relaxation of magnons in a complex magnet. Phys. Rep. **229**, 81 (1993)
27. Serga, A.A., Chumak, A.V., Hillebrands, B.: YIG magnonics. J. Phys. D. Appl. Phys. **43**, 264002 (2010)
28. Chumak, A.V., Vasyuchka, V.I., Serga, A.A., Hillebrands, B.: Magnon spintronics. Nat. Phys. **11**, 453 (2015)
29. Princep, A.J., Ewings, R.A., Ward, S., Tóth, S., Dubs, C., Prabhakaran, D., Boothroyd, A.T.: The full magnon spectrum of yttrium iron garnet. NPJ Quantum Mater. **2**, 63 (2017)
30. Brooks Harris, A.: Spin-wave spectra of yttrium and gadolinium iron garnet. Phys. Rev. **132**, 2398 (1963)
31. Plant, J.S.: 'Pseudo-acoustic' magnon dispersion in yttrium iron garnet. J. Phys. C Solid State Phys. **16**, 7037 (1983)
32. Barker, J., Bauer, G.E.W.: Thermal spin dynamics of yttrium iron garnet. Phys. Rev. Lett. **117**, 217201 (2016)
33. Rezende, S.M., Rodríguez-Suárez, R.L., Lopez Ortiz, J.C., Azevedo, A.: Thermal properties of magnons and the spin Seebeck effect in yttrium iron garnet/normal metal hybrid structures. Phys. Rev. B. **89**, 134406 (2014)
34. Sparks, M., Loudon, R., Kittel, C.: Ferromagnetic relaxation. I. Theory of the relaxation of the uniform precession. Phys. Rev. **122**, 791 (1961)
35. Sparks, M.: Theory of three-magnon ferromagnetic relaxation frequency for low temperatures and small wave vectors. Phys. Rev. **160**, 364 (1967)
36. Kasuya, T., Le Craw, R.C.: Relaxation mechanisms in ferromagnetic resonance. Phys. Rev. Lett. **6**, 223 (1961)
37. Montarroyos, E., Rezende, S.M.: Radiation damping by magnetostatic modes in YIG. Solid State Comm. **19**, 795 (1976)
38. A. N. Anisimov and A. G. Gurevich, Damping of spin waves due to four-magnon scattering, Fiz. Tverd. Tela **18**, 38 (1976); Sov. Phys. Solid State **18**, 20 (1976)
39. Rezende, S.M., Rodríguez-Suárez, R.L., Cunha, R.O., Rodrigues, A.R., Machado, F.L.A., Fonseca Guerra, G.A., Lopez Ortiz, J.C., Azevedo, A.: Magnon spin-current theory for the longitudinal spin-Seebeck effect. Phys. Rev. B. **89**, 014416 (2014)
40. Jermain, C.L., Aradhya, S.V., Reynolds, N.D., Buhrman, R.A., Brangham, J.T., Page, M.R., Hammel, P.C., Yang, F.Y., Ralph, D.C.: Increased low-temperature damping in yttrium iron garnet thin films. Phys. Rev. **95**, 174411 (2017)
41. Anderson, E.E.: Molecular field model and the magnetization of YIG. Phys. Rev. **134**, A1581 (1964)
42. Dyson, F.J.: Thermodynamic behavior of an ideal ferromagnet. Phys. Rev. **102**, 1230 (1956)
43. Yelon, W.B., Berger, L.: Magnon heat conduction and magnon scattering processes in Fe-Ni alloys. Phys. Rev. B. **6**, 1974 (1972)

## Further Reading

Akhiezer, A.I., Bar'yakhtar, V.G., Peletminskii, S.V.: Spin Waves. North-Holland, Amsterdam (1968)
Cottam, M.G., Lockwood, D.J.: Light Scattering in Magnetic Solids. Wiley, New York (1986)

Gurevich, A.G., Melkov, G.A.: Magnetization Oscillations and Waves. CRC Press, Boca Raton, FL (1994)

Kabos, P., Stalmachov, V.S.: Magnetostatic Waves and their Applications. Chapman and Hall, London (1994)

Keffer, F.: Spin waves. In: Flugge, S. (ed.) Handbuch der Physik, vol. XVIII/B. Springer, Heidelberg (1966)

Kittel, C.: Quantum Theory of Solids. Wiley, New York (1963)

Lax, B., Button, K.J.: Microwave Ferrites and Ferrimagnetics. McGraw-Hill, New York (1962)

Mattis, D.C.: The Theory of Magnetism. Harper & Row, New York (1965)

Sparks, M.: Ferromagnetic Relaxation. Mc Graw-Hill, New York (1964)

Stancil, D.D., Prabhakar, A.: Spin Waves: Theory and Applications. Springer Science, New York (2009)

White, R.M.: Quantum Theory of Magnetism, 3rd edn. Springer, Heidelberg (2007)

Wu, M., Hoffmann, A. (eds.): Recent Advances in Magnetic Insulators – From Spintronics to Microwave Applications. Academic Press, San Diego (2013)

# Magnonics in Ferromagnetic Films

**4**

In this chapter, we study spin waves in thin ferromagnetic films. Initially, we present the semiclassical theory of magnetostatic waves developed in the 1960s by Damon and Eshbach to explain microwave measurements in yttrium iron garnet bulk samples in the form of slabs. Later, with the development of techniques to fabricate thin films, the theory became the basis to study magnonic phenomena in these systems. Next, we present a quantum formulation of spin waves in thin films that applies to wave numbers from the magnetostatic to the exchange regions. The third topic is the spin wave relaxation by two-magnon scattering originating in the roughness of the surfaces and interfaces of ultrathin ferromagnetic films. Finally, we present a semiclassical calculation of the spin wave properties in coupled magnetic films.

## 4.1 Magnetostatic Spin Waves in Ferromagnetic Films

In the study of spin waves in the previous chapters, we have ignored the boundary conditions at the sample surfaces. This is a valid approximation for $k > 10^3$ cm$^{-1}$, because this range of wave numbers corresponds to wavelengths much smaller than typical sample dimensions. In this case, the variation of the magnetization at the sample surfaces is such that the surface demagnetizing fields are negligible. As we saw in Chap. 1, for the $k = 0$ magnon mode, the surface demagnetization has an important effect in the FMR frequency. Here we are interested in the effect of the surface dipolar fields on the frequency of spin waves with wavelengths comparable to the sample dimensions, such as thin films. The standard method to study these waves is to use the equation for the magnetic scalar potential in a ferromagnetic medium, first derived by Larry Walker [1].

© Springer Nature Switzerland AG 2020
S. M. Rezende, *Fundamentals of Magnonics*, Lecture Notes in Physics 969,
https://doi.org/10.1007/978-3-030-41317-0_4

## 4.1.1  Walker Equation

We consider spin waves with wave numbers in the range $10 < k < 10^3$ cm$^{-1}$, so that the effect of the exchange interaction is negligible and also $kc \gg \omega$. The last condition implies that the electric field of the wave can be neglected, so that the magnetic field can be described by Maxwell's equations in the magnetostatic limit

$$\nabla \times \vec{h} = 0, \quad \nabla \cdot \vec{h} + 4\pi \nabla \cdot \vec{m} = 0, \tag{4.1}$$

where we consider for the *rf* magnetic field $\vec{h}(\vec{r},t) = \mathrm{Re}\left[\vec{h}(\vec{r})\exp\left(-i\omega t\right)\right]$, $\vec{h}(\vec{r})$ being a complex vector, and for the *rf* magnetization $\vec{m}(\vec{r},t) = \mathrm{Re}\left[\vec{m}(\vec{r})\exp\left(-i\omega t\right)\right]$. As in the previous chapters, we consider that the ferromagnet is under a static magnetic field along the *z*-direction of a Cartesian coordinate system, so that $\vec{m}$ has components only in the *x*- and *y*-directions.

In order to solve Eq. (4.1), we need a constitutive relation between $\vec{m}$ and $\vec{h}$. This was obtained in Chap. 1 with the Landau–Lifshitz equation, and is given by Eq. (1.22), that in the CGS reads

$$4\pi\vec{m} = \overline{\chi} \cdot \vec{h}, \tag{4.2}$$

where $\overline{\chi}$ is the *rf* magnetic susceptibility tensor, that here we write as

$$\overline{\chi} = \begin{pmatrix} \kappa & -iv & 0 \\ iv & \kappa & 0 \\ 0 & 0 & 0 \end{pmatrix}, \tag{4.3}$$

where

$$\kappa = \frac{\omega_M \omega_H}{\omega_H^2 - \omega^2}, \quad v = \frac{\omega_M \omega}{\omega_H^2 - \omega^2}, \tag{4.4}$$

where $\omega_M = \gamma 4\pi M$ and $\omega_H = \gamma H_z$. Since $\nabla \times \vec{h} = 0$, we can introduce the magnetic scalar potential $\psi$, defined by $\vec{h} = -\nabla\psi$. Substituting this relation in the second equation (4.1) we obtain

$$-\nabla^2\psi + 4\pi\nabla \cdot \vec{m} = 0. \tag{4.5}$$

Finally, using Eqs. (4.2–4.4) in (4.5), we obtain an equation for the magnetic potential inside the ferromagnetic medium

$$(1 + \kappa)\left(\frac{\partial^2\psi}{\partial x^2} + \frac{\partial^2\psi}{\partial y^2}\right) + \frac{\partial^2\psi}{\partial z^2} = 0, \tag{4.6}$$

which is called Walker's equation. Note that outside the ferromagnetic sample, $\kappa = 0$ and this equation reduces to the Laplace equation

$$\frac{\partial^2 \psi}{\partial x^2} + \frac{\partial^2 \psi}{\partial y^2} + \frac{\partial^2 \psi}{\partial z^2} = 0, \tag{4.7}$$

that is used to calculate the magnetic potential in magnetostatic arrangements of currents and magnetic moments in free space. Walker used these equations to find the standing wave modes in spheroidal samples, which he called *magnetostatic modes*. These modes are denoted by a set of indices $(r, m, n)$ that represent the variations in the potential along the radius and the two angle coordinates. In a spherical sample the FMR mode is represented by (100), since the magnetization varies with the radius, but not with angle. In a FMR experiment with fixed frequency in a sample with low magnetic damping, such as YIG, as the field is scanned several magnetostatic modes are observed in the absorption or transmission spectra, as shown in Fig. 3.12.

### 4.1.2 Magnetostatic Waves in Unbounded Films Magnetized in the Plane

In this section, we shall use Walker's equation to calculate the dispersion relation of magnetostatic spin waves in ferro–ferrimagnetic films with infinite lateral dimensions, under a static magnetic field applied in the film plane. We follow the classic paper of Damon and Eshbach [2], with a slightly different notation for some quantities. Note that at the time of the Damon–Eshbach (DE) paper, spin wave phenomena were studied with bulk samples, since ferromagnetic films with good quality were not available. Thus, the DE paper refers to the sample as a slab. Figure 4.1 shows views of the film (or slab) and the coordinates system used in the calculation. The magnetic potential will be expressed by different functions in the three regions of space: $\psi_1(\vec{r})$ in region (1), $y > t/2$; $\psi_2(\vec{r})$ in region (2), $-t/2 < y < t/2$; and $\psi_3(\vec{r})$ in region (3), $y < -t/2$. In region (2), the potential satisfies Eq. (4.6) and in regions (1) and (2) it satisfies Eq. (4.7). In all regions the equations can be solved by separation of variables, so that we write the potential as

**Fig. 4.1** (a) Ferromagnetic film magnetized in the plane used to study magnetostatic spin waves. (b) Side view showing the coordinate system

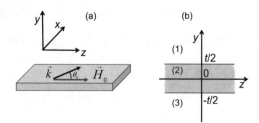

$$\psi_\lambda(x, y, z) = X_\lambda(x)Y_\lambda(y)Z_\lambda(z), \tag{4.8}$$

where each function depends on only one variable ($\lambda$ denotes the region). Consider propagating waves in the film plane, with wave vector $\vec{k} = \hat{x}k_x + \hat{z}k_z$. Thus, in order to satisfy the boundary conditions at the film surfaces, in all regions we have

$$X_\lambda(x)Z_\lambda(z) = e^{ik_x x}e^{ik_z z}. \tag{4.9}$$

The proper solutions of Eqs. (4.6 and 4.7) for the function $Y_\lambda(y)$ are $e^{\pm k_y y}$ in regions (1) and (3), because the potential has to vanish for $y \to \pm \infty$. In region (2), we will consider harmonic solutions, so that we write for $\psi_\lambda(\vec{r})$ in the three regions

$$\psi_1(\vec{r}) = C e^{-k_y^e y}e^{ik_x x}e^{ik_z z}, \tag{4.10a}$$

$$\psi_2(\vec{r}) = [A \sin (k_y^i y) + B \cos (k_y^i y)]\, e^{ik_x x}e^{ik_z z}, \tag{4.10b}$$

$$\psi_3(\vec{r}) = D e^{k_y^e y}e^{ik_x x}e^{ik_z z}, \tag{4.10c}$$

where $k_y^e$ and $k_y^i$ denote, respectively, the coefficients of the exponents for the $y$-function outside and inside the film. The four coefficients in Eqs. (4.10a–4.10c) are determined by the boundary conditions at the two surfaces of the film, namely, continuity of the fields $\vec{h}_{tan}$ and $\vec{b}_{normal}$. The first condition implies continuity of the magnetic potential, so that at $y = t/2$ we have $\psi_1 = \psi_2$, and at $y = -t/2$ we have $\psi_2 = \psi_3$. For the second condition we use the relation $b_y = -\partial\psi/\partial y + 4\pi m_y$. Thus, using $4\pi m_y = -iv\,\partial\psi/\partial x - \kappa\,\partial\psi/\partial y$, the four boundary conditions give

$$A \sin \left(k_y^i \frac{t}{2}\right) + B \cos \left(k_y^i \frac{t}{2}\right) = C\, e^{-k_y^e\, t/2}, \tag{4.11a}$$

$$-A \sin \left(k_y^i \frac{t}{2}\right) + B \cos \left(k_y^i \frac{t}{2}\right) = D\, e^{k_y^e t/2}, \tag{4.11b}$$

$$\begin{aligned}(1 + \kappa) \left[A k_y^i \cos \left(k_y^i \frac{t}{2}\right) - B k_y^i \sin \left(k_y^i \frac{t}{2}\right)\right] \\ -v k_x \left[A \sin \left(k_y^i \frac{t}{2}\right) + B \cos \left(k_y^i \frac{t}{2}\right)\right] = -C k_y^e e^{-k_y^e t/2}\end{aligned}, \tag{4.11c}$$

$$\begin{aligned}(1 + \kappa) \left[A k_y^i \cos \left(k_y^i \frac{t}{2}\right) + B k_y^i \sin \left(k_y^i \frac{t}{2}\right)\right] \\ -v k_x \left[-A \sin \left(k_y^i \frac{t}{2}\right) + B \cos \left(k_y^i \frac{t}{2}\right)\right] = D k_y^e e^{k_y^e t/2}\end{aligned}. \tag{4.11d}$$

The system of Eqs. (4.11a–4.11d) can be written in matrix form

$$\begin{bmatrix} \sin & \cos & -e^{-k_y^e t/2} & 0 \\ -\sin & \cos & 0 & -e^{k_y^e t/2} \\ (1+\kappa)k_y^i\cos -\nu k_x\sin & -(1+\kappa)k_y^i\sin -\nu k_x\cos & k_y^e e^{-k_y^e t/2} & 0 \\ (1+\kappa)k_y^i\cos +\nu k_x\sin & (1+\kappa)k_y^i\sin -\nu k_x\cos & 0 & -k_y^e e^{k_y^e t/2} \end{bmatrix} \begin{pmatrix} A \\ B \\ C \\ D \end{pmatrix} = 0 \qquad (4.12)$$

where we have used $\cos = \cos(k_y^i t/2)$ and $\sin = \sin(k_y^i t/2)$ to simplify the notation. The solution of the matrix equation (4.12) is obtained by equating to zero the main determinant. The resulting equation is

$$k_y^{e2} + 2(1+\kappa)k_y^i k_y^e \cot(k_y^i t) - (1+\kappa)^2 k_y^{i2} - (\nu k_x)^2 = 0. \qquad (4.13)$$

In order to obtain an expression involving only the frequency $\omega$, that appears in the parameter $\kappa$ as in Eq. (4.4), and the two components of the wave vector in the plane, $k_x$ and $k_z$, we need two additional relations to eliminate $k_y^e$ and $k_y^i$. One of them is obtained by substitution of either (4.10a) or (4.10c) in Laplace equation (4.7), that gives

$$k_x^2 - k_y^{e2} + k_z^2 = 0. \qquad (4.14)$$

The other relation is obtained by substitution of (4.10b) in Walker's equation (4.6), that gives

$$(1+\kappa)(k_x^2 + k_y^{i2}) + k_z^2 = 0. \qquad (4.15a)$$

This equation can be written as

$$k_y^{i2} = -\frac{(1+\kappa)k_x^2 + k_z^2}{(1+\kappa)} = -\frac{k^2 + \kappa k_x^2}{(1+\kappa)}, \qquad (4.15b)$$

where $k^2 = k_x^2 + k_z^2$. Introducing the parameters

$$\sin\theta_k = \frac{k_x}{k}, \quad \delta = \frac{1+\kappa\sin^2\theta_k}{1+\kappa}, \qquad (4.16)$$

we can write Eq. (4.15b) as

$$k_y^i = k(-\delta)^{1/2} = k\left(-\frac{1+\kappa\sin^2\theta_k}{1+\kappa}\right)^{1/2}. \qquad (4.17)$$

Using this result and Eq. (4.14) in Eq. (4.13) one can show that [3]

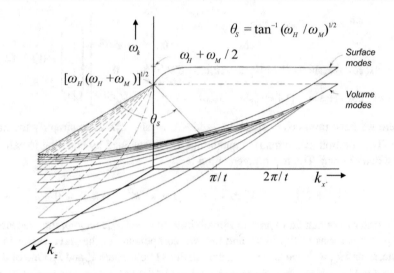

**Fig. 4.2** (a) Magnetostatic spin wave dispersion relations in a ferromagnetic film magnetized in the plane. After Damon and Eshbach [2], with the notation of the quantities used here. Reproduced with kind permission from Elsevier Inc.

$$1 + 2(1 + \kappa)(-\delta)^{1/2} \cot \left[k t (-\delta)^{1/2}\right] + \delta(1 + \kappa)^2 - v^2 \sin^2 \theta_k = 0. \qquad (4.18)$$

This equation represents the dispersion relation for magnetostatic spin waves in a ferromagnetic film magnetized in the plane. Note that for a given wave vector in the plane, characterized by a pair of values $k_x$, $k_z$, or equivalently $k$, $\theta_k$, Eq. (4.18) has several solutions for the frequency $\omega_k$, each corresponding to a different transverse mode pattern characterized by a discrete $k_y^i$. The solutions of Eq. (4.18) are shown in Fig. 4.2 as a family of surfaces in the first quadrant of $\omega_k$, $k_x$, $k_z$ space, as in Damon and Eshbach [2], with the notation of the quantities as used here. From Eq. (4.15b) and the parameters (4.4), one can see that $k_y^i$ can be real or imaginary, depending on the range of frequency. Real values of $k_y^i$ correspond to the so-called *volume magnetostatic spin waves* (VMSW), for which the magnetization components have a dependence on the transverse coordinate $y$ of the type $\cos k_y^i$, $\sin k_y^i$. Imaginary values of $k_y^i$ correspond to *the surface magnetostatic spin waves* (SMSW), which have an exponential dependence on $y$ decaying away from one of the film surfaces. The properties of these modes will be studied in detail in the next subsections.

### 4.1.3   Volume Magnetostatic Modes

For frequencies in the range

$$\omega_H < \omega < [\omega_H(\omega_H + \omega_M)]^{1/2}, \qquad (4.19)$$

the parameter $\kappa$ in Eq. (4.4) is such that $(1 + \kappa) < 0$. Notice that for films magnetized in the plane, with a field $H_0$ applied in a direction of crystalline anisotropy, the parameter $\omega_H$ in Eq. (4.19) becomes $\omega_H = \gamma(H_0 + H_A)$. For frequencies in the range of Eq. (4.19), we see in (4.16) that for any $\theta_k < \pi/2$, the parameter $\delta$ is negative. This means that, according to Eq. (4.17), $k_y^i$ is real. Thus, the potential given by Eq. (4.10b) fills the whole thickness of the film, characterizing a volume mode. The upper limit in Eq. (4.19) corresponds to the *top* of the bulk spin wave band in the limit of low wave number $k$, as in Eq. (2.36) for $\theta_k = \pi/2$. On the other hand, the lower limit coincides with the *bottom* of the bulk spin wave at low $k$ for $\theta_k = 0$. These spin wave band frequency limits, therefore, also represent the band of volume modes in the Damon–Eshbach theory. Note in Fig. 4.2, that as the propagation angle varies from zero to $\pi/2$, the bottom frequency of the VMSW band moves up to join the top of the band. That is, the volume mode band *collapses to zero width* as the mode propagation angle approaches $\pi/2$ [3].

Let us study in more detail the volume mode waves that propagate in the direction of the applied field, that is, $\theta_k = 0$. In this case, from Eq. (4.16) we have $\delta = 1/(1 + \kappa)$, and Eq. (4.18) becomes

$$1 + 2(1 + \kappa)(-\delta)^{1/2} \cot\left[kt(-\delta)^{1/2}\right] + (1 + \kappa) = 0. \tag{4.20a}$$

Introducing the parameter $\alpha = [-1/(1 + \kappa)]^{1/2}$, this equation can be written as

$$2\cot(\alpha k_z t) = \alpha - 1/\alpha. \tag{4.20b}$$

This equation has multiple roots because the function cot is periodic in the argument. Notice that for fixed frequency $\omega_k$ and field $H_z = H_0$, $\alpha$ is fixed, so that the solutions of Eqs. (4.20a, 4.20b) are equally spaced in $k_z$. This can be seen in Fig. 4.3 showing the variation of $\omega_k$ with $k_z$, for $k_x = 0$. Each curve in the figure

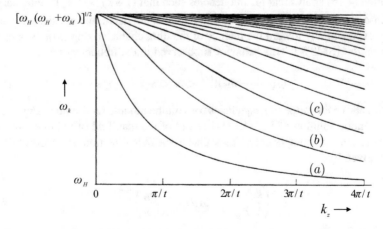

**Fig. 4.3** Magnetostatic volume spin wave dispersion relations in a ferromagnetic film magnetized in the plane for $k_x = 0$ [2]. Reproduced with kind permission from Elsevier Inc.

corresponds to a different root of (4.20a, 4.20b), and thus to a different mode pattern in the film characterized by the value of $k_y^i$, given by Eq. (4.17). Notice that for $k_z = 0$, all modes are degenerate, with frequency given by the same value as in Kittel equation (1.38) with $N_x = N_z = 0$, $N_y = 4\pi$, namely $\omega_0 = [\omega_H(\omega_H + \omega_M)]^{1/2}$. As $k_z$ increases, the dipolar interaction decreases and so does the frequency. The result is that the group velocity $v_g = \partial\omega_k/\partial k$ of all volume modes is negative. For this reason, these modes are also called *backward volume modes*. For $k_z \to \infty$ all frequencies coalesce at the bottom of the band, $\omega_H$. Of course, for $k_z > 10^5\text{cm}^{-1}$, the exchange interaction comes into play and the frequencies increase with increasing $k_z$, as will be studied in the next section. It has been shown [4] that, for very thin films, Eqs. (4.20a, 4.20b) can be solved approximately to give an explicit expression for the frequency. For the lowest-order mode (a), the dependence of the frequency on $k_z$ is given by [4]

$$\omega_k^2 = \omega_H[\omega_H + \omega_M(1 - e^{-k_z t})/k_z t]. \tag{4.21}$$

This result will also be demonstrated in the next section. Figure 4.4 shows the variation of the function $Y(y)$ for the potential in Eq. (4.8) for the three lowest frequency modes, corresponding to the curves labeled (a), (b), and (c) in Fig 4.3. For each mode the potentials are shown for three different ranges of $k_z$. The patterns for $k_z = 0$ correspond to the uniform precession, or FMR, mode. In this case the potential vanishes outside the film, in other words, there is no *rf* field outside. Figure 4.4 also shows that for $k_z > 0$, the potential function outside the film decays rapidly away from the surface, as in Eqs. (4.10a and 4.10c).

### 4.1.4  Surface Magnetostatic Modes

Equation (4.16) shows that for frequencies such that $(1 + \kappa) > 0$, $k_y^i$ is imaginary. In this case, the amplitudes of the magnetic potential are maximum at each surface and decay exponentially toward the interior of the film, characterizing *surface magnetostatic modes* (SMSW). This behavior is observed in the frequency range

$$[\omega_H(\omega_H + \omega_M)]^{1/2} < \omega < \omega_H + \omega_M/2, \tag{4.22}$$

that lies above the range of frequencies for volume modes. As shown in Fig. 4.2, the surface mode solution of Eq. (4.18) rises out of the manifold of volume modes for wave vectors with component $k_x$ transverse to the field direction larger than a critical value, given by

$$\left(\frac{k_x}{k_z}\right)_{cr} = \tan\theta_S = \left(\frac{\omega_H}{\omega_M}\right)^{1/2}. \tag{4.23}$$

Let us study in more detail the surface waves that propagate in the direction perpendicular to the applied field, that is, $\theta_k = \pi/2$, or $k_z = 0$. In this case we have

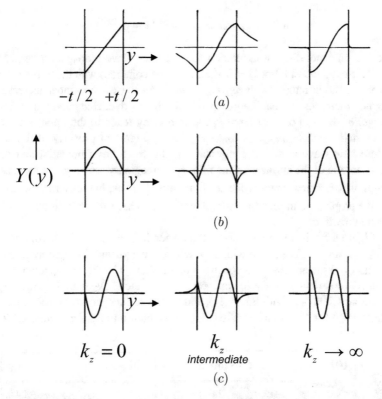

**Fig. 4.4** Magnetic scalar potential function corresponding to the three lowest-order magnetostatic volume modes, viewed in the cross section of the ferromagnetic film magnetized in the plane. After Damon and Eshbach [2]. Reproduced with kind permission from Elsevier Inc.

from Eq. (4.14) that $k_y^e = k_x$, and from (4.15a, 4.15b) that $k_y^i = \pm i k_x$. Thus, the amplitude in the $y$-direction is attenuated exponentially away from each surface toward the interior and exterior of the film, with a characteristic length $1/k_x$. Of course, as the value of $k_x$ increases, the magnetic excitation becomes more concentrated close to the film surface, characteristic of a surface wave. One important property of the surface waves is that, since the potential function is not periodic through the thickness of the film, they do not exhibit multiple modes. The dispersion relation for the surface wave can be obtained from Eq. (4.18) with $k_y^i = i k_x$, $\theta_k = \pi/2$, and $\delta = 1$. The result is

$$1 + 2(1 + \kappa)\coth(k_x t) + (1 + \kappa)^2 - \nu^2 = 0.$$

Using in this expression the relations for the parameters $\kappa$ and $\nu$ in Eq. (4.4), one can show that

$$\omega_k^2 = \omega_H^2 + \omega_H\omega_M + \frac{\omega_M^2}{4}\left(1 - e^{-2k_x t}\right). \tag{4.24}$$

This is an exact dispersion relation for a surface wave propagating perpendicularly to the applied field. For $k_x = 0$, the frequency coincides with the upper bound of the volume mode range, $\omega_0^2 = \omega_H(\omega_H + \omega_M)$. As the wave number increases the frequency increases, so that the group velocity is positive, characterizing a *forward propagating wave*. For very large $k_x$ the frequency tends to the upper limit of all magnetostatic modes, $\omega_{k\max} \to \omega_H + \omega_M/2$. An important property of the surface magnetostatic wave is the nonreciprocity. If the wave has a potential that decays away from one of the surfaces of the film, $k_y^i$ is imaginary with a given sign, so that the sign of $k_x$ is determined by the relation with $k_y^i$. Thus, in one surface of the film the wave propagates in one direction, while in the other surface it propagates in the opposite direction.

In Figure 4.5a, b we show plots of the dispersion curves for both volume mode with $\theta_k = 0$ and surface mode with $\theta_k = \pi/2$ for two yttrium iron garnet films with thickness $t = 3\,\mu$m and $28\,\mu$m magnetized in the plane. The dispersion relations were calculated numerically for the lowest order volume mode (V) with the exact expression (4.20a, 4.20b) and with Eq. (4.21), and for the surface mode (S) with Eq. (4.24). We have used for YIG $4\pi M = 1.76\,$kG, corresponding to $\omega_M/2\pi = 4.93\,$GHz, and

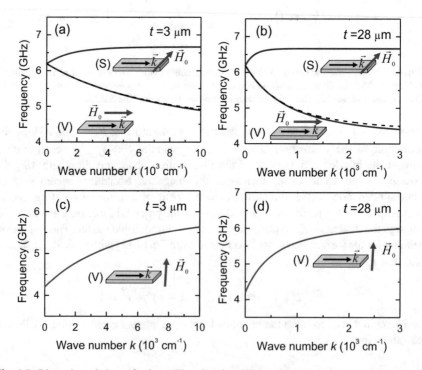

**Fig. 4.5** Dispersion relations of volume (V) and surface (S) magnetostatic spin waves for two films of yttrium iron garnet with thickness $t$ indicated. In (**a**) and (**b**), the films are magnetized in the plane with $H_0 = 1.5\,$kOe. In (**c**) and (**d**), the films are magnetized perpendicularly to the plane with $H_0 = 3.26\,$kOe, for which the internal field is $H_z = 1.5\,$kOe

$H_0 = 1.5$ kOe, that gives $\omega_H/2\pi = 4.2$ GHz. Using these numbers, we find $f_0 = [4.2 \times (4.2 + 4.93)]^{1/2} = 6.19$ GHz for the $k = 0$ mode, as in Fig. 4.5. As the wave number increases, the variation of the frequencies for both modes is more pronounced for the thicker film, as expected from Eqs. (4.21 and 4.24).

The solid curves were calculated with the exact dispersion (4.20a, 4.20b) and the dashed curves were calculated with the approximate expression (4.21) for the volume mode. We see that the agreement with the exact calculation is quite good over the whole range of $k$ for the thinner film, but for the thicker film there is some departure at larger $k$. The negative slopes of the curves for the volume mode in Fig. 4.5a, b show clearly the negative group velocities characteristic of backward waves, while the positive slopes for the surface modes show that they are forward propagating waves.

### 4.1.5 Volume Magnetostatic Waves in Perpendicularly Magnetized Films

The discussions of the previous sections were restricted to films magnetized in the plane. Actually, volume and surface magnetostatic waves can be supported by magnetic films under a static field in any direction relative to the film plane. Let us consider here a magnetic film with magnetization $4\pi M$ under an external field $\vec{H}_0$ applied perpendicularly to its plane. As long as $H_0 > 4\pi M$, the internal magnetic field is also perpendicular to the film and has intensity $H_z = H_0 - 4\pi M$. The properties of the magnetostatic waves can be calculated with a procedure similar to the one presented in the previous sections. The dependence of the magnetic potential on the coordinate normal to the film is considered to have different forms in the three regions of space, as in Fig. 4.1b, and the dependence along a coordinate in the plane has the form $\exp(i\vec{k}.\vec{r})$, where $k$ is the in-plane wave number. The solutions of the Walker equation (4.6) and Laplace equation (4.7) that satisfy the electromagnetic boundary conditions give the wave dispersion relations. Similarly to the configuration studied in Sect. 4.1.2, for each in-plane wave number there are several modes with different transverse functions, each with different frequency [See Stancil and Prabhakar]. The lowest-order mode has an approximate analytical dispersion relation, given by [4]

$$\omega_k^2 = \omega_H^2 + \omega_H\omega_M\left(1 - \frac{1 - e^{-kt}}{kt}\right), \tag{4.25}$$

where $\omega_H = \gamma H_z$ and $\omega_M = \gamma 4\pi M$. One can see that for $k = 0$ the frequency is $\omega_0 = \omega_H$, that coincides with the lower bound of the volume mode in films magnetized in the plane. As the wave number increases the frequency increases, so that the group velocity is positive, characterizing a *forward propagating wave*. For very large $k$ the frequency tends to $\omega_0 = [\omega_H(\omega_H + \omega_M)]^{1/2}$, which is the same expression for the frequency of the lowest-order $k = 0$ volume and surface magnetostatic modes in films magnetized in the plane. The dispersion curves for the lowest-order volume mode in YIG films with thickness $t = 3$ μm and 28 μm, magnetized perpendicular to the plane are shown in Fig. 4.5c, d.

### 4.1.6  Experiments with Magnetostatic Spin Waves

In this section, we present two typical experimental techniques to study magneto-static spin waves. The first is based on the FMR technique, described in Sect. 1.5.1, and is used to measure the absorption spectra of standing waves. The second one uses the pulsed microwave techniques presented in Sect. 2.5.1 to study propagating spin wave packets.

The absorption spectra of magnetostatic modes can be measured with the same experimental setup used for FMR studies. As illustrated by the schematic diagram in Fig. 1.7a, the setup consists basically of a CW microwave generator, transmission lines and components to carry the radiation to the sample housing, and the components to detect the radiation reflected from the sample. The sample placed in a shorted microwave waveguide is subject to a *rf* magnetic field applied perpen-dicularly to the static field $\vec{H}_0$ produced by an electromagnet. If the microwave frequency is kept fixed, the absorption spectrum is measured by scanning the applied field. In order to increase the sensitivity of the system, one uses a pair of Helmholtz coils to modulate the applied field and phase-sensitive (phase lock-in) amplification, so that the measured signal is proportional to the derivative of the absorption line relative to the magnetic field.

Figure 4.6 shows the absorption spectra measured in a YIG film with thickness 28 μm, grown by liquid-phase epitaxy on a 0.5-mm thick substrate of (111) gadolinium gallium garnet (GGG), cut with rectangular shape with lateral dimensions of $2 \times 4$ mm$^2$ [5]. The sample is introduced through a small hole in the back wall of a rectangular shorted X-band microwave waveguide, fed with

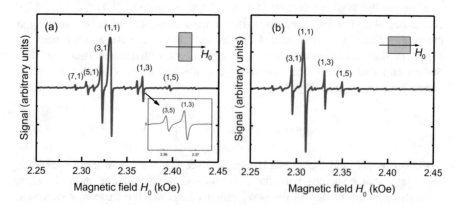

**Fig. 4.6** Field scan microwave absorption spectra of a 28-μm thick YIG film with lateral dimensions $2 \times 4$ mm$^2$, with the static field orientation as indicated, measured at a frequency 8.6 GHz. The inset in (**a**) shows details of the lines of modes (3,5) and (1,3) [5]. Reproduced from L. H. Vilela-Leão *et. al.*, Appl. Phys. Lett. **99**, 102505 (2011), with the permission of AIP Publishing

radiation of frequency $f = 8.6$ GHz and power 100 mW, and placed between the poles of an electromagnet. The static magnetic field with intensity $H_0$ and the microwave magnetic field are kept perpendicular to each other and in the film plane as the sample is rotated. With this configuration, one can investigate the angular dependence of the absorption spectra by scanning the applied field.

The various peaks in the spectra of Fig. 4.6 correspond to resonating standing-wave magnetostatic modes. The standing waves are formed by the travelling spin waves described by Eqs. (4.10a–4.10c), that due to the boundary conditions at the edges of the film are reflected producing counter propagating waves. Denoting the lateral film dimensions by $L_x$ and $L_z$, $z$ being the direction of the static field, the standing waves have quantized in-plane wave numbers $k_x = n_x \pi / L_x$ and $k_z = n_z \pi / L_z$. The values of the resonance fields for 8.6 GHz, calculated numerically using the dispersion relation (4.18), allow a precise identification of the modes represented by $n_x$, $n_z$. The strongest line corresponds to the FMR mode with $n_x = 1$, $n_z = 1$, that has frequency close to the one for the spin wave with $k = 0$, given by $f_0 = \gamma (H_0 + H_A)^{1/2} (H_0 + H_A + 4\pi M)^{1/2}$. The calculated field positions for the modes agree very well with the measured values using for YIG $\gamma = 2.8$ GHz/kOe, $4\pi M = 1.76$ kG, and $H_A = 0.01$ kOe. The lines to the left of the FMR mode (1,1) correspond to hybridized surface modes, whereas those to the right are hybridized volume magnetostatic modes. The two spectra in Fig. 4.6a, b are different because when the sample is rotated by 90°, the values of the dimensions $L_x$ and $L_z$ are switched, since the field direction changes from parallel to perpendicular to the short dimension of the film.

Experiments with propagating magnetostatic waves consist basically of exciting wave packets in the sample with pulsed microwave radiation in a launching antenna, and detecting the signal generated by the dipolar field of the packet in a receiving antenna at another position in the sample. Figure 4.7a shows the schematic diagram of an experimental setup used to study either volume or surface waves, depending on the orientation of the applied field relative to the direction of propagation [6].

The results in Fig. 4.7 were obtained with a YIG film strip with thickness 8 μm and lateral dimensions $2 \times 12$ mm$^2$ [6]. The YIG film is magnetized to saturation by a uniform static magnetic field with intensity $H_0$ applied in the film plane, in the direction transverse to the length of the strip. In this configuration, the film supports the propagation of surface magnetostatic spin waves (SMSW) along its length. Two copper fine wire antennas close to the ends of the strip are used to excite and detect spin wave packets.

Microwave radiation pulses with frequency 3.4 GHz, peak power of 20 mW, duration 25 ns, and repetition rate 10 kHz are directed to one of the antennas. The component of the *rf* magnetic field of the antenna perpendicular to the static field excites coherent spin wave packets with low wave number that propagate along the YIG film in the configuration of the lowest-order surface magnetostatic mode. The packet propagates with a certain group velocity and reaches the other end of the strip, where it produces a *rf* pulse in the pickup antenna with a time delay that changes with the intensity of $H_0$.

One can see in Fig. 4.7b that as the field decreases, the group velocity given by the slope of the curve, decreases, resulting in longer time delays. This is confirmed by

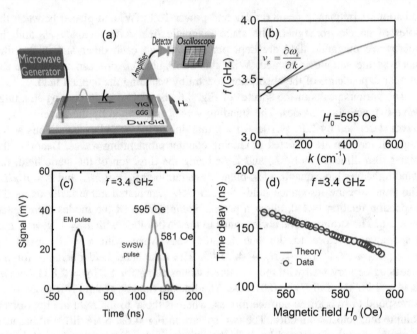

**Fig. 4.7** Experiments with propagating surface magnetostatic spin waves in a YIG film strip. Adapted from Fig. 1 in Holanda et al. [6]. (**a**) Schematic diagram of the microwave setup. (**b**) Dispersion relation for surface waves showing the slope of the curve for $f = 3.4$ GHz and $H_0 = 595$ Oe that gives the group velocity of the wave packet. (**c**) Signal pulses detected at the receiving antenna. The pulse with no delay corresponds to the direct electromagnetic coupling between the launching and receiving antennas. The pulses with a delay of 130 ns and 142 ns correspond to the signal generated by the spin wave packet in the pickup antenna for two field values. (**d**) Time delay versus field intensity, measurements and calculations. Reproduced from J. Holanda et al., Nat. Phys. **14**, 500 (2018), with permission of Springer Nature

the oscilloscope traces of the time-resolved measurements shown in Fig. 4.7c for a frequency of 3.4 GHz. The large initial pulse on the left corresponds to the direct electromagnetic coupling between excitation and pickup antennas. The delayed pulses correspond to the signal transmitted by the propagating spin wave packet that has group velocity several orders of magnitude smaller than that of electromagnetic waves in vacuum. The symbols in Fig. 4.7d represent the measured time delay for 3.4 GHz versus field intensity, while the solid curve represents the delay calculated as follows.

Since the static field is uniform, the wave packet travels in the film strip with constant group velocity, given by $v_g = \partial \omega_k / \partial k$, where $\omega_k = 2\pi f_k$. Thus, the time delay can be calculated with $\tau = L_a / v_g$, where $L_a$ is the distance between the two antennas. Figure 4.7b shows the spin wave dispersion in the region of small wave numbers, $k < 10^3$ cm$^{-1}$, for $H_0 = 595$ Oe, calculated with Eq. (4.24). The group velocity of the SMSW obtained from Eq. (4.24) is

$$v_g = \frac{\omega_M^2 t}{4\omega_k} e^{-2k_x t},$$
(4.26)

For each value of the field, the wave number $k_x$ is calculated with Eq. (4.24) for the magnon frequency of 3.4 GHz and the group velocity is determined by (4.26). The calculated time delay versus field shown in Fig. 4.7d agrees quite well with the experimental data. The small discrepancy in larger fields is due to the fact that Eq. (4.24) is exact for an infinite film. As the field increases, the wave number decreases and the finite width of the film strip results in discrete modes with slight shifts in frequency, as discussed previously. Finally, note that as the field decreases, the time delay increases and the pulse amplitude decreases. This is due to the spin wave damping during propagation, resulting in a peak amplitude that decays in time as $A(t) \propto \exp(-2\eta_k t)$, where $\eta_k$ is the spin wave relaxation rate.

## 4.2 Quantum Theory of Spin Waves in Thin Ferromagnetic Films

As we saw in Sect. 3.1.2, the quantum theory of spin waves in the HP formalism requires knowledge of the wave function of the spin excitation for the transformation from the spin deviation operators into magnon operators. One difficulty of the quantum theory for magnetostatic waves in films is that, due to the complex form of the dipolar field, we do not have an analytical expression for the spin wave function in the coordinate perpendicular to the film plane. In this section, we present a quantum theory for spin waves in thin films, that applies to wave numbers in the magnetostatic and exchange regions. The theory is based on the approach of Arias and Mills [7], except that here we use the formalism of second quantization [8]. The theory is valid for the lowest-order modes in thin films, with thickness typically less than 10 μm in YIG. The main results shown here agree with those of more sophisticated calculations for dipole-exchange spin waves in films [9].

We consider for the magnetic Hamiltonian contributions from Zeeman, dipolar, and exchange interaction energies. As in Sects. 3.1 and 3.2, we treat the quantized excitations of the magnetic system with the approach of Holstein–Primakoff, which consists of three transformations that allow the spin operators to be expressed in terms of boson operators that create or destroy magnons. In the first transformation, the components of the local spin operator are related to the creation and annihilation operators of spin deviation at site $i$, denoted, respectively, by $a_i^\dagger$ and $a_i$, which satisfy the boson commutation rules $[a_i, a_j^\dagger] = \delta_{ij}$ and $[a_i, a_j] = 0$.

We use a coordinate system with $\widehat{z}$ along the equilibrium direction of the spins and the raising and lowering spin operators, defined by $S_i^+ = S_i^x + iS_i^y$ and $S_i^- = S_i^x - iS_i^y$. Then we employ the linearized transformations $S_i^+ \cong (2S)^{1/2}a_i$, $S_i^- \cong (2S)^{1/2}a_i^\dagger$, and $S_i^z = S - a_i^\dagger a_i$ in order to express the magnetic Hamiltonian in a quadratic form containing only lattice sums of products of two boson operators. In the second transformation from the localized field operators to collective boson operators $a_k^\dagger$ and $a_k$, we use the Fourier transform in Eqs. (3.30–3.32), where the wave vector $\overrightarrow{k} = \widehat{x}k_x + \widehat{z}k_z$ is considered to be in the film plane, as in Sect. 4.1.

For very thin films, the various contributions to the Hamiltonian can be obtained with approximations valid for the nearly uniform transverse mode, which corresponds to the lowest-lying exchange branch of Kalinikos and Slavin [9], with $k_y \approx 0$. Following [7] we neglect the variation of the magnetization on the transverse coordinate and work with the averages over $y$,

$$m_{x,y}(x,z;t) = \int_{-t/2}^{t/2} \frac{1}{t} m_{x,y}(x,y,z;t) dy. \tag{4.27}$$

The magnetic potential $\psi$ created by the spatial variation of the small-signal transverse components of the magnetization is written in the form

$$\psi(x,y,z) = \frac{1}{V^{1/2}} \sum_k \psi_k(y) e^{i\vec{k}\cdot\vec{r}}, \tag{4.28}$$

where $V$ is the volume of the film, and $\vec{k}$ and $\vec{r}$ denote, respectively, the wave vector and the position vector in the plane. The Fourier transform of the potential $\psi_k(y)$ can be obtained from the solution of $\nabla^2 \psi = 4\pi \nabla \cdot \vec{m}$ derived from Maxwell's equations (4.1), subject to the magnetic boundary conditions at $y = \pm d/2$ [7],

$$\psi_k(y) = 4\pi i [e^{-kt/2} \cosh(ky) - 1] \frac{k_x}{k^2} m_x(k) + 4\pi e^{-kt/2} \sinh(ky) \frac{1}{k} m_y(k), \tag{4.29}$$

where the Fourier components of the magnetization can be expressed in terms of the collective boson operators. Using the relation between the transverse components of the magnetization and the spin operators $m_{x,y} = g\mu_g (N/V) S_i^{x,y}$, where $N$ is the number of spins, we have

$$m_x(k) = \hbar\gamma \left(\frac{NS}{2V}\right)^{1/2} (a_k + a_{-k}^\dagger), \tag{4.30a}$$

$$m_k(k) = -i\hbar\gamma \left(\frac{NS}{2V}\right)^{1/2} (a_k - a_{-k}^\dagger). \tag{4.30b}$$

The contributions from the Zeeman and exchange energies to the quadratic Hamiltonian in boson operators are obtained as in Sect. 3.1.2. Equations (3.34) and (3.35) give, for wave numbers $ka \ll 1$

$$H_Z + H_{exc} = \hbar \sum_k \gamma (H_0 + Dk^2) a_k^\dagger a_k, \tag{4.31}$$

where $D = 2JSa^2/\gamma\hbar$. The contribution from the magnetic dipolar energy is calculated with

$$H_{dip} = -\frac{1}{2} \int dxdydz(m_x h_x^{dip} + m_y h_y^{dip}), \tag{4.32}$$

where the transverse components of the dipolar field can be obtained from the magnetic potential with $\vec{h} = -\nabla \psi$. The integration in (4.32) is performed by expressing the magnetization and the dipolar field in terms of their Fourier transforms and using the orthonormality relations. One can show that

$$H_{dip} = \hbar\gamma 2\pi M \sum_k [(1 - F_k)\sin^2\theta_k + F_k]a_k^\dagger a_k$$
$$+ \frac{1}{2}\{[(1 - F_k)\sin^2\theta_k - F_k]\,a_k a_{-k} + H.c.\} \tag{4.33}$$

With (4.31) and (4.33) one can write the total Hamiltonian for the free magnon system as,

$$H_0 = \hbar\sum_k A_k\, a_k^\dagger a_k + \frac{1}{2}B_k\, a_k a_{-k} + \frac{1}{2}B_k\, a_k^\dagger a_{-k}^\dagger, \tag{4.34}$$

where

$$A_k = \gamma[H_0 + Dk^2 + 2\pi M(1 - F_k)\sin^2\theta_k + 2\pi M F_k], \tag{4.35a}$$

$$B_k = \gamma[2\pi M(1 - F_k)\sin^2\theta_k - 2\pi M F_k], \tag{4.35b}$$

$$F_k = (1 - e^{-kt})/kt. \tag{4.35c}$$

Following the same procedures used in Sect. 3.2, we diagonalize the Hamiltonian (4.34) with a Bogoliubov transformation to the magnon creation and annihilation operators $c_k^\dagger$ and $c_k$ given by

$$a_k = u_k c_k + v_k c_{-k}^\dagger, \tag{4.36a}$$

$$a_k^\dagger = u_k c_k^\dagger + v_k c_{-k}, \tag{4.36b}$$

where $u_k^2 - v_k^2 = 1$, as appropriate for a unitary transformation. The coefficients of this transformation must be such that the Hamiltonian acquires the diagonal form

$$H_0 = \hbar\sum_k \omega_k c_k^\dagger c_k. \tag{4.37}$$

It can be shown that the coefficients of the transformations (4.36a, 4.36b) are given by

$$u_k = \left(\frac{A_k + \omega_k}{2\omega_k}\right)^{1/2},\tag{4.38a}$$

$$v_k = \pm(u_k^2 - 1)^{1/2} = \pm\left(\frac{A_k - \omega_k}{2\omega_k}\right)^{1/2},\tag{4.38b}$$

where the sign of $v_k$ is the opposite one of the parameter $B_k$, and the frequency $\omega_k$ of the eigenmodes is

$$\omega_k = (A_k^2 - |B_k|^2)^{1/2}.\tag{4.39}$$

Using the expressions for the parameters in (4.35a–4.35c) we obtain from (4.39) an explicit equation for the frequency of magnons with wave vector in the plane [8]

$$\omega_k^2 = \gamma^2 [H_0 + Dk^2 + 4\pi M(1 - F_k)\sin^2\theta_k][H_0 + Dk^2 + 4\pi M F_k].\tag{4.40}$$

This equation is the same as the one obtained for the lowest-lying branch of the "dipole-exchange" modes with more rigorous treatment of the exchange interaction [9]. Note that for $k = 0$ it gives the same expression as the one for the magnetostatic modes. For $kt \gg 1$, $F_k \to 0$, and (4.40) reduces to the magnon dispersion relation (2. 36) for an infinite medium. Finally, notice that for $\theta_k = 0$ and $D = 0$, Eq. (4.40) coincides with the approximate dispersion relation for VMSW, Eq. (4.21), while for $\theta_k = \pi/2$ and $D = 0$, it is the same as Eq. (4.24) for SMSW.

Figure 4.8 shows the magnon dispersion relations calculated for two YIG films with thicknesses $t = 0.1$ μm and $t = 5$ μm, for the field $H_0$ applied in the film plane, and wave vector in the plane at an angle $\theta_k$ with the field. The plots are made with the horizontal axes in a logarithmic scale so as to cover several decades of wave number to show the magnetostatic region and also the effect of exchange. The solid lines

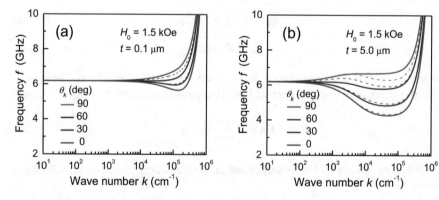

**Fig. 4.8** Magnon dispersion relations for two YIG films with thickness $t$ as indicated, with the field $H_0$ applied in the film plane, and wave vector in the plane, at an angle $\theta_k$ with the field. The solid lines were calculated with Eq. (4.18) including exchange, and the dashed lines were calculated with Eq. (4.40)

were calculated with Eq. (4.18), with the exchange interaction introduced in an *ad hoc* manner by making $\omega_H = \gamma(H_0 + H_A + Dk^2)$. The dashed lines were calculated with Eq. (4.40). The agreement between the two calculations is quite good in both films for angles $\theta_k$ smaller than a certain value, and is better for the thinner film.

The dispersion relations in Fig. 4.8 reveal an important feature of magnons in thin ferromagnetic films, namely, the minimum energy for wave vector along the magnetic field in the range $10^4 < k < 10^5$ cm$^{-1}$. The energy minimum away from $k = 0$, due to the combined effects of the exchange and magnetic dipolar interactions, produces a peak in the density of states that makes possible for magnons to undergo Bose–Einstein condensation, as will be studied in Chap. 7. Another important consequence of the shape of the magnon dispersion relations in films is that the manifold of modes degenerate with the $k = 0$ magnon creates a new relaxation channel based on two-magnon scattering. This will be studied in the next section.

## 4.3   Spin Wave Relaxation in Ultrathin Films by Two-magnon Scattering

As briefly mentioned in Sects. 1.5.1 and 3.4.3, metallic ferromagnets have magnetic damping much larger than in YIG. The large damping observed in metals is due to the existence of various intrinsic relaxation mechanisms involving conduction electrons. The details of these mechanisms are outside the scope of this introductory book and can be found in several references [10–13]. In general, they predict a relaxation rate that is proportional to the spin precession frequency, so that they are usually referred to as Gilbert damping processes. In this section, we will present the concepts of an extrinsic relaxation mechanism that is important in ultrathin magnetic films, namely the two-magnon scattering process induced by roughness at the interfaces or surfaces of the film.

One of the first hints of the importance of extrinsic relaxation processes in ultrathin films was provided by the FMR data of Celinski and Heinrich [14] on Fe metal films grown with several different conditions. By measuring the FMR in the range 10–35 GHz, they found that the linewidth increased proportionally to the frequency. They also found that as the frequency was extrapolated to zero, the residual linewidth was sensitive to film quality; the highest quality films exhibited the smallest residual values. The FMR linewidth measurements in films are commonly fitted by the empirical expression

$$\Delta H = \Delta H(0) + \lambda \omega_0 / \gamma, \tag{4.41}$$

where $\lambda$ is proportional to the Gilbert damping parameter $\alpha$, and $\Delta H(0)$ is the residual linewidth that has an extrinsic origin. The data of [14] also showed that in ultrathin films, the value of the Gilbert damping constant extracted from the fit with (4.41) was substantially larger than that for single-crystal bulk materials [14–16]. Many years later, Hurben and Patton [17] proposed a simple theory for the two-magnon scattering relaxation in magnetic thin films, that was subsequently confirmed

qualitatively by experiments with thin magnetic metallic films deposited on antiferromagnetic substrates [18]. Following these developments, Arias and Mills [7]
proposed a formal theory for the two-magnon relaxation in ultrathin magnetic
films. Their theory is the base of the presentation here.

## 4.3.1   Two-magnon Relaxation in Single Ferromagnetic Films

As discussed in Sect. 3.7.2, the two-magnon scattering mechanism is an important
source of FMR line broadening in bulk YIG samples. A key condition for the
two-magnon mechanism to be operative is the existence of a manifold of modes
degenerate with the magnon to be relaxed. This is so because the scattering by static
imperfections does not conserve momentum but must conserve energy. As we saw in
the previous section, in thin films magnetized in the plane, the $k = 0$ magnon
is degenerate with many modes with wave vectors $k > 0$ and $\theta_k$ below a certain
value. We consider magnons in an ultrathin film with thickness $t$, magnetized in the
plane by a field $\vec{H}_0$, with wave vector $\vec{k}$ in the plane, at an angle $\theta_k$ with the field. For
$kt \ll 1$, the factor $F_k$ in Eq. (4.40) is $F_k \approx 1 - kt/2$, and the magnon dispersion
relation is given approximately by the same expression used in [7], namely

$$\omega_k^2 = \omega_0^2 - \gamma^2 2\pi M kt[H_0 - (H_0 + 4\pi M_{eff})\sin^2\theta_k]$$
$$+ \gamma^2(2H_0 + 4\pi M_{eff})Dk^2, \qquad (4.42)$$

where

$$\omega_0 = \gamma[H_0(H_0 + 4\pi M_{eff})]^{1/2} \qquad (4.43)$$

is the frequency of the uniform mode, and $4\pi M_{eff}$ is the effective magnetization. In
the CGS this is given by $4\pi M_{eff} = 4\pi M - H_s$, where $H_s = 2K_s/Mt$ is the surface
anisotropy field. This was introduced in the magnon frequency as in Sect. 1.5.2, and
since it varies inversely with the thickness, it has an important role in ultrathin films.

Equation (4.42) shows that in ultrathin films the dipolar energy generates a term
in the dispersion relation linear in the wave number that has a negative slope for a
certain range of propagation angles. Thus, with the positive exchange contribution
that increases with the square of the wave number, the dispersion exhibits a range of
modes degenerate with the uniform FMR mode. Examination of the second term in
Eq. (4.42) shows that the condition for the initial negative slope is

$$\sin^2\theta_k < \sin^2\theta_c = \frac{H_0}{H_0 + 4\pi M_{eff}}, \qquad (4.44)$$

where $\theta_c$ is the critical propagation angle above which the modes do not contribute to
the scattering of the $k = 0$ magnon. In many commonly used metallic films

**Fig. 4.9** Dispersion relations for magnons propagating in a 20-nm thick film of permalloy, under an in-plane static field $H_0 = 1$ kOe, with the wave vector in the plane, at an angle $\theta_k$ with the field

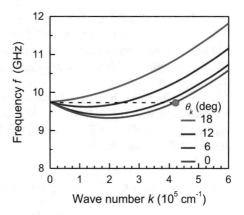

$H_0 \ll 4\pi M_{eff}$, so that the critical angle is approximately $\theta_c \approx (H_0/4\pi M_{eff})^{1/2}$. Using for permalloy $H_0 = 1$ kOe and $4\pi M_{eff} = 10$ kG, this gives $\theta_c \approx 18$ deg.

Figure 4.9 shows the magnon dispersion relations calculated with Eq. (4.40) for a 20-nm thick film of permalloy under an in-plane static field $H_0 = 1$ kOe, with wave vector $\vec{k}$ in the plane, at an angle $\theta_k$ with the field. The parameters used for Py are $D = 2 \times 10^{-12}$ kOe cm$^2$ and $4\pi M_{eff} = 10$ kG. Figure 4.9 shows clearly that the initial slope of the $\omega_k(k)$ curves rises with increasing propagation angle $\theta_k$, and becomes positive for $\theta_k > 18°$, as determined with Eq. (4.40). Notice that the dispersion relations calculated with the approximate expression (4.42) are nearly identical to those in Fig. 4.9 for wave vectors with $k < 5 \times 10^5$ cm$^{-1}$ and $\theta_k < 10$ deg, which is the important range for the two-magnon scattering relaxation of the $k = 0$ magnon.

As shown by Arias and Mills [7], the two-magnon scattering mechanism in ultrathin ferromagnetic films originates in the roughness of the surfaces and interfaces. We recall that, as the film thickness decreases, the fraction of the magnetic ions in surface or interface sites increases. Thus, defects at the surfaces and interfaces produce local perturbations in the magnetic interaction energies that are significant in the film volume, resulting in the coupling of magnons with different wave vectors. In Arias and Mills [7] it is assumed that the defects, pits, or bumps, in the shape of platelets, have sizes small compared to the wavelength of the spin waves involved in the two-magnon scattering event. Thus, their contributions to the matrix element can be calculated considering that the spin wave amplitudes do not vary in their vicinity.

Of the three possible energy contributions to the matrix element, Zeeman, dipolar, and surface anisotropy, Arias and Mills [7] shows that the latter dominates. The matrix element $V_{0,k}^{(2)}$ for scattering of the $k = 0$ mode into a magnon with wave vector $\vec{k}$, due to the perturbation of the surface anisotropy by defects with lateral rectangular shape with dimensions $a$ and $c$, and height or depth $b$, is [7]

$$V_{0,k}^{(2)} = \hbar \frac{\gamma^2 H_s b}{\omega_0} \left( \frac{pac}{A} \right)^{1/2} \left[ (2H_0 + 4\pi M_{eff}) \frac{1}{c} - (H_0 + 4\pi M_{eff}) \frac{1}{a} \right], \quad (4.45)$$

where $H_S = 2K_S/(M\,t)$ is the mean surface anisotropy field and $p$ is the fraction of the film surface $A$ covered by the defects. Equation (4.45) would be valid if all defects had identical sizes and topology. Actually, one has to perform an average appropriate to an ensemble of defects of diverse topology. That is, on a typical surface, one may have some defects with $c > a$, and others with $c < a$, and the ratio $c/a$ is expected to fluctuate about unity in value. Also, in the calculation of the two-magnon relaxation rate with Eq. (3.190), one uses the modulus of (4.45) squared. Thus, upon expanding the bracket in Eq. (4.45) squared and averaging over this aspect of the defect topology, one obtains

$$\langle|V_{0k}^{(2)}|\rangle = \hbar^2 \frac{\gamma^4 H_S^2 b^2 p}{\omega_0^2 A}[H_0^2 + (2H_0 + 4\pi M_{eff})^2(\langle a/c\rangle - 1)$$
$$+(H_0 + 4\pi M_{eff})^2(\langle c/a\rangle - 1)] \tag{4.46}$$

Using Eq. (3.190) we have an expression for the relaxation rate of the $k = 0$ mode due to two-magnon scattering

$$\eta_0 = \frac{2\pi}{\hbar^2}\sum_k \langle|V_{0k}^{(2)}|^2\rangle\delta(\omega_0 - \omega_k). \tag{4.47}$$

The first step for the evaluation of (4.47) consists of replacing the summation over wave vectors by an integration. Since we consider the wave vectors lying in the film plane, instead of using Eq. (3.88) valid for a 3D space we have

$$\sum_k \rightarrow \frac{A}{(2\pi)^2}\int k\,dk\,d\theta_k, \tag{4.48}$$

where $A$ is the area of the film. The next step consists in transforming the argument of the Dirac delta function in Eq. (4.47) into one with the wave number. For this we use the property

$$\delta[f(x)] = \frac{1}{|df/dx|_{x_0}}\delta(x - x_0), \tag{4.49}$$

where $x_0$ is the value for which $f(x_0) = 0$. Since we have an expression for $\omega_k^2$, the transformation is done in two steps. Using Eqs. (4.49 and 4.42) we have

$$\delta(\omega_0 - \omega_k) = 2\omega_0\delta(\omega_0^2 - \omega_k^2) = \frac{2\omega_0}{\gamma^2(2H_0 + 4\pi M_{eff})Dk_c}\delta(k - k_c), \tag{4.50}$$

where $k_c$ is the zero of the derivative of $(\omega_0^2 - \omega_k^2)$ with respect to $k$, that is easily obtained from Eq. (4.42)

$$k_c = \frac{2\pi M t[H_0 - (H_0 + 4\pi M_{eff})\sin^2\theta_k]}{(2H_0 + 4\pi M_{eff})D}. \tag{4.51}$$

With Eqs. (4.48 and 4.50), the summation that appears in (4.47) becomes

$$\sum_k \delta(\omega_0 - \omega_k) = \frac{2\omega_0}{\gamma^2(2H_0 + 4\pi M_{eff})Dk_c} \frac{A}{(2\pi)^2} \int kdk\delta(k - k_c)2 \int\limits_{-\theta_c}^{\theta_c} d\theta_k,$$

where the integral in $\theta_k$ is limited to the ranges that contribute to the relaxation, given by Eq. (4.44), $-\theta_c < \theta_k < \theta_c$ and $\pi - \theta_c < \theta_k < \pi + \theta_c$. Using this result in Eq. (4.47), we obtain for the two-magnon relaxation rate

$$\eta_0 = \frac{4\gamma^2 H_S^2 b^2 p\theta_c}{\omega_0\pi D(2H_0 + 4\pi M_{eff})}[H_0^2 + (2H_0 + 4\pi M_{eff})^2(\langle a/c\rangle - 1) + (H_0 + 4\pi M_{eff})^2(\langle c/a\rangle - 1)]. \tag{4.52}$$

Finally, it is necessary to obtain an expression for the quantity that is measured experimentally, namely the linewidth of the FMR absorption curve. At this point we notice that the relation $\Delta H = \eta_0/\gamma$, given in Eq. (1.32), is valid only if the FMR frequency is linearly proportional to the applied field, because the relaxation rate corresponds to the width $\Delta\omega_0$ in the frequency spectrum. In typical FMR experiments, the driving frequency is kept fixed and the applied field is swept through the resonance. The field linewidth can be calculated with the relation $\Delta H = \eta_0/(\partial\omega_0/\partial H_0)$ for the specific shape of the sample and field direction. In the case of a film magnetized in the plane, with this relation and Eq. (4.43), we obtain for the half-width at half-maximum

$$\Delta H = \frac{2\omega_0\eta_0}{\gamma^2(2H_0 + 4\pi M_{eff})}. \tag{4.53}$$

Substituting Eq. (4.52) in (4.53), and using expression (4.44) for the critical angle, we obtain for the two-magnon linewidth [7]

$$\Delta H = \frac{8H_S^2 b^2 p}{\pi D(2H_0 + 4\pi M_{eff})^2}[H_0^2 + (2H_0 + 4\pi M_{eff})^2(\langle a/c\rangle - 1)$$
$$+ (H_0 + 4\pi M_{eff})^2(\langle c/a\rangle - 1)] \times \sin^{-1}\left(\frac{H_0}{H_0 + 4\pi M_{eff}}\right)^{1/2}. \tag{4.54}$$

Assuming further that there is no anisotropy in the two directions of the rectangular defects, i.e., $\langle a/c\rangle = \langle c/a\rangle$, considering that for transition metal films $H_0 \ll 4\pi M_{eff}$, and using $H_S = 2K_S/(Mt)$, the linewidth can be written approximately as [19]

$$\Delta H = \frac{32K_S^2 s}{\pi DM^2 t^2} \sin^{-1} \left( \frac{H_0}{H_0 + 4\pi M_{eff}} \right)^{1/2}, \qquad (4.55)$$

where $s$ is a geometrical factor given by $s = pb^2(<a/c> - 1)$. This result shows that as the film thickness increases, the two-magnon scattering linewidth decreases with $1/t^2$.

The experimental verification of the $1/t^2$ thickness dependence of the FMR linewidth of ultrathin in $Ni_{50}Fe_{50}$ films was reported in Azevedo et al. [19]. The FMR measurements were carried out as described in Sects. 1.5.2 and 1.5.3 in the X-band microwave frequency band, using various TE102 rectangular microwave cavities appropriate for the operating frequency. The sample was mounted on the tip of an external goniometer and introduced through a hole in the shorted end of the cavity so that it could be rotated in the plane to allow measurements of the in-plane resonance field and linewidth as a function of the angle. The measurements were done with the static magnetic field modulated by an 1.1-kHz $ac$ component, so that the field for resonance and the linewidth were determined by fitting the derivative of a Lorentzian line shape to the measured field spectrum.

The FMR linewidth measured in Azevedo et al. [19] was the peak-to-peak value, related to the linewidth in Eq. (4.55) by $\Delta H_{pp} = (2/\sqrt{3})\Delta H$, as discussed in Sect. 1.5.1. In each sample, the FMR spectrum was measured as a function of the in-plane angle. The angular variations of both the resonance field and linewidth displayed a small uniaxial anisotropy, and the reported values are averages over the angular variations. The symbols in Fig. 4.10 represent the linewidths measured at 8.53 and 10.83 GHz as a function of the film thickness. Clearly, the linewidth increases markedly with decreasing thickness below 50 Å. Thicker films exhibit a thickness independent linewidth approaching 14 and 20 Oe at 8.53 and 10.83 GHz, respectively. These values are approximately proportional to the measuring frequencies and are attributed to the intrinsic Gilbert damping. The curves in Fig. 4.10 represent fits to data with Eq. 4.55 plus a constant term. By measuring the variation of the resonance field with thickness, the authors of Azevedo et al. [19] obtain for the $Ni_{50}Fe_{50}$ films, $4\pi M = 13.2$ kG, and $H_S = -(82/t)$ kOe Å$^{-1}$. With these values and using for the exchange parameter $D = 2 \times 10^{-9}$ Oe cm$^2$, the fitted curves yield

**Fig. 4.10** Thickness dependence of the FMR peak-to-peak linewidth in $Ni_{50}Fe_{50}$ $(t)$/Si (100), measured at two frequencies, 8.53 GHz and 10.83 GHz. The lines are fits with Eq. (4.55) plus a constant term [19]. Reprinted with permission from A. Azevedo et. al., Phys. Rev. B **62**, 5331 (2000). Copyright (2000) by the American Physical Society

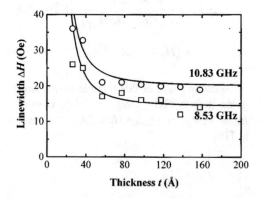

for the geometrical factor $s \approx 16 \, \text{Å}^2$. This value is compatible with realistic estimates for the geometry of the defects: $b = 8 \, \text{Å}$; $p = 0.5$; and $<a/c> = 1.5$. The quantitative agreement of the FMR data with theory provides a clear demonstration that the two-magnon scattering processes, resulting from the presence of defects on the film surfaces, accounts for the relaxation of the $k = 0$ magnon in ultrathin ferromagnetic films.

Additional evidence of the two-magnon relaxation mechanism in ultrathin films and its separation from the intrinsic damping processes is provided by FMR experiments carried out in a wide range of microwave frequencies [20]. In order to obtain the frequency dependence of the two-magnon linewidth, we need to relate the applied field with the operating frequency $\omega$. Since the experiments are made with fixed frequency, and the FMR is detected at the field for which $\omega_0 = \omega$, we use Eq. (4.43) to express $H_0$ in terms of $\omega$. Substitution of the expression in (4.55) gives for the frequency dependence of the two-magnon linewidth

$$\Delta H = \frac{8H_S^2 s}{\pi D} \sin^{-1} \left[ \frac{\left(\omega^2 + \omega_M^2/4\right)^{1/2} - \omega_M/2}{\left(\omega^2 + \omega_M^2/4\right)^{1/2} + \omega_M/2} \right]^{1/2}, \tag{4.56}$$

where the parameter $\omega_M$ is defined in terms of the effective magnetization $\omega_M = \gamma 4\pi M_{eff}$. Clearly, the linewidth vanishes in the limit $\omega \to 0$, so the term zero-frequency residual linewidth does not apply to the two-magnon mechanism. The physical origin of the frequency dependence of the extrinsic contribution to the linewidth is clearly in the fact that as the frequency is decreased, the field for resonance decreases, and so does the critical angle $\theta_c$ given by Eq. (4.44). Thus, the two magnon mechanism "shuts off" as the frequency goes to zero.

Figure 4.11 shows the various contributions to the linewidth in an ultrathin film as a function of frequency [20]. The intrinsic Gilbert damping contribution $\Delta H_G$ is characterized by a linear increase of the FMR linewidth with frequency, as in the

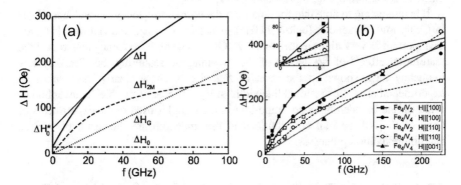

**Fig. 4.11** (a) Schematic representation of the frequency dependence of the various FMR linewidth contributions. (b) Measured linewidths of two Fe/V multilayer samples as a function of microwave frequency. The inset is a magnification of the low-frequency regime [20]. Reprinted with permission from K. Lenz *et. al.*, Phys. Rev. B **73**, 144424 (2006). Copyright (2006) by the American Physical Society

dotted line in Fig. 4.11a. The dashed line shows the linewidth $\Delta H_{2M}$ due to two-magnon scattering, as in Eq. (4.56). The dashed-dotted line represents a residual linewidth $\Delta H_0$, attributed to some inhomogeneous or impurity-driven process. The sum of the three contributions is indicated by the full line $\Delta H$. Clearly, a limited set of low-frequency experimental data can always be fitted with a linear frequency dependence and a constant inhomogeneous term, as in Celinski and Heinrich [14].

The FMR measurements of Lenz et al. [20], carried out over a large frequency range, from 1 to 225 GHz, with the field applied either parallel or perpendicular to the film plane, made possible the elucidation of the roles of the several mechanisms. The authors used two multilayer samples $[Fe_4/V_2]_{60}$ and $[Fe_4/V_2]_{45}$ with 60 and 45 repetitions of the Fe/V layers, respectively, grown on MgO (001). The subscripts in $Fe_4/V_2$ and $Fe_4/V_4$ denote the number of atomic monolayers. The authors used multilayer samples to increase the volume because the films were very thin. Figure 4.11b shows the frequency dependencies of the FMR linewidths measured in the two samples with the field applied in three different directions. The solid and dashed curves for the field in the [100] and [110] in-plane directions are fits using a sum of Eqs. (4.41 and 4.56). The dotted line for the [001] out-of-plane direction corresponds to a linear frequency dependence added to an inhomogeneous contribution $\Delta H_0$.

The two key results of the FMR experiments of Lenz et al. [20] are the following: (i) FMR experiments with in-plane applied fields, in a frequency range of more than two orders of magnitude, show an unambiguous nonlinear frequency dependence of the linewidth $\Delta H(\omega)$, characteristic of the two-magnon mechanism; (ii) Measurements with the field perpendicular to the film plane, fit with the dotted line in Fig. 4.11b, show a linear frequency dependence up to more than 200 GHz. This is also in agreement with theory, since when the field is perpendicular to the film plane, the magnon dispersion relation has a positive slope for all wave numbers [21]. In this case, there are no modes degenerate with the $k = 0$ magnon, so the two-magnon scattering is not operative [22]. Thus, the only remaining mechanism is the intrinsic Gilbert damping, represented by Eq. (4.41).

It is interesting to note that for obtaining the correct frequency dependence of $\Delta H$ $(\omega)$, very low frequencies (below 9 GHz), as used in Celinski and Heinrich [14], are as important as very high frequencies (225 GHz). The high frequency data reveal the "saturation effect" of Eq. (4.56) of the two-magnon scattering, but the very low frequency data, shown in the inset of Fig. 4.11b, show clearly that within an experimental error the "residual linewidth" $\Delta H_0$ is almost negligible. Thus, inhomogeneous line broadening is not important in the samples used in the experiments of Lenz et al. [20], and all contributions to the linewidth are the result of various magnon relaxation processes.

## 4.3.2  Enhanced Two-magnon Relaxation in Exchange-biased Films

As mentioned at the beginning of this section, one of the first clear evidence of the two-magnon scattering relaxation in magnetic thin films was demonstrated by FMR measurements in metallic ferromagnetic films deposited on substrates of antiferro-magnetic NiO [18]. In order to test the role of the contact with the AF material, the same films were deposited on 2 nm thick layers of Ta separating the magnetic films from the NiO. The linewidths measured in the films in direct contact with NiO were substantially larger than in the films deposited on the Ta layer [18]. Similar results were obtained by measuring the linewidths of the magnon peaks detected with Brillouin light scattering [22], providing unequivocal demonstration of the enhanced two-magnon relaxation in exchange-biased films. A systematic experimental study of FMR and BLS linewidths in exchange-biased films and their quantitative inter-pretation were reported in Rezende et al. [23].

In the exchange bias phenomenon, that occurs when a thin ferromagnetic (FM) film is in direct contact with an antiferromagnet (AF), it is known that the field shifts of the hysteresis loops are roughly two orders of magnitude smaller than those associated with the direct exchange between moments at the two sides of the AF/FM interface [24]. The most accepted explanation for this is that, due to the roughness of the interface, the spins in the FM side are subject to random exchange fields created by the spins in small randomly arranged AF grains at the interface [25]. Since the AF/FM samples are prepared under an applied magnetic field, there is a slight imbalance in the number of spins on each sublattice within the grain, so that there is an effective bias field acting on the FM film much smaller than the exchange field in an atomically perfect interface.

As explained in Rezende et al. [23], the enhancement of the magnon damping in exchange-biased films has its origin in the microscopic roughness of the AF/FM interface. The roughness gives rise to a large fluctuating field because the FM spins interact alternatively with one or the other AF sublattice via the atomic exchange coupling. As a result, the perturbation potential is much larger than the one due to the surface anisotropy in the single FM film, producing an enhanced two-magnon scattering potential. Here we extend the two-magnon scattering calculation to exchange-biased films, both for the $k = 0$ magnon, appropriate for the FMR linewidth, and for $k \neq 0$ magnons, as probed in BLS experiments, following Rezende et al. [23].

Initially, we have to obtain the magnon dispersion relation for an exchange-biased FM film. For this, we need to find the contribution to the quadratic Hamilto-nian (4.34) from the AF/FM interface coupling. Differently than in Sect. 1.5, here we consider a Cartesian coordinate system with the $x$, $z$ axes in the film plane, with $z$ in the direction of the exchange bias field. The local interface exchange interaction energy per unit volume is written as

$$E_I = -J_I \frac{\vec{M} \cdot \hat{z}}{Mt},  \tag{4.57}$$

where $\vec{M}$ is the magnetization of the FM film, and $J_I$ represents the local interfacial exchange energy per unit area. Note that the exchange bias field measured by the shift in the hysteresis loop in AF/FM bilayers represents an average over the defects at the interface and over the grains of the polycrystalline AF layer. In order to find the equilibrium direction of the FM magnetization we use for exchange bias energy per unit volume, in CGS units,

$$E_{eb} = -\vec{H}_{eb} \cdot \vec{M}, \tag{4.58}$$

where $H_{eb}$ is expected to be much smaller than the local exchange interaction field given by $H_I = J_I/Mt$. Since in most experiments the FM film also exhibits uniaxial anisotropy, we also consider a contribution to the energy in the form

$$E_u = -\frac{K_u}{M^2} (\vec{M} \cdot \widehat{u}_u)^2, \tag{4.59}$$

where $\widehat{u}_u$ is the unit vector in the direction of symmetry, with polar angle $\theta_u$, and $K_u$ is the first-order anisotropy constant. Adding to the contributions from exchange bias and uniaxial anisotropy, given by Eqs. (4.57–4.59), the ones from Zeeman, demagnetizing and surface anisotropy as in Eq. (1.67), we obtain for the total energy per unit volume, in the CGS system

$$E = -\vec{H}_0 \cdot \vec{M} + 2\pi(n \cdot \vec{M})^2 - \frac{K_S}{M^2 t}(n \cdot \vec{M})^2 - \vec{H}_{eb} \cdot \vec{M} - \frac{K_u}{M^2}(\vec{M} \cdot u_u)^2, \tag{4.60}$$

where $\widehat{n} = \widehat{y}$ is the unit vector normal to the film. With the external field applied in the film plane, at an angle $\theta_H$ with the z-axis, we find the angle $\theta_M$ of the equilibrium direction of the magnetization in the plane by writing the energy as a function of the polar angles

$$E(\theta_M) = -H_0 M \cos(\theta_H - \theta_M) - M H_{eb} \cos \theta_M - K_u \cos^2(\theta_M - \theta_u). \tag{4.61}$$

Thus, with $\partial E/\partial \theta_M = 0$ we obtain the equation for the angle of equilibrium

$$-H_0 \sin(\theta_H - \theta_M) + H_{eb} \sin \theta_M - H_u \sin 2(\theta_M - \theta_u) = 0, \tag{4.62}$$

where $H_u = 2K_u/M$ is the uniaxial anisotropy field. This equation is usually solved numerically to find the magnetization angle $\theta_M$ in terms of the field angle $\theta_H$ for each value of the applied field.

In order to find the Hamiltonian in terms of the collective spin deviation operators, we consider the terms arising from Zeeman, exchange, and dipolar energies, as in Eqs. (4.34 and 4.35a–4.35c), and add the contributions from exchange bias and surface and uniaxial anisotropies. For this, we express the magnetization in Eq. (4.60) in terms of the spin operator components in a coordinate system with the z'-axis along the equilibrium direction of the magnetization, and use the transformations (3.22) and (3.30). The quadratic Hamiltonian becomes

$$H_0 = \hbar \sum_k A_k a_k^\dagger a_k + \frac{1}{2} B_k a_k a_{-k} + \frac{1}{2} B_k a_k^\dagger a_{-k}^\dagger, \tag{4.63}$$

where

$$A_k = \gamma[H_0 \cos(\theta_H - \theta_M) + Dk^2 + 2\pi M(1 - F_k)\sin^2\theta_k + 2\pi M F_k + H_S$$
$$+ H_{eb}\cos\theta_M + H_u \cos^2(\theta_M - \theta_u) - (1/2)H_u \sin^2(\theta_M - \theta_u)], \tag{4.64a}$$

$$B_k = \gamma[2\pi M(1 - F_k)\sin^2\theta_k - 2\pi M F_k - (1/2)H_u \sin^2(\theta_M - \theta_u)], \tag{4.64b}$$

where, for $kt \ll 1$, $F_k \approx 1 - kt/2$. This Hamiltonian can be diagonalized with the transformations (4.36a, 4.36b) to give the magnon frequency as in Eq. (4.39), that we write in the form

$$\omega_k = \gamma[H_1(k)H_3(k)]^{1/2}, \tag{4.65}$$

where the new field parameters are

$$H_1 = (A_k - B_k)/\gamma, \quad H_3 = (A_k + B_k)/\gamma, \tag{4.66}$$

With Eqs. (4.64a, 4.64b) these parameters become

$$\begin{aligned} H_1(k) = & H_0\cos(\theta_H - \theta_M) + Dk^2 + 4\pi M(1 - kt/2) \\ & + H_S + H_u\cos^2(\theta_M - \theta_u) + H_{eb}\cos\theta_M \end{aligned}, \tag{4.67a}$$

$$\begin{aligned} H_3(k) = & H_0\cos(\theta_H - \theta_M) + Dk^2 + 2\pi M kt \sin^2\theta_k \\ & + H_S + H_u\cos2(\theta_M - \theta_u) + H_{eb}\cos\theta_M \end{aligned}. \tag{4.67b}$$

Clearly, the general shapes of the dispersion curves of magnons in exchange-biased FM films are similar to the ones studied in Sect. 4.2. The introduction of the contributions from the exchange bias and uniaxial anisotropy is necessary for a quantitative comparison with experimental data.

The scattering of a magnon with wave vector $\vec{q}$ into another magnon $\vec{k}$ by a process that does not conserve momentum is described by the perturbation Hamiltonian

$$H_{pert}^{(2)} = \sum_{q,k} V_{q,k}^{(2)}(c_q c_k^\dagger + c_q^\dagger c_k), \tag{4.68}$$

where $c_k$ and $c_k^\dagger$ are the creation and annihilation magnon operators and $V_{q,k}^{(2)}$ represents the scattering perturbation (the vector sign is dropped for simplicity). In the case of a FM/AF bilayer, the main source of scattering is the fluctuation in the exchange coupling due to the interface roughness. Following the ideas of Arias and Mills [7], we assume a simple model for the roughness. The interface is considered to be atomically flat, with randomly distributed defects in the form of bumps and pits, with the shape of parallelepipeds with large faces of area $A_d$ parallel to the film plane, and height, or depth, $b$. Using Eq. (4.57) one can see that a defect with $b$ equal to an odd multiple of the AF inter-sublattice distance results in a local perturbation of the exchange coupling that has energy $-2J_IM_{z'}\cos\theta_M/M$ per unit area. Here $z'$ refers to the coordinate axis along the magnetization equilibrium direction. Thus, with Eqs. (3.22c and 3.30), the perturbation Hamiltonian for a single defect labeled $j$ is

$$H_j = J_I \cos\theta_M A_d (NS)^{-1} \sum_{q,k} e^{-i(\vec{q}-\vec{k})\cdot\vec{r}_i}(a_q^\dagger a_k + a_q a_k^\dagger). \tag{4.69}$$

The total perturbation Hamiltonian, obtained by summing (4.69) overall defects becomes

$$H_d^{(2)} = H_I \cos\theta_M \frac{g\mu_B A_d}{A} \sum_{q,k} S_d(\vec{q}-\vec{k}) a_q^\dagger a_k, \tag{4.70}$$

where

$$S_d(\vec{q}-\vec{k}) = \sum_j e^{-i(\vec{q}-\vec{k})\cdot\vec{r}_j} \tag{4.71}$$

is the defect structure factor, and we have used $H_I = J_I/(Mt) = J_I A(NSg\mu_B)^{-1}$, where $A$ is the film area. Using the transformations (4.36a, 4.36b), Eq. (4.70) can be expressed in terms of magnon operators. Assuming that the transformation coefficients for $q$ and $k$ are approximately the same, and using the relation obtained with Eqs. (4.38a, 4.38b) and (4.66), we have $u_q^2 + v_q^2 = A_q/\omega_q = \gamma(H_1 + H_3)/2\omega_q$. With this relation, and disregarding terms that do not conserve energy, the Hamiltonian (4.70) is cast in the form of Eq. (4.68), where the matrix element is

$$V_{q,k}^{(2)} = \hbar\gamma^2 \frac{H_I \cos\theta_M A_d}{2\omega_q A} S_d(\vec{q}-\vec{k})[H_1(q) + H_3(q)]. \tag{4.72}$$

With Eq. (3.180) we obtain an expression for the relaxation rate of the $\vec{q}$ magnon due to the exchange-bias driven two-magnon scattering

$$\eta_q = \frac{2\pi}{\hbar^2} \sum_k \langle |V_{q,k}^{(2)}|^2 \rangle \delta(\omega_q - \omega_k), \tag{4.73a}$$

where, with the averaging overall defects, the coefficient in the summation is

$$\frac{1}{\hbar^2}\langle |V_{q,k}^{(2)}|^2 \rangle = \gamma^4 H_I^2 \langle\cos^2\theta_M\rangle p \frac{A_d}{A} \frac{[H_1(q) + H_3(q)]^2}{4\omega_q^2}, \tag{4.73b}$$

where $p = A_d N_d / A$ is the fraction of the film area covered by $N_d = \langle |S_d|^2 \rangle$ defects with depth or height corresponding to odd multiples of AF lattice spacings. Notice that we have also averaged over $\cos^2 \theta_M$ to account for the fact that most experiments are made with polycrystalline AF layers, so that the angle between the FM and AF moments varies from grain to grain.

The evaluation of the relaxation rate (4.73a, 4.73b) depends on the wave vector $\vec{q}$ of the magnon mode under study. For the $q = 0$ uniform mode, the calculation is done as in Sect. 4.3.1. Considering $H_u, H_{eb} \ll H_0$ one obtains

$$\eta_0 = \frac{\gamma^2 H_I^2 \langle \cos^2 \theta_M \rangle p \langle A_d \rangle}{\pi D \omega_0} (2H_0 + 4\pi M_{eff}) \sin^{-1} \left( \frac{H_0}{H_0 + 4\pi M_{eff}} \right)^{1/2}. \quad (4.74)$$

For $H_0 \ll 4\pi M_{eff}$ this expression reduces approximately to the result of [23]

$$\eta_0 = \frac{\gamma H_I^2 \langle \cos^2 \theta_M \rangle p \langle A_d \rangle}{\pi D}. \quad (4.75)$$

Using Eq. (4.53) we obtain for the approximate FMR linewidth [23]

$$\Delta H = \frac{2H_I^2 \langle \cos^2 \theta_M \rangle p \langle A_d \rangle H_0^{1/2}}{\pi D (H_0 + 4\pi M_{eff})^{1/2}}. \quad (4.76)$$

The FMR linewidths in exchange-biased films were investigated in detail in Rezende et al. [23]. The experiments were carried out with films of $Ni_{50}Fe_{50}$ of various thickness, with the same apparatus as described in Sect. 4.3.1. The FMR linewidth data, obtained at 8.53 GHz, are shown in Fig. 4.12. The circles represent angular averages of the peak-to-peak linewidths measured in the NiFe ($t$)/NiO

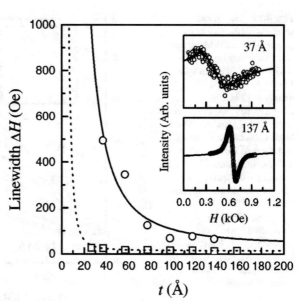

**Fig. 4.12** Thickness dependence of the FMR peak-to-peak linewidths measured at 8.53 GHz in NiFe($t$)/NiO (circles) and in NiFe($t$)/Si (100) (squares). The solid line is a fit with Eq. (4.76) plus a constant term of 40 Oe, as described in the text. Insets show the absorption derivative spectra exhibiting the broadening with decreasing FM film thickness [23]. Reprinted with permission from S. M. Rezende, et. al., Phys. Rev. B **63**, 214418 (2001). Copyright (2001) by the American Physical Society

(860 Å) samples, which fluctuate randomly by up to 20% as the sample is rotated. The square symbols represent the linewidths measured in NiFe films deposited directly on Si(100) substrates. In both cases, the linewidths increase with decreasing sample thickness, but the films deposited on the AF layer show a dramatic 20-fold increase in the linewidth compared to those on Si. The insets in Fig. 4.12 illustrate the line broadening of the exchange-biased film with decreasing thickness. The dashed line represents a fit to the data for the NiFe/Si samples with Eq. (4.55) (times $2/\sqrt{3}$) [19], while the solid line is the fit to the data for the NiFe/NiO samples with Eq. (4.76) (times $2/\sqrt{3}$) [23]. Both curves correspond to a $t^{-2}$ dependence, predicted by the two-magnon relaxation mechanism, plus a small constant term.

As described in Sect. 4.3.1, the fit to the data for the NiFe/Si samples was made with a surface anisotropy field of $H_S = -(82/t)$ kOe Å$^{-1}$, that for $t = 37$ Å corresponds to a value of $-2.2$ kOe [19]. On the other hand, the fit to the data for the exchange-biased sample, NiFe/NiO, was made with Eq. (4.76) using the following parameters [23]: $\langle \cos^2 \theta_M \rangle = 0.5$, $\langle A_d \rangle = 20$ Å$^2$, $p = 0.3$, and $J_I = 11.6$ erg/cm$^2$. Using for the $Ni_{50}Fe_{50}$ films, $4\pi M = 13.2$ kG, this value of the interaction parameter gives for the $t = 37$ Å film in contact with the NiO layer an interaction field of $H_I = J_I/Mt \approx 30$ kOe. This is over one order of magnitude larger than the surface anisotropy field, which is the reason for the large enhancement of the two-magnon relaxation in exchange-biased films. Note also that this interaction field is more than two orders of magnitude larger than the macroscopic exchange bias field $H_{eb} \approx 70$ Oe for the same sample [23].

Additional evidence of the enhancement of the two-magnon relaxation in exchange-biased films is provided by studies of the linewidths of magnon peaks observed by Brillouin light scattering (BLS). In order to illustrate the calculation of the relaxation rate of magnons with $k \sim 10^5$ cm$^{-1}$, we indicate in Fig. 4.13 a point that corresponds to a magnon with wave number similar to the ones observed in

**Fig. 4.13** Dispersion relations for magnons propagating in the film plane calculated with Eqs. (4.65–4.67b) for a NiFe film with $t = 50$ Å, $H_0 = 0.8$ kOe, $4\pi M_{eff} = 12$ kG, $D = 2 \times 10^{-9}$ Oe cm$^2$, and neglecting $H_{eb}$ and $H_u$. The angle between the wave vector and the field varies in steps of 10°. The inset shows the region around $k \sim 0$ expanded to illustrate the modes degenerate with the FMR mode [23]. Reprinted with permission from S. M. Rezende, et. al., Phys. Rev. B **63**, 214418 (2001). Copyright (2001) by the American Physical Society

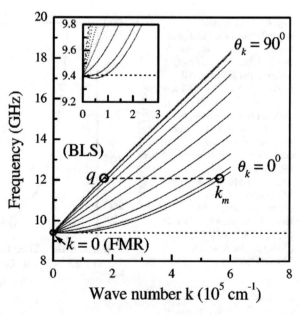

BLS, propagating perpendicularly to the magnetic field. Notice that in the backscattering configuration with the film magnetized in the plane, the magnon wave vector is approximately normal to the plane. Here we assume that the magnons with large wave numbers have relaxation rate that does not depend on the azimuthal angle, and perform the calculation for wave vectors in the plane.

The evaluation of the relaxation rate of the $q$ magnon, carried out with Eqs. (4.73a, 4.73b) using Eq. (4.48) to replace the sum over wave vectors by an integral gives

$$\eta_q = \int k\,dk\,d\theta_k \gamma^4 H_I^2 \langle \cos^2\theta_M \rangle p A_d \frac{[H_1(q) + H_3(q)]^2}{8\pi\omega_q^2} \delta(\omega_q - \omega_k). \qquad (4.77)$$

Here, in contrast to the case of the FMR mode, we see in Fig. 4.13 that the $q$ magnon is scattered into degenerate modes with wave vectors in the whole range of angle $\theta_k = 0 - 2\pi$. Thus, to evaluate Eq. (4.77), it is more convenient to replace the delta function in frequency by one in the angle $\theta_k$. Using the property (4.49) and the approximate dispersion relation (4.21), we have

$$\delta(\omega_q - \omega_k) = \frac{2\omega_q}{\gamma^2 2\pi M k t (H_0 + 4\pi M_{eff}) \sin 2\theta_k} \delta(\theta_k - \theta_d), \qquad (4.78)$$

where

$$\sin^2\theta_d = \frac{q}{k} - \frac{Dk}{2\pi M t}. \qquad (4.79)$$

Then the integral in $k$ runs from $q$ to $k_m = (2\pi M t\,q/D)^{1/2}$, the wave number of the degenerate mode with $\theta_k = 0$, indicated in Fig. 4.13. Using Eq. (4.78) in (4.77), and considering that $H_0 \ll 4\pi M$, one can show that the frequency linewidth of the $q$ magnon, which is equal to the relaxation rate, becomes approximately

$$\Delta\omega_q = \eta_q \approx \int\limits_{q}^{k_m} dk\,\gamma^2 H_I^2 \langle \cos^2\theta_M \rangle p A_d \frac{1}{\pi\omega_q t \sin 2\theta_d}. \qquad (4.80)$$

Finally, using Eq. (4.79), we write this result as

$$\Delta\omega_q = \frac{\gamma^2 p \langle A_d \rangle \langle \cos^2\theta_M \rangle 4\pi M \xi}{2\pi D\omega_q} H_I^2, \qquad (4.81)$$

where $\xi$ is a dimensionless number of order unity, given by the integral

$$\xi = \int\limits_{x_q}^{x_m} \left(\frac{x_0}{x} - x\right)^{1/2} \left(1 + x - \frac{x_0}{x}\right)^{-1/2} dx, \qquad (4.82)$$

where $x$ is a dimensionless variable defined by $x = k/k_0$, and $k_0 = (2\pi M t/D)^{1/2}$.

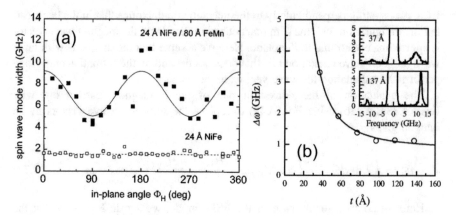

**Fig. 4.14** (a) Spin wave BLS linewidths as a function of the angle of the in-plane applied field, with the in-plane [001] direction for the 24-Å thick $Ni_{80}Fe_{20}$ layer, with and without a $Fe_{50}Mn_{50}$ cover layer. The full (open) squares are the data of the covered (uncovered) layer. The full and dashed lines are guides to the eye [22]. Reproduced from C. Mathieu et. al., J. Appl. Phys. **83**, 2863 (1998), with the permission of AIP Publishing. (b) Thickness dependence of the BLS linewidths in NiFe (*t*)/NiO (860 Å) and fit with Eq. (4.81) plus a constant term of 0.69 GHz, as described in the text. Insets illustrate the broadening of the BLS magnon peaks with decreasing FM film thickness [23]. Reprinted with permission from S. M. Rezende, et. al., Phys. Rev. B **63**, 214418 (2001). Copyright (2001) by the American Physical Society

The enhancement of the two-magnon relaxation in exchange-biased films has been confirmed by measurements of the frequency linewidths of the magnon peaks in Brillouin light scattering (BLS) experiments. Experimental data from Mathieu et al. [22] and Rezende et al. [23] are shown in Fig. 4.14. The measurements were carried out in the backscattering geometry as described in Sect. 2.6.2. The experiments of Mathieu et al. [22] were carried with films of $Ni_{80}Fe_{20}$ (Py) covered by a layer of the metallic antiferromagnet $Fe_{50}Mn_{50}$ and also with no cover layer. The linewidths of the covered and uncovered 24 Å thick NiFe films were measured as a function of the in-plane field angle, as shown in Fig. 4.14a. Two important results are observed in the data. The first is that the averaged magnon linewidth increases by a factor of about five upon covering the FM film with the AF layer. The second is that the linewidth of the exchange-biased NiFe film varies within a factor of two as a function of the field angle. This provides an experimental demonstration of the angle dependence of the linewidth, as predicted in Eq. (4.81), since the magnetization angle $\theta_M$ follows the field as it is rotated in the plane.

The BLS experiments reported in Rezende et al. [23] were carried out in the same $Ni_{50}Fe_{50}$ samples used in the FMR experiments described earlier. The FMR linewidths were measured in NiFe films grown directly on Si substrates, as well as on layers of the AF insulator NiO. In the BLS experiments, the linewidths of the Si/NiFe samples were smaller than the instrument width, and could not be measured. However, with the pronounced broadening due to the AF/FM interface exchange, the linewidths of the NiO/NiFe samples were investigated in detail. The insets in Fig. 4.14b show the increase in the linewidth of the magnon peaks with decreasing thickness. The main panel of the figure shows a fit of the linewidth data with a $1/t^2$ function of the film thickness plus a constant term of 0.69 GHz. Considering the same parameters for the interface roughness used to fit the FMR data, $\langle \cos^2\theta_M \rangle = 0.5$, $\langle A_d \rangle = 20^2 \text{ Å}^2$, $p = 0.3$, and using the value $\xi = 0.65$ calculated numerically with Eq. (4.82), the fit of the data in Fig. 4.14b with Eq. (4.81) yields $J_I = 11.2$ erg/cm$^2$. This is in very close agreement with the value obtained from the FMR data, $J_I = 11.6$ erg/cm$^2$. A key point in the analysis of the experimental data in Rezende et al. [23] is the ability to account for the FMR and BLS linewidth data with the use of the same set of parameters. This represents an unambiguous confirmation of the two-magnon relaxation mechanism in ultrathin magnetic films [26].

## 4.4 Spin Waves in Coupled Magnetic Films

As briefly mentioned in Sect. 2.6.2, the study of spin waves by Brillouin light scattering in the 1980s led to the discovery of the coupling between thin ferromagnetic (FM) films separated by nonmagnetic metallic (NM) layers. By analyzing the frequencies of the spin wave modes in Fe/Cr/Fe and other trilayer systems, Grünberg and collaborators [27, 28] showed that the magnetizations in the two FM films were coupled. This discovery triggered investigations of the properties of magnetic multilayers worldwide that led to the discovery of the giant magnetoresistance by Fert and Grünberg and propelled the development of the field of Spintronics [29, 30]. In this section, we present a calculation of the spin wave dispersion relations in magnetic trilayers, which can readily be used to interpret experimental data obtained with ferromagnetic resonance and Brillouin light scattering techniques.

### 4.4.1 Interlayer Coupling in Magnetic Trilayers

Initially thought to originate in the magnetic dipolar interaction between the two FM layers [31], it was soon understood that the interlayer coupling has a quantum origin and arises from the exchange interaction between electrons in localized magnetic moments and conduction electrons. The form of this interaction was proposed by Ruderman and Kittel to explain the interaction between nuclear moments mediated by conduction electrons [32]. Then Kasuya investigated the interaction between electrons $s$ and $d$ in ferromagnetic metals and calculated its effect on the electrical resistivity [33]. Nearly at the same time, Yosida made use of the $s$–$d$ interaction to explain some magnetic properties of Cu–Mn alloys [34]. These developments led the indirect coupling between magnetic moments by conduction electrons to be called Ruderman–Kittel–Kasuya–Yosida (RKKY) interaction.

It is known that the RKKY interaction oscillates in sign and decays in amplitude as the distance between magnetic impurities increases. A similar type of oscillation was identified by Parkin et. al. in the exchange coupling of Co/Ru, Co/Cr, and Fe/Cr superlattices, by means of magnetization and magnetoresistance measurements [35]. The coupling oscillates in sign and decays in amplitude with increasing thickness of the spacer layers, with an oscillation period on the order of a few atomic layers [35]. It is now established that the indirect coupling between FM layers separated by an NM layer has the same origin as the mechanism of the RKKY interaction [36, 37].

In a trilayer system made of two FM layers with magnetizations $\vec{M}_1$ and $\vec{M}_2$, separated by an NM layer, there is an interlayer exchange coupling that can be expressed phenomenologically by an energy per unit area in the form

$$E_{ex1} = -J_1 \frac{\vec{M}_1 \cdot \vec{M}_2}{M_1 M_2}, \tag{4.83}$$

where $J_1$ is called the bilinear coupling constant. This coupling constant has amplitude and sign that depend on the spacer layer thickness and material composition. For $J_1 > 0$ the energy is minimum for $\vec{M}_1$ parallel to $\vec{M}_2$, that is, ferromagnetic alignment. On the other hand, for $J_1 < 0$ the energy is minimum in the antiferromagnetic alignment. The resistance measured across a multilayer structure depends on the sign and amplitude of $J_1$, as well as on the applied field. When the magnetizations of the layers are aligned in parallel, the electrons with one of the spin directions can go easily through all the magnetic layers and the short circuit through this channel leads to a small resistance. However, in the antiparallel configuration, the electrons of each channel are slowed down every second magnetic layer and the resistance is high. In this case, if a magnetic field is applied so as to align the magnetizations in parallel, the resistance changes from high to low, resulting in a giant magnetoresistance [29, 30].

In a trilayer with spacer layer thickness close to the one at which $J_1$ changes sign, $J_1 \approx 0$ and the bilinear coupling is small. Then another interaction comes into play, the so-called biquadratic coupling, represented by an energy per unit area in the form

$$E_{ex2} = J_2 \left( \frac{\vec{M}_1 \cdot \vec{M}_2}{M_1 M_2} \right)^2, \tag{4.84}$$

where $J_2$ is the biquadratic coupling constant [37–39]. For $J_2 > 0$, this coupling favors an arrangement with $\vec{M}_1$ perpendicular to $\vec{M}_2$, resulting in interesting effects [39–41]. The biquadratic coupling is usually attributed to extrinsic effects, such as interface roughness [37, 38]. With a spacer layer such that $J_1 \approx 0$, the interface roughness produces a local variation of the thickness, so that the interlayer exchange coupling fluctuates in sign. The frustration between parallel and antiparallel alignments results in a perpendicular configuration. Both bilinear and biquadratic couplings will be used in the calculation of the equilibrium magnetization configuration of the FM layers and of the spin wave properties.

## 4.4.2 Equilibrium Configuration of the Magnetizations

In order to study the spin excitations, initially we have to determine the equilibrium directions of the magnetizations in the two FM layers of the trilayer system. The geometry and the coordinate systems employed are shown in Fig. 4.15. We consider two magnetic thin single-crystal films, 1 and 2, having a cubic lattice structure. They have thicknesses $d_1$ and $d_2$ and are separated by a nonmagnetic spacer layer with thickness $s$. The coordinate systems are chosen so that their $xz$-planes are parallel to the film surface, with the $x$- and $z$-axes along [100] and [001] crystal directions, respectively. We study only the situation where the external static magnetic field $\vec{H}_0$ is applied in the plane of the films, at an arbitrary angle $\theta_H$ with respect to the [001] direction. In this case, the equilibrium directions of the magnetizations of the two films, $\vec{M}_1$ and $\vec{M}_2$, are also in the $xz$-plane, characterized by the polar angles $\theta_1$ and $\theta_2$. As is well known, the static and dynamic responses result from the competition of several interactions. In general, each interaction tries to align the magnetizations of the two magnetic films along different directions.

The equilibrium directions of $\vec{M}_1$ and $\vec{M}_2$ are determined by the minima of the total free energy. We consider a free energy per unit area with four contributions

$$E = E_Z + E_A + E_{ex} + E_{dip}, \tag{4.85}$$

where the subscripts denote, in order, Zeeman, anisotropy, exchange, and dipolar terms. The Zeeman energy per unit area is

$$E_Z = -d_1 \vec{M}_1 \cdot \vec{H}_0 - d_2 \vec{M}_2 \cdot \vec{H}_0. \tag{4.86}$$

The anisotropy energy has three contributions, a cubic magnetocrystalline anisotropy energy $E_c$, an in-plane uniaxial term $E_u$ due to distortions introduced by mismatches between the lattices of the films and the substrate, and a surface energy $E_s$ due to the broken cubic symmetry at the film surfaces. These three contributions to the energy per unit area are

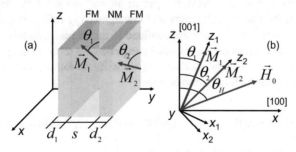

**Fig. 4.15** (a) Trilayer structure used to calculate the spin wave modes with the respective equilibrium magnetizations. (b) Front view of the coordinate systems used to represent the fields and magnetizations in the two magnetic layers. The $x$–$z$ plane is parallel to the plane of the films and the axes $z_1$ and $z_2$ are chosen to coincide with the equilibrium magnetization directions in the films

$$E_C = \sum_i \frac{K_c^{(i)} d_i}{M_i^4} (M_{ix}^2 M_{iy}^2 + M_{iy}^2 M_{iz}^2 + M_{iz}^2 M_{ix}^2), \qquad (4.87a)$$

$$E_u = -\sum_i \frac{K_u^{(i)} d_i}{M_i^2} (\vec{M}_i \cdot \hat{\theta}_{ui})^2, \qquad (4.87b)$$

$$E_S = \sum_i \frac{K_S^{(i)}}{M_i^2} M_{iy}^2, \qquad (4.87c)$$

where $K_c^{(i)}$ is the first-order cubic anisotropy constant of film $i$, $K_u^{(i)}$ is the uniaxial in-plane anisotropy constant in a direction defined by a polar angle unit vector $\hat{\theta}_{ui}$, and $K_S^{(i)}$ is the uniaxial surface anisotropy energy constant. Note that the anisotropy constants $K_c^{(i)}$ and $K_u^{(i)}$ have units of erg/cm$^3$ and each one is associated with an effective anisotropy field $H_{(i)}^A = 2K^{(i)}/M_i$ with units of Oe. We will consider that the two FM layers are made of the same material, so that the anisotropy parameters are the same for both films, and the superscript $(i)$ will be dropped for simplicity.

The exchange energy in (4.85) has the usual volume intralayer exchange contribution plus the interlayer exchange coupling. The first will be considered in the next section to study spin waves, but vanishes for the uniform magnetizations considered here. The second is composed of the bilinear and biquadratic terms given in Eqs. (4.83 and 4.84). Finally, the dipolar energy has surface and volume contributions. The surface contribution is the demagnetizing energy given by

$$E_{demag} = \sum_i 2\pi d_i M_{iy}^2, \qquad (4.88)$$

which has the same form as the surface anisotropy energy (4.87c). The volume contribution is associated with the magnetic field created by the spatial variations of the small-signal magnetization. Since it contributes to the dynamics of the system, but not to the equilibrium configuration, we leave its discussion for the next section.

If the external magnetic field is applied in the plane of the films and the combination of the surface anisotropy and demagnetizing effects has an easy-plane character, the two magnetizations are confined to the $xz$-plane. In this case, $M_{iy} = 0$ so that all contributions to the energy can be expressed in terms of the polar angles of the magnetizations. Using Eqs. (4.83–4.88), we write the total energy per unit area to determine the equilibrium configuration as

$$E = \sum_{i=1}^2 d_i [M_i H_0 \cos(\theta_i - \theta_H) + (1/4) K_c^{(i)} \sin^2 2\theta_i - K_u^{(i)} \cos^2(\theta_i - \theta_u^{(i)})] \\ - J_1 \cos(\theta_1 - \theta_2) + J_2 \cos^2(\theta_1 - \theta_2) \qquad (4.89)$$

In simple situations, the equilibrium configuration can be obtained analytically by equating to zero the derivatives of the energy in Eq. (4.89) with respect to $\theta_1$ and $\theta_2$. However, in more general situations this leads to transcendental equations that cannot be solved analytically. In this case, a simple numerical solution can be obtained by varying $\theta_1$ and $\theta_2$ in small steps to find the values that minimize (4.89). Once $\theta_1$ and $\theta_2$ are found, the total magnetization component in the field direction can be calculated with

$$\frac{M(H_0)}{M_s} = \frac{M_1 \cos(\theta_1 - \theta_H) + M_2 \cos(\theta_2 - \theta_H)}{M_1 + M_2}. \tag{4.90}$$

This equation will be used later to fit magnetization data for specific samples.

### 4.4.3   Derivation of the Spin Wave Dispersion Relation

The spin wave dispersion relations for trilayer structures have been calculated by several authors. The earlier calculations, made before the exchange coupling was discovered, considered only the effect of the dipolar interaction on the surface and volume magnetostatic modes in the coupled films [27]. The simultaneous presence of dipolar and exchange interactions complicates the problem considerably. The equations of motion involve three components of the magnetization and of the dipolar magnetic field in each layer, matched by the appropriate boundary conditions at the interfaces. The full solutions including the bilinear exchange coupling, but restricted to ferromagnetic alignment, were worked out by Hillebrands [42] for an arbitrary number of magnetic layers. For a trilayer structure, formed by two magnetic films separated by a nonmagnetic spacer layer, the dispersion relation is obtained from a system of 16 linear equations. This requires the use of appropriate numerical tools that complicate the interpretation of the observed spectra. This fact, added to the need to interpret data in systems with antiferromagnetic coupling, led several authors to develop alternative calculations, each with its own limitations and simplifying assumptions [28, 42–48].

Here we follow Rezende et al. [48] and derive the spin wave dispersion relations for a trilayer structure with several types of anisotropy and interlayer coupling. In the following section, we shall use the results to interpret BLS and FMR data in specific samples. The calculation is based on the torque equations of motion for the continuous magnetizations of the two magnetic films. Compared to earlier calculations, Rezende et al. [48] introduced the biquadratic exchange coupling and used an improved approximation for the dipolar coupling, making the results valid for relatively thick magnetic layers. We use the Landau–Lifshitz equation (1.10) to describe the motion of the magnetization of film $i$, written as

$$\frac{d\vec{M}_i}{dt} = -\gamma_i \vec{M}_i \times \vec{H}_{eff}^{(i)}. \tag{4.91}$$

where $\vec{H}_{eff}^{(i)}$ is the effective field acting on $\vec{M}_i$. All fields and magnetizations are decomposed in a static part and a small-signal dynamic component. For each film, we use a Cartesian coordinate system $x_i$ $y_i$ $z_i$, obtained from the one with the axes in <100> directions, by rotation about the $y$-axis so that the $z_i$-axis coincides with the equilibrium direction of the magnetization $\vec{M}_i$, as shown in Fig. 4.15b. Hence, the magnetization in film $i$ can be written as

$$\vec{M}_i = \hat{x}_i m_{ix_i} + \hat{y}_i m_{iy_i} + \hat{z} M_{iz_i}, \tag{4.92}$$

where it is assumed that $m_{ix_i}, m_{iy_i} << M_{iz_i}$. The transformations from the original variables are given by

$$M_{ix} = M_{iz_i} \sin \theta_i + m_{ix_i} \cos \theta_i, \tag{4.93a}$$

$$M_{iy} = m_{iy}, \tag{4.93b}$$

$$M_{iz} = M_{iz_i} \cos \theta_i - m_{ix_i} \sin \theta_i. \tag{4.93c}$$

Likewise, the effective field is written as

$$\vec{H}_{eff}^{(i)} = \hat{x}_i h_{ix_i} + \hat{y}_i h_{iy_i} + z_i H_{iz_i}. \tag{4.94}$$

The effective fields corresponding to the Zeeman, anisotropy, bilinear, and biquadratic exchange energy contributions $E_\lambda$ (per unit area) are given by

$$\vec{H}_\lambda^{(i)} = -\nabla_{M_i}(E_\lambda/d_i). \tag{4.95}$$

The calculation of the volume dipolar magnetic field is more involved and requires some approximations in order to be carried out analytically, as will be shown later. Since the spin-wave frequencies are determined by the linearized equations, in the transformation of the energy expressions (4.83–4.89) to the new variables (4.93a–4.93c and 4.94), only terms quadratic in small-signal components have to be kept. Furthermore, in the calculation of the effective fields, only terms linear in small quantities in the $x_i$ and $y_i$ components and constant in the $z_i$ components need to be retained. Note also that it is not necessary to expand the $M_{iz_i}$ components into $m_{ix}$, and $m_{iy}$. It is simpler to evaluate the derivatives of $E_\lambda$ with respect to $M_{iz_i}$ and enter the equation of motion with the corresponding $z_i$ components of the effective fields.

From the Zeeman energy (4.86), we obtain only one relevant field component

$$H_{iz_i} = -\frac{1}{d_i} \frac{\partial E_Z}{\partial M_{iz_i}} = H_0 \cos(\theta_i - \theta_H). \tag{4.96}$$

The contributions from the cubic anisotropy energy to the effective field are

$$h^c_{ix_i} \approx \frac{K^{(i)}_c}{M^2_i}(3\sin^2 2\theta_i - 2)m_{ix_i}, \tag{4.97a}$$

$$h^c_{iy} \approx -\frac{2K^{(i)}_c}{M^2_i}m_{iy}, \tag{4.97b}$$

$$H^c_{iz_i} \approx -\frac{K^{(i)}_c}{M_i}\sin^2 2\theta_i. \tag{4.97c}$$

The relevant components of the uniaxial anisotropy field are

$$h^u_{ix_i} \approx \frac{2K^{(i)}_u}{M^2_i}\sin^2(\theta_i - \theta^{(i)}_u)m_{ix_i}, \tag{4.98a}$$

$$H^u_{iz_i} \approx \frac{2K^{(i)}_u}{M_i}\cos^2(\theta_i - \theta^{(i)}_u), \tag{4.98b}$$

and from the surface anisotropy energy, the only component is

$$h^s_{iy} \approx \frac{2K^{(i)}_s}{d_i M^2_i}m_{iy}. \tag{4.99}$$

Finally, from the bilinear and biquadratic exchange energies we obtain the following components

$$h^{ex}_{ix_i} \approx \frac{J_1}{d_i M_1 M_2}\cos(\theta_1 - \theta_2)m_{jx_j} - \frac{2J_2}{d_i M_1 M_2}\cos[2(\theta_1 - \theta_2)]m_{jx_j}$$
$$- \frac{2J_2}{d_i M^2_i}\sin^2(\theta_1 - \theta_2)m_{ix_i}, \tag{4.100a}$$

$$h^{ex}_{iy} \approx \frac{J_1}{d_i M_1 M_2}m_{jy} - \frac{2J_2}{d_i M_1 M_2}\cos(\theta_1 - \theta_2)m_{jy}, \tag{4.100b}$$

$$H^{ex}_{iz_i} \approx \frac{J_1}{d_i M_i}\cos(\theta_1 - \theta_2) - \frac{2J_2}{d_i M_i}\cos^2(\theta_1 - \theta_2). \tag{4.100c}$$

The treatment of the dipolar magnetic field is more involved and requires approximations to be carried out analytically. This field does not exist when the magnetization is uniform and lies in the film plane, therefore it was not considered in the calculation of the equilibrium configuration. However, as we saw in Sects. 4.1 and 4.2, the spatial variation of the magnetization in a spin wave creates a dipolar magnetic field that obeys Maxwell's equations. This field can be calculated from the magnetic potential $\psi$, which, as shown earlier, can be obtained from the solutions of Walker equation (4.6) inside the film and Laplace equation (4.7) outside, subject to the appropriate boundary conditions at the surfaces. For a thin unbounded ferromagnetic film under an in-plane magnetic field in the $z$-direction, the magnetic potential is given by Eq. (4.29). Assuming a spin wave propagating in the $x$-direction, with transverse magnetization components of the form

$$M_x(x) = M_x e^{ikx}, \quad \text{and} \quad M_y(x) = M_y e^{ikx}, \tag{4.101}$$

the dipolar field is calculated with $\vec{h} = -\nabla\psi$, with the potential given by Eqs. (4.29 and 4.101). One can show that the components of the dipolar field inside an unbounded film of thickness $d$, with the coordinate $y$ perpendicular to the film plane, and $y = 0$ in the middle of the film, are [Problem 4.7]

$$h_x^{dip} = [2\pi M_x e^{-kd/2}(e^{ky} + e^{-ky}) - 4\pi M_x - i2\pi M_y e^{-kd/2}(e^{ky} - e^{-ky})]e^{ikx}, \tag{4.102a}$$

$$h_y^{dip} = [-i2\pi M_x e^{-kd/2}(e^{ky} - e^{-ky}) - 2\pi M_y e^{-kd/2}(e^{ky} + e^{-ky})]e^{ikx}. \tag{4.102b}$$

However, we also have to consider the dipolar field outside the film, because the field of one film can drive the magnetization in the other film. As shown in Cochran et al. [43], for $y \geq d/2$ the field components are

$$h_x^{dip} = 2\pi(M_x + iM_y)(e^{-kd/2} - e^{kd/2})e^{-ky}e^{ikx}, \tag{4.103a}$$

$$h_y^{dip} = 2\pi(iM_x - M_y)(e^{-kd/2} - e^{kd/2})e^{-ky}e^{ikx}, \tag{4.103b}$$

and for $y \leq -d/2$

$$h_x^{dip} = 2\pi(M_x - iM_y)(e^{-kd/2} - e^{kd/2})e^{ky}e^{ikx}, \tag{4.104a}$$

$$h_y^{dip} = -2\pi(iM_x + M_y)(e^{-kd/2} - e^{kd/2})e^{ky}e^{ikx}, \tag{4.104b}$$

In order to simplify these equations, as in Sect. 4.2 we assume that the dipolar fields inside the films do not vary along $y$ and replace expressions (4.102a, 4.102b) by suitable averages

$$\langle \vec{h}_1^{dip} \rangle = \frac{1}{d_1} \int_{-d_1/2}^{d_1/2} \vec{h}^{dip}(y)\, dy, \tag{4.105a}$$

$$\langle \vec{h}_2^{dip} \rangle = \frac{1}{d_1} \int_{-d_1/2+s}^{d_1/2+s+d_2} \vec{h}^{dip}(y)\, dy. \tag{4.105b}$$

Considering the field and the magnetization at the angles shown in Fig. 4.15, the average dipolar field in film 1 has components

$$\langle h_{1x}^{dip} \rangle = -4\pi M_{1x}[1 - (1 - e^{-kd_1})/kd_1]$$
$$+ 2\pi(iM_{2y} - M_{2x})(1 - e^{-kd_1})[(1 - e^{-kd_2})e^{-ks}/kd_1]\cos\theta_1 \tag{4.106a}$$

$$\langle h_{1y}^{dip} \rangle = -4\pi M_{1y}(1 - e^{-kd_1})/kd_1$$
$$+ 2\pi(iM_{2x} + M_{2y})(1 - e^{-kd_1})(1 - e^{-kd_2})e^{-ks}/kd_1 \tag{4.106b}$$

The final expressions for the components of the dipolar field are obtained by introducing the coordinate transformations in (4.93a–4.93c) and expanding the exponential functions in the parameters $kd_1 \ll 1$. Similar expressions follow for the dipolar field in film 2.

Using the equation of motion (4.91), the dipolar fields and the other fields given by Eqs. (4.96–4.100c), we obtain the equations for the small-signal magnetization components in the two films in their respective coordinate systems. Assuming the time variation $\exp(-i\omega t)$, where $\omega$ is the angular frequency, and retaining only terms to first order in small quantities we obtain [48]

$$\begin{pmatrix} i\omega/\gamma_1 & H_1 & iH_5 & H_2 \\ -H_3 & i\omega/\gamma_1 & H_4 & iH_6 \\ -iG_5 & G_2 & i\omega/\gamma_2 & G_1 \\ G_4 & -iG_6 & -G_3 & i\omega/\gamma_2 \end{pmatrix} \begin{pmatrix} m_{1x_1} \\ m_{1y} \\ m_{2x_2} \\ m_{2y} \end{pmatrix} = 0, \tag{4.107}$$

where

$$H_1 = H_0\cos(\theta_1 - \theta_H) + \frac{H_c^{(1)}}{4}(3 + \cos 4\theta_1) + H_u^{(1)}\cos^2(\theta_1 - \theta_u^{(1)}) + D^{(1)}k^2$$
$$+ 4\pi M_1\left(1 - \frac{kd_1}{2}\right) - H_s^{(1)} + H_{ex1}^{(1)}\cos(\theta_1 - \theta_2) - 2H_{ex2}^{(1)}\cos^2(\theta_1 - \theta_2) \tag{4.108a}$$

$$H_2 = -H_{ex1}^{(2)} - 2\pi M_1 kd_1\left(1 - \frac{kd_2}{2}\right)e^{-ks} + 2H_{ex2}^{(2)}\cos(\theta_1 - \theta_2), \tag{4.108b}$$

$$H_3 = H_0\cos(\theta_1 - \theta_H) + H_c^{(1)} + \cos 4\theta_1 + H_u^{(1)}\cos 2(\theta_1 - \theta_u^{(1)}) + D^{(1)}k^2$$
$$+ 2\pi M_1 kd_1\cos^2\theta_1 + H_{ex1}^{(1)}\cos(\theta_1 - \theta_2) - 2H_{ex2}^{(1)}\cos[2(\theta_1 - \theta_2)] \tag{4.108c}$$

$$H_4 = -H_{ex1}^{(2)}\cos(\theta_1 - \theta_2) - 2H_{ex2}^{(1)}\cos[2(\theta_1 - \theta_2)]$$
$$- 2\pi M_1 kd_1\left(1 - \frac{kd_2}{2}\right)e^{-ks}\cos\theta_1\cos\theta_2 \tag{4.108d}$$

$$H_5 = -2\pi M_1 kd_1\left(1 - \frac{kd_2}{2}\right)e^{-ks}\cos\theta_2, \tag{4.108e}$$

$$H_6 = 2\pi M_1 kd_1 \left(1 - \frac{kd_2}{2}\right) e^{-ks} \cos\theta_1, \qquad (4.108f)$$

and the coefficients $G_1$–$G_6$ are given by similar expressions, replacing the film label 1 by 2, and vice versa. The effective fields in (4.108a–4.108f) are defined by:
$H_c^{(i)} = 2K_c^{(i)}/M_i$ ; $H_u^{(i)} = 2K_u^{(i)}/M_i$ ; $H_S^{(i)} = 2K_S^{(i)}/(d_iM_i)$ ; $H_{ex1,2}^{(i)} = J_{1,2}/(d_iM_i)$ .
Note that $D^{(i)}$ is the intralayer exchange parameter for film $i$, which was introduced in the usual manner.

The solutions of Eq. (4.107) are obtained by requiring that the secular determinant vanishes. This leads to spin wave frequencies that are given by the zeroes of the following equation

$$\frac{\omega^4}{\gamma_1^2\gamma_2^2} + a\omega^2 + b\omega + c = 0, \qquad (4.109)$$

where

$$a = (G_2H_4 + G_4H_2 + G_5H_5 + G_6H_6)/\gamma_1\gamma_2 - H_1H_3/\gamma_2^2 - G_1G_3/\gamma_1^2, \qquad (4.110a)$$

$$b = (G_3G_5H_2 + G_1G_4H_5 - G_1G_6H_4 - G_2G_3H_6)/\gamma_1$$
$$+ (G_6H_2H_3 + G_4H_1H_6 - G_5H_1H_4 - G_2H_3H_5)/\gamma_2 , \qquad (4.110b)$$

$$c = G_1G_6H_3H_5 + G_1G_3H_1H_3 + G_5G_6H_5H_6 + G_5G_6H_2H_4$$
$$+ G_3G_5H_1H_6 + G_2G_4H_2H_4 - G_2G_3H_2H_3 - G_1G_4H_1H_4 + G_2G_4H_5H_6 , \qquad (4.110c)$$

For any value and direction of the applied in-plane field, Eq. (4.109) with the coefficients in (4.110a–4.110c) has two real solutions, corresponding to the acoustic and optic modes. Of course, in order to find the frequencies one must first determine the equilibrium angles, as described earlier. Note that the FMR technique probes the $k = 0$ modes, and since they are not influenced by the dipolar coupling, the calculation simplifies considerably. In this case $H_5 = H_6 = G_5 = G_6 = 0$, so that the term linear in $\omega$ in Eq. (4.109) vanishes. This allows the FMR frequencies to be determined analytically. For $\gamma_1 = \gamma_2 = \gamma$ we have

$$(\omega/\gamma)^2 = -(a_0/2) \pm [(a_0/2)^2 - c_0]^{1/2}, \qquad (4.111)$$

where

$$a_0 = (G_2H_4 + G_4H_2 - H_1H_3 - G_1G_3)_{k=0}, \qquad (4.112a)$$

$$c_0 = (G_1G_3H_1H_3 + G_2G_4H_2H_4 - G_2G_3H_2H_3 - G_1G_4H_1H_4)_{k=0}. \qquad (4.112b)$$

Let us apply this result to the simple case of a trilayer having two identical magnetic films, coupled through a bilinear antiferromagnetic exchange. Consider further, that the in-plane anisotropy is uniaxial with easy axis in the $z$-direction and that the external field is applied in this direction. In this case there are three equilibrium phases, depending on the field value. By minimizing Eq. (4.89) with $\theta_H = \theta_1 = \theta_2 = 0$ and $K_c^{(i)} = J_2 = 0$, one can see that, in the field range

$$0 \leq H_0 \leq H_{SF} = [H_u(2H_{ex} + H_u)]^{1/2}, \tag{4.113}$$

where $H_{ex}$ is the absolute value of the AF bilinear exchange field and $H_u$ is the uniaxial anisotropy field for both films, the magnetizations in the two films are aligned antiferromagnetically, i.e., $\theta_1 = 0$ and $\theta_2 = \pi$. In this case, the FMR frequencies for the optic and acoustic modes obtained from (4.111) are given by

$$
\begin{aligned}
(\omega/\gamma)^2 &= H_0^2 - H_u^2 + (H_u + H_{ex})(2H_u + 4\pi M_{eff}) \\
&\pm \left[ H_0^2(2H_u + 4\pi M_{eff})(2H_u + 4\pi M_{eff} + 4H_{ex}) + (H_{ex} + 4\pi M_{eff})^2 \right]^{1/2},
\end{aligned} \tag{4.114}
$$

where $4\pi M_{eff} = 4\pi M - H_S$. As the field is increased beyond the spin–flop critical field $H_{SF}$, the magnetizations become canted with angles $\theta_1 = \theta$ and $\theta_2 = \pi - \theta$, where $\sin\theta = H_0/(2H_{ex} - H_u)$. As the field increases further in the spin–flop phase, the magnetizations rotate toward the $z$-axis and the FMR frequencies vary with field as

$$
\begin{aligned}
(\omega/\gamma)^2 &= [2H_{ex}^2 + H_u(H_{ex} + 4\pi M_{eff})\sin^2\theta + (H_{ex} + 4\pi M_{eff})(H_{ex} - H_u) \\
&\quad -H_{ex}^2 \pm H_{ex}[(2H_{ex} + 8\pi M_{eff} + H_u)\sin^2\theta - (H_u + 4\pi M_{eff})] .
\end{aligned} \tag{4.115}
$$

Finally, if the field is increased beyond the critical value $H_c = 2H_{ex} - H_u$, the magnetizations become aligned in the $z$-direction. In this phase, called saturated or ferromagnetic, the FMR frequencies become

$$(\omega/\gamma)_+^2 = (H_0 + H_u)(H_0 + H_u + 4\pi M_{eff}), \tag{4.116a}$$

$$(\omega/\gamma)_-^2 = (H_0 + H_u - 2H_{ex})(H_0 + H_u - 2H_{ex} + 4\pi M_{eff}). \tag{4.116b}$$

These expressions show that in an FMR experiment, the difference between the fields for resonance of the acoustic and optic modes in the ferromagnetic phase, observed at a fixed frequency, yields a direct measurement of the exchange coupling field. In the case of the BLS technique the field is fixed, and one observes the frequency shifts of the two modes. One difficulty encountered in both FMR and BLS is that the optic mode intensity is much smaller than that of the acoustic mode. In fact, as is well known, for films of identical materials the intensity of the optic mode in the ferromagnetic phase is theoretically zero. Actually, experimental films are always somewhat different from each other and present various magnetic phases, so that although the optic mode is weak it can often be observed.

### 4.4.4  FMR and BLS Experiments in Fe/Cr/Fe Trilayers

To close this section, we shall use the theory presented in the previous sections to analyze experimental data in (100) Fe(40 Å)/Cr($s$)/Fe (40 Å) trilayers investigated in Azevedo et al. [40] and Rezende et al. [48]. The films were grown by dc magnetron sputtering in an ultrahigh vacuum chamber onto polished, chemically cleaned single-crystal MgO (100) substrates. In order to determine the overall behavior of the coupling as a function of Cr spacer thickness, one sample was fabricated with a wedged trilayer structure, in which the thickness of the spacer layer varies continuously from zero to a certain value [49, 50]. Then, by scanning the position of the laser beam of a MOKE magnetometer along the direction of the wedge, one can measure the total magnetization as a function of the spacer layer thickness. Figure 4.16 shows a plot of the measured saturation field as a function of the Cr layer thickness. The saturation field is defined arbitrarily as the field at which the magnetization reaches 75% of the saturation value. The field is applied along the [100] axis. The coupling reaches the first antiferromagnetic maximum at about $s \approx 10$ Å and crosses to ferromagnetic at $s \approx 15$ Å. The magnetic behavior was studied in detail in two trilayer samples having uniform Cr spacers with thickness $s = 11$ Å and $s = 15$ Å.

Consider initially the (100) Fe/Cr/Fe sample with Cr thickness $s = 11$ Å, a value close to the first AF maximum. The initial characterization is made with MOKE magnetometry. Figure 4.17a shows data (open circles) measured with the field along the hard [101] axis. The solid line represents a theoretical fit with the calculation described in Sect. 4.4.2 using the following parameters: $4\pi M = 19.5$ kG; $H_c = 0.57$ kOe; $H_u = 0$; $H_{ex1} = -0.89$ kOe, corresponding to $J_1 = -0.55$ erg/cm$^2$; $H_{ex2} = 0.07$ kOe, corresponding to $J_2 = 0.044$ erg/cm$^2$ and $J_2/J_1 = -0.08$. The field dependencies of the equilibrium angles $\theta_1$ and $\theta_2$ calculated with these parameters are shown by the solid lines in Fig. 4.17b. The dashed line represents the calculation with $J_2 = 0$, indicating that the existence of a small positive biquadratic

**Fig. 4.16** Saturation field measured with MOKE magnetometry in a (100) Fe/Cr/Fe wedged trilayer structure as a function of the Cr spacer layer thickness. The solid line is a guide to the eyes [40]. Reprinted with permission from A. Azevedo et. al., Phys. Rev. Lett. **76**, 4837 (1996). Copyright (1996) by the American Physical Society

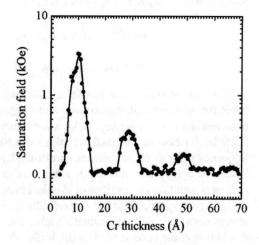

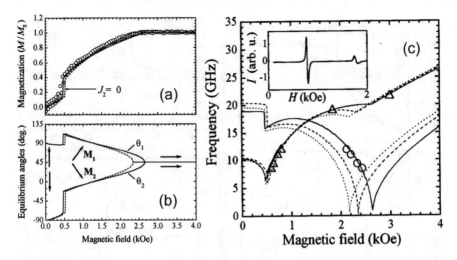

**Fig. 4.17** (**a**) Normalized magnetization versus external field applied along the [101] axis ($\theta_H = 45°$) in the (100) Fe(40 Å )/Cr(11 Å )/Fe(40 Å) sample. Open circles are MOKE data. The solid line represents the calculation with the parameters obtained by the least square deviation fit. Dashed line is the calculation with $J_2 = 0$. (**b**) Calculated equilibrium magnetization angles. (**c**) FMR frequencies measured with the sweeping magnetic field at several frequencies. Solid, dashed, and dotted lines are the results of calculations. Inset shows the FMR spectrum at 9.5 GHz with the acoustic and optic mode lines [48]. Reproduced from S. M. Rezende et. al., J. Appl. Phys. **84**, 958 (1998), with the permission of AIP Publishing

exchange coupling does not change much the behavior of the magnetizations in the sample with large bilinear coupling. Figures 4.17a and b show clearly that for the field in the [101] direction, the system exhibits three distinct phases, illustrated by the arrows indicating the magnetizations. At fields below the critical value of 0.5 kOe, the magnetizations remain close to an antiferromagnetic alignment along the [001] axis. At this field there is a sudden transition to a spin–flop state. As the field increases above the critical value, the spins rotate toward the field and only at 2.63 kOe the system reaches saturation.

Of course, we expect the dependence of the magnon frequencies on the applied field to change with spin configuration. Indeed, this is clearly observed in measurements carried out with both FMR and BLS techniques. Figure 4.17c shows FMR data obtained by measuring the field for resonance at constant microwave frequency. The symbols represent measurements made with several microwave cavities to vary the frequency discretely. The inset shows the FMR spectrum at 9.5 GHz where one can see both the acoustic and optic mode lines [48]. Note that at higher frequencies the optic mode resonance is not observed because its intensity in the FM phase is theoretically zero. The solid lines represent the calculation with Eqs. (4.111–4.112b) using the same parameters obtained from the fit to the MOKE data. The dashed line is obtained with $H_{ex2} = 0$, keeping the same values for the other parameters. As expected, this produces a large departure of the optic mode line from the data because the critical field for the SF–FM transition is very sensitive to

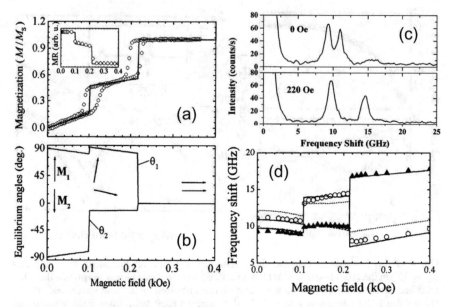

**Fig. 4.18** (a) Normalized magnetization versus external field applied along the [100] axis (easy axis) in the (100) Fe(40 Å)/Cr(15 Å)/Fe(40 Å) sample. Open circles are MOKE data. The solid line represents the calculation with the parameters obtained by the least square deviation fit. (b) Calculated equilibrium magnetization angles. (c) BLS spectra for two field values showing the peaks for both acoustic and optic magnon modes. (d) BLS frequency shift as a function of the field. Symbols represent data and lines are results of calculations [48]. Reproduced from S. M. Rezende et. al., J. Appl. Phys. **84**, 958 (1998), with the permission of AIP Publishing

the value of $H_{ex2}$. One may ask what happens if we keep $H_{ex2} = 0$ and vary the other parameters to obtain a least-square fit to the data. The result is shown by the dotted line in Fig. 4.17c, obtained with $4\pi M_{eff} = 19.5$ kG, $H_c = 0.53$ kOe, $H_{ex1} = 0.82$ kOe, and $g = 2.1$. This shows that although $H_{ex2}$ is small, it is not possible to fit the data with theory without the inclusion of the biquadratic exchange energy in the calculation. The dispersion relations calculated with Eqs. (4.109–4.110c) for the magnon wave number probed with BLS, using the parameters of the FMR fit, are also in excellent agreement with the BLS data for this sample [48].

Finally, let us look to some results obtained in the sample with Cr thickness $s = 15$ Å, expected to have a small bilinear coupling, as indicated in Fig. 4.16 and correspondingly small critical fields for all magnetic phase transitions. Figure 4.18 shows data and calculated curves for this sample. Interestingly, besides the expected low critical fields, the data furnished by the various techniques displayed sudden jumps as the field varied. This is a result of the relatively large biquadratic exchange coupling. Figure 4.18a shows MOKE measurements in the field range $0 < H_0 < 0.4$ kOe, with the field applied along the easy axis [100]. The circles in the inset are magnetoresistance data, obtained with a standard four-probe technique. Both data without the coercive field are very well fitted by the theoretical lines shown in the figure, obtained with Eqs. (4.89 and 4.90) with the following parameters: $4\pi M = 19.0$ kG; $H_c = 0.55$ kOe; $H_u = 0$; $H_{ex1} = -0.15$ kOe; $H_{ex2} = 0.05$ kOe. Note that the biquadratic exchange field is similar to the one of the $s = 11$ Å sample. However, since the bilinear exchange is quite smaller, the role of the biquadratic coupling becomes important. In fact, it is responsible for the two first-order phase transitions that produce the jumps in the data. As shown in Fig. 4.18b, the calculated equilibrium angles reveal that at $H_0 \approx 0.1$ kOe the configuration changes from AF to nearly $90°$, and at $H_0 \approx 0.22$ kOe it changes from $90°$ to FM alignment.

Signatures of the first-order phase transitions are also observed in the spin-wave frequency versus field data obtained with BLS. Figure 4.18c shows the anti-Stokes spectra obtained at two values of the field applied along the [001] axis in the backscattering geometry with an incidence angle of $30°$, corresponding to a magnon wave number $k = 1.22 \times 10^5$ cm$^{-1}$. Note that the wave vector is nearly perpendicular to the film plane and to the field. Even though the calculations presented in this section considered the wave vector in the plane, for large $k$ we assume that the magnon frequency depends on the direction relative to the field, but not relative to the film plane. The symbols in Fig. 4.18d represent the measured frequency shifts as a function of the field (triangles for the acoustic mode and circles for the optic mode) and the curves calculated with Eqs. (4.109–4.110c) with the same parameters used to fit the MOKE data.

As shown in Figs. 4.18c, d, in the AF phase, at zero field, the two magnon modes are separated in frequency by about 2 GHz, the optic mode having higher frequency than the acoustic one. As the field is increased above 0.1 kOe, the system acquires the near $90°$ alignment and the separation between the acoustic and optic modes suddenly increases. The two peaks at 0.220 kOe are 5 GHz apart. At 0.230 kOe the system reaches the FM state, and the acoustic mode frequency becomes larger than the optic mode frequency. The dotted lines in Fig. 4.18d represent the frequencies calculated without the contribution of the biquadratic exchange field. The presence of $H_{ex2}$ is not very important for the values of the frequencies, but it is essential to determine the correct ground states. The most pronounced effect of the biquadratic coupling, in this case, is to shift the frequency of the optic mode, downward in the AF and FM phases and upward in the $90°$ central region.

## Problems

4.1. Show that for frequencies in the range $\omega_H < \omega < [\omega_H(\omega_H + \omega_M)]^{1/2}$, the parameter $\kappa$ in Eq. (4.4) is such that $(1 + \kappa) < 0$.

4.2. Using the dispersion relation Eq. (4.18) for magnetostatic waves in a film magnetized in the plane, show that for the wave vector perpendicular to the magnetic field, the dispersion relation is given by Eq. (4.24).

4.3. Using Eq. (4.24) for a surface magnetostatic spin wave in a ferromagnetic film, demonstrate Eq. (4.26) for the group velocity.

4.4. Show that the frequency dependence of the two-magnon linewidth for an ultrathin ferromagnetic film given by Eq. (4.56) can be obtained from (4.55) by expressing the applied field in terms of the frequency.

4.5. In the two-magnon relaxation process of a spin wave with wave number $q$ and $\theta_k = \pi/2$, illustrated in Fig. 4.13, show that $k_m = (2\pi M t q/D)^{1/2}$ is the wave number of the degenerate mode with $\theta_k = 0$.

4.6. From Eqs. (4.83–4.88), show that the total energy per unit area in the magnetic trilayer illustrated in Fig. 4.15 is given by expression (4.89).

4.7. Show that the magnetic dipolar field in a ferromagnetic film with the magnetic potential given by Eq. (4.29) has the components given by Eqs. (4.102a, 4.102b).

4.8. Show that in a magnetic trilayer with bilinear antiferromagnetic coupling and uniaxial anisotropy, with an applied field such that the magnetizations are aligned in the field direction, the FMR frequencies are given by Eq. (4.115).

# References

1. Walker, L.R.: Resonant modes of ferromagnetic spheroids. J. Appl. Phys. **2**, 318 (1958)
2. Damon, R.W., Eshbach, J.R.: Magnetostatic modes of a ferromagnet slab. J. Phys. Chem. Solids. **19**, 308 (1961)
3. Hurben, M.J., Patton, C.E.: Theory of magnetostatic waves for in-plane magnetized isotropic films. J. Mag. Mag. Mat. **139**, 263 (1995)
4. Kalinikos, B.A.: Excitation of propagating spin waves in ferromagnetic films. IEE Proc. **127**, 4 (1980)
5. Vilela-Leão, L.H., Salvador, C., Azevedo, A., Rezende, S.M.: Unidirectional anisotropy in the spin pumping voltage in yttrium iron garnet/platinum bilayers. Appl. Phys. Lett. **99**, 102505 (2011)
6. Holanda, J., Maior, D.S., Azevedo, A., Rezende, S.M.: Detecting the phonon spin in magnon-phonon conversion experiments. Nat. Phys. **14**, 500 (2018)
7. Arias, R., Mills, D.L.: Extrinsic contributions to the ferromagnetic resonance response of ultrathin films. Phys. Rev. B. **60**, 7395 (1999)
8. Rezende, S.M.: Theory of coherence in Bose-Einstein condensation phenomena in a microwave-driven interacting magnon gas. Phys. Rev. B. **79**, 174411 (2009)
9. Kalinikos, B.A., Slavin, A.N.: Excitation theory of dipole-exchange spin wave spectrum for ferromagnetic films with mixed exchange boundary conditions. J. Phys. C: Solid State Phys. **19**, 7013 (1986)
10. Korenman, V., Prange, R.E.: Anomalous damping of spin waves in magnetic metals. Phys. Rev. B. **6**, 2769 (1972)
11. Mills, D.L.: Ferromagnetic resonance relaxation in ultrathin metal films: The role of the conduction electrons. Phys. Rev. B. **68**, 014419 (2003)
12. Kamberský, V.: Spin-orbital Gilbert damping in common magnetic metals. Phys. Rev. B. **76**, 134416 (2007)
13. Gilmore, K., Idzerda, Y.U., Stiles, M.D.: Spin-orbit precession damping in transition metal ferromagnets. J. Appl. Phys. **103**, 07D303 (2008)
14. Celinski, Z., Heinrich, B.: Ferromagnetic resonance linewidth of Fe ultrathin films grown on a bcc Cu substrate. J. Appl. Phys. **70**, 5935 (1991)
15. Frait, Z., Heinrich, B.: Intrinsic ferromagnetic resonance linewidth in metal single crystals. J. Appl. Phys. **35**, 904 (1964)
16. Bhagat, S.M., Lubitz, P.: Temperature variation of ferromagnetic relaxation in the 3d transition metals. Phys. Rev. B. **10**, 179 (1974)
17. Hurben, M.J., Patton, C.E.: Theory of two magnon scattering microwave relaxation and ferromagnetic resonance linewidth in magnetic thin films. J. Appl. Phys. **83**, 4344 (1998)

18. McMichael, R.D., Stiles, M.D., Chen, P.J., Egelhoff Jr., W.F.: Ferromagnetic resonance linewidth in thin films coupled to NiO. J. Appl. Phys. **83**, 7037 (1998)
19. Azevedo, A., Oliveira, A.B., de Aguiar, F.M., Rezende, S.M.: Extrinsic contributions to spin-wave damping and renormalization in thin $Ni_{50}Fe_{50}$ films. Phys. Rev. B. **62**, 5331 (2000)
20. Lenz, K., Wende, H., Kuch, W., Baberschke, K., Nagy, K., Jánossy, A.: Two-magnon scattering and viscous Gilbert damping in ultrathin ferromagnets. Phys. Rev. B. **73**, 144424 (2006)
21. Landeros, P., Arias, R.E., Mills, D.L.: Two magnon scattering in ultrathin ferromagnets: The case where the magnetization is out of plane. Phys. Rev. B. **77**, 214405 (2008)
22. Mathieu, C., Bauer, M., Hillebrands, B., Fassbender, J., Güntherodt, G., Jungblut, R., Kohlhepp, J., Reinders, A.: Brillouin light scattering investigations of exchange biased (110)-oriented NiFe/FeMn bilayers. J. Appl. Phys. **83**, 2863 (1998)
23. Rezende, S.M., Azevedo, A., Lucena, M.A., de Aguiar, F.M.: Anomalous spin-wave damping in exchange-biased films. Phys. Rev. B. **63**, 214418 (2001)
24. Nogués, J., Schuller, I.K.: Exchange bias. J. Magn. Magn. Mater. **192**, 203 (1999)
25. Malozemoff, A.P.: Random-field model of exchange anisotropy at rough ferromagnetic-antiferromagnetic interfaces. Phys. Rev. B. **35**, 3679 (1987)
26. Mills, D.L., Rezende, S.M.: In: Hillebrands, B., Ounadjela, K. (eds.) Spin Dynamics in Confined Magnetic Structures II. Springer, Heidelberg (2002)
27. Grünberg, P., Schreiber, R., Pang, Y., Brodsky, M.B., Sowers, H.: Layered magnetic structures: evidence for antiferromagnetic coupling of Fe layers across Cr interlayers. Phys. Rev. Lett. **57**, 2442 (1986)
28. Vohl, M., Barnás, J., Grünberg, P.: Effect of interlayer exchange coupling on spin wave spectra in magnetic double layers: theory and experiments. Phys. Rev. B. **39**, 12003 (1989)
29. Fert, A.: Nobel lecture: origin, development, and future of spintronics. Rev. Mod. Phys. **80**, 1517 (2008)
30. Grünberg, P.: Nobel lecture: from spin waves to giant magnetoresistance and beyond. Rev. Mod. Phys. **80**, 1531 (2008)
31. Grünberg, P.: Layered magnetic structures: history, facts and figures. J. Mag. Mag. Mater. **226**, 1688 (2001)
32. Ruderman, M.A., Kittel, C.: Indirect exchange coupling of nuclear magnetic moments by conduction electrons. Phys. Rev. **96**, 99 (1954)
33. Kasuya, T.: Electrical resistance of ferromagnetic metals. Prog. Theor. Phys. **16**, 45–58 (1958)
34. Yosida, K.: Magnetic properties of Cu-Mn alloys. Phys. Rev. **106**, 99 (1957)
35. Parkin, S.S.P., More, N., Roche, K.P.: Oscillations in exchange coupling and magnetoresistance in metallic superlattice structures: Co/Ru, Co/Cr, and Fe/Cr. Phys. Rev. Lett. **64**, 2304 (1990)
36. Edwards, D.M., Mathon, J., Muniz, R.B., Phan, M.S.: Oscillations in the exchange coupling of ferromagnetic layers separated by a non-magnetic metallic layer. J. Phys. : Condens. Matter. **3**, 4941 (1991)
37. Stiles, M.D.: Interlayer exchange coupling. J. Mag. Mag. Mater. **200**, 322 (1999)
38. Slonczewski, J.C.: Fluctuation mechanism for biquadratic exchange coupling in magnetic multilayers. Phys. Rev. Lett. **67**, 3172 (1991)
39. Heinrich, B., Celinski, Z., Cochran, J.F., Arrott, A.S., Myrtle, K.: Bilinear and biquadratic exchange coupling in bcc Fe/Cu/Fe trilayers: Ferromagnetic-resonance and surface magneto-optical Kerr-effect studies. Phys. Rev. B. **47**, 5077 (1993)
40. Azevedo, A., Chesman, C., Rezende, S.M., de Aguiar, F.M., Bian, X., Parkin, S.S.P.: Biquadratic exchange coupling in sputtered (100) Fe/Cr/Fe. Phys. Rev. Lett. **76**, 4837 (1996)
41. Demokritov, S.O.: Biquadratic interlayer coupling in layered magnetic systems. J. Phys. D. **31**, 925 (1998)
42. Hillebrands, B.: Spin-wave calculations for multilayered structures. Phys. Rev. B. **41**, 530 (1990)
43. Cochran, J.F., Rudd, J., Muir, W.B., Heinrich, B., Celinski, Z.: Brillouin light-scattering experiments on exchange-coupled ultrathin bilayers of iron separated by epitaxial copper (001). Phys. Rev. B. **42**, 508 (1990)

44. Layadi, A., Artman, J.O.: Ferromagnetic resonance in a coupled two-layer system. J. Magn. Magn. Mater. **92**, 143 (1990)
45. Heinrich, B., Cochran, J.F.: Ultrathin metallic magnetic films: magnetic anisotropies and exchange interactions. Adv. Phys. **42**, 523 (1993)
46. Stamps, R.L.: Spin configurations and spin-wave excitations in exchange-coupled bilayers. Phys. Rev. B. **49**, 339 (1994)
47. Zhang, Z., Zhou, L., Wigen, P.E., Ounadjela, K.: Angular dependence of ferromagnetic resonance in exchange-coupled Co/Ru/Co trilayer structures. Phys. Rev. B. **50**, 6094 (1994)
48. Rezende, S.M., Chesman, C., Lucena, M.A., Azevedo, A., de Aguiar, F.M., Parkin, S.S.P.: Studies of coupled metallic magnetic thin-film trilayers. J. Appl. Phys. **84**, 958 (1998)
49. Unguris, J., Celotta, R.J., Pierce, D.T.: Observation of two different oscillation periods in the exchange coupling of Fe/Cr/Fe (100). Phys. Rev. Lett. **67**, 140 (1991)
50. Rührig, M., Schäfer, R., Huber, A., Mosler, R., Wolf, J.A., Demokritov, S., Grünberg, P.: Domain observations on Fe-Cr-Fe layered structures. Phys. Status Solidi (a). **125**, 635 (1991)

## Further Reading

Guimarães, A.P.: Principles of Nanomagnetism, 2nd edn. Springer, Cham (2017)
Gurevich, A.G., Melkov, G.A.: Magnetization Oscillations and Waves. CRC Press, Boca Raton, FL (1994)
Heinrich, B., Bland, J.A.C. (eds.): Ultrathin Magnetic Structures II. Springer, Heidelberg (1994)
Hillebrands, B., Ounadjela, K. (eds.): Spin Dynamics in Confined Magnetic Structures II. Springer, Heidelberg (2002)
Kabos, P., Stalmachov, V.S.: Magnetostatic Waves and their Applications. Chapman and Hall, London (1994)
Lax, B., Button, K.J.: Microwave Ferrites and Ferrimagnetics. McGraw-Hill, New York (1962)
Stancil, D.D., Prabhakar, A.: Spin Waves: Theory and Applications. Springer Science, New York (2009)
White, R.M.: Quantum Theory of Magnetism, 3rd edn. Springer, Heidelberg (2007)
Wu, M., Hoffmann, A. (eds.): Recent Advances in Magnetic Insulators – From Spintronics to Microwave Applications. Academic Press, San Diego (2013)

# Magnons in Antiferromagnets

**5**

In this chapter we study the basic concepts and properties of magnons in antiferro-magnetic (AF) materials. *Initially we present a brief* introduction to AF materials and a semiclassical calculation of the equilibrium spin configuration in simple systems with two types of magnetic anisotropy: uniaxial and biaxial. Then we study the antiferromagnetic resonance and apply the results to three important AF insulators: $MnF_2$, $FeF_2$, and NiO. Finally, we present a quantum theory of magnons for these materials and compare the results with experimental data.[1]

## 5.1 Antiferromagnetic Materials

Antiferromagnets are magnetic materials that have no net macroscopic magnetization and thus are almost insensitive to external magnetic fields. The existence of antiferromagnetism was proposed by Louis Néel [1] to account for the magnetic behavior of some salts with complicated lattice structures. Néel first assumed that the ordered magnetic arrangements could be described in terms of *sublattices*, each having all spins aligned in the same direction. Then, he considered that the interaction between the spins in different sublattices could be negative, so that they would be aligned antiparallel. Now we know that the interaction between spins originates mainly in the exchange interaction, with magnitude represented by an exchange parameter $J$. The intra-sublattice $J$ is positive because it is due to direct exchange, while the inter-sublattice $J$ is negative because it arises from the super-exchange that is mediated by ligand ions between the magnetic ions. If the sublattices have different magnetizations, the materials are called *ferrimagnets*. They have spontaneous net magnetization below the ordering temperature $T_c$ and in many aspects

---

[1]This chapter is largely based on the tutorial by S. M. Rezende, A. Azevedo, and R. L. Rodríguez-Suárez, *Introduction to antiferromagnetic magnons*, J. Appl. Phys. **126**, 151101 (2019), with the permission of AIP Publishing.

© Springer Nature Switzerland AG 2020                                                    187
S. M. Rezende, *Fundamentals of Magnonics*, Lecture Notes in Physics 969,
https://doi.org/10.1007/978-3-030-41317-0_5

behave like ferromagnets. Materials with $J < 0$ that have sublattices with the same magnetization are called *antiferromagnets*. In the ordered phase, below a critical temperature called Néel temperature $T_N$, they are magnetic, but have no net macroscopic magnetization.

In his 1970 Nobel lecture, Louis Néel stated that "Antiferromagnetic materials are extremely interesting from the theoretical viewpoint, but do not seem to have any application." This pessimistic view of antiferromagnets would change dramatically a few decades later. Today, these materials have some practical applications, and have the potential for many others. The debut of antiferromagnets in technology was made possible with the discovery of the giant magnetoresistance effect (GMR) in 1968, by Albert Fert [2] and Peter Grünberg [3]. They showed that in magnetic structures with nanometer thick multilayers, the electron transport can be controlled by the spin. This discovery triggered research in magnetic multilayers and gave birth to the field of spintronics, that has revolutionized magnetic recording technologies and promises new functionalities to electronic devices [4].

Antiferromagnetic materials proved to be essential in GMR sensors for pinning the magnetization of a reference ferromagnetic layer by means of the exchange bias phenomenon [5]. Since late 1990s the disk-drive magnetic read-heads are based on GMR sensors, much more sensitive to changes in magnetic fields than the traditional induction heads. Although important, AF materials have had a passive role in spintronic devices, the active role is played by ferromagnetic materials. However, in recent years this scenario began to change with new experimental and theoretical results showing that antiferromagnets have several advantages over ferromagnets in spintronics phenomena. One of them is the fact that AF materials are insensitive to external magnetic perturbations. Another one is the ultrafast dynamics of antiferromagnets that might have application in devices for the terahertz frequency range. The late developments gave rise to the field of antiferromagnetic spintronics [6–8]. Antiferromagnets are known to exist with a great variety of crystalline structures and physical properties, and new materials continue to be discovered [8]. Here we will restrict attention to simple AF insulators with only two sublattices, one class with easy-axis anisotropy, also called uniaxial anisotropy, such as the fluorides $MnF_2$ and $FeF_2$, and the other with easy-plane anisotropy, or hard-axis anisotropy, such as the oxides NiO and MnO.

We consider initially two simple antiferromagnets with uniaxial anisotropy: $MnF_2$ and $FeF_2$. Both have the rutile crystal structure, a body-centered tetragonal lattice with the magnetic ions occupying the corner and body-centered positions, as shown in Fig. 5.1a. Below the Néel temperature, 66.5 K for $MnF_2$ and 78.4 K for $FeF_2$, and in the absence of an external magnetic field, the spins are arranged in two oppositely directed sublattices, pointing along the easy-anisotropy direction ($c$-axis). The magnetic properties are described very well by a Hamiltonian consisting of contributions from Zeeman, exchange and magnetic anisotropy energies in the form

$$H = -\gamma\hbar\sum_i \vec{S}_i \cdot \vec{H}_0 - \sum_{i,i'\neq i} 2J_{ii'}\vec{S}_i \cdot \vec{S}_{i'} - D\sum_i (S_i^z)^2, \qquad (5.1)$$

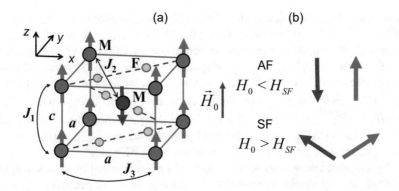

**Fig. 5.1** (a) Crystal structure of the rutile antiferromagnets $MnF_2$ and $FeF_2$. The ions M represent $Mn^{2+}$ or $Fe^{2+}$. (b) Spin configurations in the antiferromagnetic (AF) and spin-flop (SF) phases for an external field applied along the easy axis

where $\vec{H}_0$ is the applied magnetic field, $J_{ii'}$ is the exchange constant of the interaction between spins $\vec{S}_i$ and $\vec{S}_{i'}$ ($i$ and $i'$ denote generic spin sites), and $D$ is the uniaxial anisotropy constant (not to be confused with the exchange parameter in Eq. (2.19)). The important exchange interactions are illustrated in Fig. 5.1a. In $MnF_2$ the exchange parameters determined by inelastic neutron scattering measurements [9] are $J_1 = 0.028$ meV, $J_2 = -0.152$ meV, and $J_3 = -0.004$ meV. Thus, the negative inter-sublattice $J_2$ is the dominant interaction and determines the antiferromagnetic arrangement of the sublattices.

In order to find the spin equilibrium configuration, we use a macrospin approximation and associate to $\vec{S}_i$ and $\vec{S}_j$ uniform sublattice magnetizations $\vec{M}_{1,2} = \gamma \hbar N \vec{S}_{i,j}$, where $N$ is the number of spins per unit volume. For $T = 0$ the static sublattice magnetizations have the same value, $M_1 = M_2 = M$. Consider a magnetic field $H_0$ applied along the $c$-axis, the $z$-direction of the system shown in Fig. 5.1a, and use the energy per unit volume obtained from Eq. (5.1)

$$E = -H_0(M_{1z} + M_{2z}) + \frac{H_E}{M}\vec{M}_1 \cdot \vec{M}_2 - \frac{H_A}{2M}(M_{1z}^2 + M_{2z}^2), \qquad (5.2)$$

where $H_E$ and $H_A$ are the effective exchange and anisotropy fields defined by

$$H_E = 2Sz|J_2|/\gamma\hbar, \quad H_A = 2SD/\gamma\hbar, \qquad (5.3)$$

where we have considered only the negative inter-sublattice exchange interaction $J_2$ between the $z$ nearest neighbors. Due to the symmetry, the sublattice magnetizations and the applied field are in the same plane, so that the energy in Eq. (5.2) becomes

$$\frac{E(\theta_1, \theta_2)}{M} = -H_0(\cos\theta_1 + \cos\theta_2) + H_E\cos(\theta_1 + \theta_2)$$
$$- \frac{H_A}{2}(\cos^2\theta_1 + \cos^2\theta_2), \qquad (5.4)$$

where $\theta_1$ and $\theta_2$ are the polar angles of the sublattice magnetizations. The conditions $\partial E/\partial \theta_1 = \partial E/\partial \theta_2 = 0$ give two possible equilibrium configurations. As illustrated in Fig. 5.1b, $\theta_1 = 0$, $\theta_2 = \pi$ corresponds to the antiferromagnetic (AF) phase, while $\theta_1 = \theta_2 = \cos^{-1}[H_0/(2H_E - H_A)]$ corresponds to the spin-flop (SF) phase. Substitution of these angles in Eq. (5.4) shows that the AF phase has lower energy for fields $H_0 < H_{SF}$, while the SF phase has lower energy for $H_0 > H_{SF}$, where $H_{SF} = (2H_E H_A - H_A^2)^{1/2}$. For small fields applied in the $z$-direction, the spins are along the $c$-axis in the AF phase. As the field intensity increases and reaches $H_{SF}$ the system undergoes a first-order phase transition to the SF phase.

The magnetic interactions in $MnF_2$ and $FeF_2$ are dominated by nearest-neighbor exchange, having similar inter-sublattice exchange fields, respectively $H_E = 526$ kOe and $H_E = 540$ kOe. In $MnF_2$ the ground state configuration of the magnetic $Mn^{2+}$ ions is $3d^5(^6S_{5/2})$, with no orbital angular momentum and thus very small single-ion anisotropy. The origin of the magnetic anisotropy of $MnF_2$ lies mainly in the dipolar interaction, with a relatively small effective anisotropy field $H_A = 8.2$ kOe [10], so that $H_{SF} \approx \sqrt{2H_E H_A} = 93$ kOe. The temperature dependence of the spin-flop field has been measured by ultrasonic techniques [11], antiferromagnetic resonance [12], and spin Seebeck effect [13]. In $FeF_2$ the ground state configuration of the magnetic $Fe^{2+}$ ions is $3d^5(^5D_4)$, which has a finite orbital angular momentum and consequently a large effective anisotropy field $H_A = 200$ kOe, arising from the spin–orbit coupling. Thus, the spin-flop field is large, $H_{SF} = 505$ kOe, and can be attained only with pulsed magnetic fields [14].

The other material we shall study is NiO, a prototypical room temperature antiferromagnetic insulator because of its simple structure and spin interactions. Its magnetic structure and spin interactions are similar to those in MnO [15, 16]. In the paramagnetic phase, NiO has the face-centered cubic structure of sodium chloride. Below the Néel temperature $T_N \approx 523K$, the $Ni^{2+}$ spins are ordered ferromagnetically in $\{111\}$ planes, lying along $< 11\bar{2} >$ axes, with adjacent planes oppositely magnetized due to a super-exchange AF interaction, illustrated in Fig. 5.2. It is characterized by two distinct anisotropies, a negative one (hard) along $<111>$ axes that forces the spins to lie in $\{111\}$ planes, and a positive in-plane one (easy) along $< 11\bar{2} >$ axes. The spin Hamiltonian with Zeeman energy, exchange interaction and out-of-plane ($x$) and in-plane ($z$) anisotropy energies can be written as

$$\mathrm{H} = -\gamma\hbar\sum_{i,j}H_0 S_{i,j}^{z'} - \sum_{i,j}2J_{ij}\vec{S}_i \cdot \vec{S}_j + \sum_{i,j}D_x(S_{i,j}^x)^2 - D_z(S_{i,j}^{z'})^2, \qquad (5.5)$$

where the subscripts $i$ and $j$ refer to the spin sites in sublattices 1 and 2, respectively, and we have considered the field applied in the easy $z'$-direction along one of the three $< 11\bar{2} >$ axes in the $\{111\}$ plane, and only inter-sublattice exchange. The difference to Eq. (5.1) is that the Hamiltonian now contains two anisotropy terms, with anisotropy constants $D_x$ and $D_z$. Both are taken to be positive, so that the signs in Eq. (5.5) imply that $x$ is a hard direction along a $<111>$ axis. Since there are three

**Fig. 5.2** Crystal structure
and spin arrangements in the
antiferromagnetic phase of
NiO. The small yellow circles
represent $O^{2-}$ ions and the
large blue and red circles
represent the $Ni^{2+}$ ions

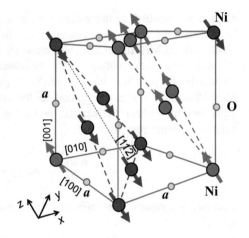

equivalent directions in the easy-plane and four {111} planes, in bulk NiO crystals
there are AF domains with spins in 12 different directions which complicate the
interpretation of some magnetic measurements.

From Eq. (5.5) we obtain for the energy per unit volume

$$E = -H_0(M_{1z'} + M_{2z'}) + \frac{H_E}{M}\vec{M}_1 \cdot \vec{M}_2 + \frac{H_{Ax}}{2M}(M_{1x}^2 + M_{2x}^2)$$
$$- \frac{H_{Az}}{2M}(M_{1z'}^2 + M_{2z'}^2), \tag{5.6}$$

where the effective exchange field is the same as in Eq. (5.3) (with the relevant inter-
sublattice exchange) and now we have two anisotropy fields

$$H_{Ax} = 2SD_x/\gamma\hbar, \quad H_{Az} = 2SD_z/\gamma\hbar. \tag{5.7}$$

In order to simplify calculations, we assume a monodomain sample with the field
applied along an easy direction. In this case, $M_{1x} = M_{2x} = 0$ and Eq. (5.6) reduces to
Eq. (5.2), so that the result previously obtained is valid here. In NiO the effective
fields are $H_E = 9\,684$ kOe, $H_{Ax} = 6.35$ kOe, $H_{Az} = 0.11$ kOe, so that the field for the
spin-flop transition $H_{SF} = \sqrt{2H_EH_{Az}}$ is 46.3 kOe [17].

## 5.2   Antiferromagnetic Resonance: The $k = 0$ Magnons

As we saw in Chap. 1, in ferromagnets the exchange interaction does not play a
direct role in the ferromagnetic resonance. Its role is to align the spins in the same
direction so that the response to a microwave field is much stronger than in
paramagnetic materials. But since the exchange field is parallel to the magnetization,
$\vec{M} \times \vec{H}_E = 0$, it does not enter in the expression for the resonance frequency. In
ferrimagnets and antiferromagnets the situation is quite different, because the
exchange field of one sublattice acts on the magnetizations of the other sublattices.
In ferri- and antiferromagnets the resonance phenomenon involves the coupled

motion of the sublattice magnetizations, so that the exchange interaction is expected to influence the resonance frequency. The first derivation of the antiferromagnetic resonance (AFMR) frequencies was made by Charles Kittel [18] for a two-sublattice AF with easy-axis anisotropy. Here we basically follow that paper to calculate the resonance frequencies of the uniform precession modes of the sublattice magnetizations, which correspond to the magnons with zero wave vector. The equations of motion for the magnetization components are obtained from the Landau–Lifshitz equation

$$\frac{d\vec{M}_{1,2}}{dt} = \gamma \vec{M}_{1,2} \times \vec{H}_{eff1,2},\tag{5.8}$$

where $\vec{H}_{eff1,2}$ represent the effective fields that act on the sublattice magnetizations, given by

$$\vec{H}_{eff1,2} = -\nabla_{\vec{M}1,2}[E(\vec{M}_{1,2})].\tag{5.9}$$

Note that the sign in (5.8) is positive, different than in (1.10), because here we consider that the magnetizations and the spins have the same direction. We treat separately the cases of the uniaxial (easy-axis) and biaxial (easy-plane) antiferromagnets.

## 5.2.1 Easy-Axis Antiferromagnets

With the expression (5.2) for the magnetic energy in easy-axis AFs, we obtain with Eq. (5.9) the Cartesian components of the effective fields that act on the magnetization of sublattice 1

$$H_{eff1x} = -\frac{\partial E}{\partial M_{1x}} = -\frac{H_E}{M}M_{2x} - \frac{H_{Ax}}{M}M_{1x},\tag{5.10a}$$

$$H_{eff1y} = -\frac{\partial E}{\partial M_{1y}} = -\frac{H_E}{M}M_{2y},\tag{5.10b}$$

$$H_{eff1z} = -\frac{\partial E}{\partial M_{1z}} = H_0 - \frac{H_E}{M}M_{2z} + \frac{H_{Az}}{M}M_{1z}.$$

With $M_{1z} = M$, and $M_{2z} = -M$, that are valid at low temperatures, the last equation becomes

$$H_{eff1z} = H_0 + H_E + H_{Az}.\tag{5.10c}$$

The expressions for the components of $H_{eff2}$ are given by Eqs. (5.10a)–(5.10c) with the interchange $1 \leftrightarrow 2$. Writing for the sublattice magnetizations

$$\vec{M}_{1,2} = \hat{z}M_{1,2z} + (\hat{x}m_{1,2x} + \hat{y}m_{1,2y})e^{-i\omega t}, \qquad (5.11)$$

substitution of Eqs. (5.10a)–(5.10c) and (5.11) in Eq. (5.8) leads to the equations of motion for the transverse components of the magnetizations

$$i\omega m_{1x} + \gamma m_{1y}(H_0 + H_E + H_A) + \gamma H_E m_{2y} = 0, \qquad (5.12a)$$

$$-\gamma m_{1x}(H_0 + H_E + H_A) + i\omega m_{1y} - \gamma H_E m_{2x} = 0, \qquad (5.12b)$$

$$-\gamma H_E m_{1y} + i\omega m_{2x} + \gamma m_{2y}(H_0 - H_E - H_A) = 0, \qquad (5.12c)$$

$$+\gamma H_E m_{1x} - \gamma m_{2x}(H_0 - H_E - H_A) + i\omega m_{2y} = 0. \qquad (5.12d)$$

The symmetry of Eqs. (5.12a–5.12d) indicates that $|m_{1x}| = |m_{1y}|$ and $|m_{2x}| = |m_{2y}|$. Thus, we define circularly polarized transverse components of the magnetizations, $m_1^+ = m_{1x} + im_{1y}$, $m_2^+ = m_{2x} + im_{2y}$, and from Eqs. (5.12a–5.12d) we obtain

$$-\omega m_1^+ + m_1^+ \gamma(H_E + H_A + H_0) = -\gamma H_E m_2^+, \qquad (5.13a)$$

$$\gamma H_E m_1^+ = -m_2^+ \gamma(H_E + H_A - H_0) - \omega m_2^+. \qquad (5.13b)$$

The solution of these equations gives for the AFMR frequencies

$$\omega_{\alpha 0, \beta 0} = \pm\gamma H_c + \gamma H_0, \qquad (5.14a)$$

where

$$H_c = (2H_E H_A + H_A^2)^{1/2}. \qquad (5.14b)$$

This result shows that in a two-sublattice AF, there are two modes, which are degenerate in the absence of an external field. In both modes each sublattice magnetization precesses circularly, clockwise in one mode and counterclockwise in the other. The application of an external field in the direction of anisotropy lifts the degeneracy. In mode $\alpha$, that has $\omega_{\alpha 0} > 0$, the frequency increases with increasing $H_0$, while in mode $\beta$, $\omega_{\beta 0} < 0$, and the frequency decreases in absolute value with increasing $H_0$.

Note that the effect of the anisotropy field is enhanced by the interplay with exchange, and this gives rise to the gap in the spectrum. The opposite directions of the anisotropy fields for the two sublattice magnetizations make it impossible for them to process in the same direction, and this brings the exchange field into play. The interplay between the anisotropy and the large exchange interaction results in a magnetization precession with frequencies that are much larger in antiferromagnets than in ferromagnets. This fast magnetization dynamics constitutes one of the motivations for pushing research in antiferromagnetic spintronics. From one of the Eqs. (5.13a) or (5.13b) we can obtain the configurations of the sublattice magnetizations in each mode. With $H_0 = 0$, Eq. (5.13b) gives for the ratio between the precession amplitudes

**Fig. 5.3** Illustration of the precessions of the sublattice magnetizations of the two modes of antiferromagnetic resonance in an easy-axis antiferromagnet

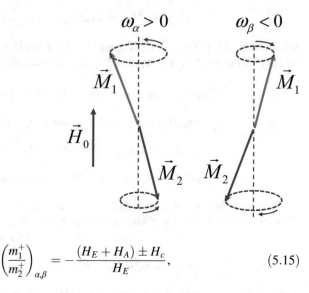

$$\left(\frac{m_1^+}{m_2^+}\right)_{\alpha,\beta} = -\frac{(H_E + H_A) \pm H_c}{H_E}, \tag{5.15}$$

where the + sign applies to the $\alpha$-mode and $-$ to the $\beta$-mode. The precessions of the sublattice magnetizations of the two modes are illustrated in Fig. 5.3. Note that Eq. (5.15) gives only the ratio of the magnetization precession amplitudes in the two sublattices. The actual amplitudes depend on the intensity of the driving field.

Equation (5.14a) shows that the frequency of the $\beta$-mode decreases with increasing field and vanishes for $H_0 = H_c = (2H_E H_A + H_A^2)^{1/2}$, the field value that defines the limit of stability of the AF phase. Note that this value is larger than the one obtained before for the thermodynamic transition, but for small anisotropies the two expressions give approximately the same value $H_{SF} \approx H_c \approx (2H_E H_A)^{1/2}$. With no applied field, the AFMR frequency is $\omega_0 = \gamma H_c$. Using for MnF$_2$ $g = 2$, $\gamma = 2.8$ GHz/kOe and $H_c = 93$ kOe, this gives for the AFMR frequency 260 GHz. Using for FeF$_2$ $g = 2.22$ and $H_c = 505$ kOe, we obtain 1.57 THz.

The AFMR frequencies have been measured in MnF$_2$ and FeF$_2$ with several techniques. The first measurements in MnF$_2$, made in zero field by Johnson and Nethercot [19] with a millimeter microwave spectrometer, showed very broad lineshapes, but served to give the temperature dependence of the AFMR frequency. By applying a large magnetic field along the $c$-axis, it is possible to lower the frequency of the $\beta$-mode in MnF$_2$ to microwave frequencies. This technique was used in Rezende et al. [12] to measure the limit of stability of the AF phase in MnF$_2$. As shown in Fig. 5.4a, measurement of the AFMR frequency versus field and extrapolation to zero-frequency yields a precise value for the field $H_c(T)$. The complete picture of the field dependence of the two AFMR frequencies in MnF$_2$ at low temperatures was measured in Hagiwara et al. [20] using a combination of microwave millimeter and submillimeter sources, as shown in Fig. 5.4b. According to Eqs. (5.14a) and (5.14b), as the field increases, the frequency of the mode with counterclockwise precession ($\alpha$) increases, as in a ferromagnet. On the other hand, the frequency of the clockwise precession mode ($\beta$) decreases with increasing field.

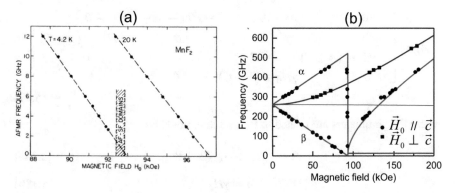

**Fig. 5.4** Experimental data for MnF$_2$. (**a**) Measurements of the AFMR frequency vs. applied field at two temperatures [12]. Reprinted with permission from S. M. Rezende et al., Phys. Rev. B **16**, 1126 (1977). Copyright (1977) by the American Physical Society. (**b**) Symbols represent the measured magnetic field dependence of the AFMR frequencies in MnF$_2$ at 1.8–5.0 K, with the field applied parallel and perpendicular to the $c$-axis [20]. The solid lines represent the field dependencies of the frequencies calculated with Eqs. (5.14a) and (5.14b) for the AF phase, with Eqs. (5.75a) and (5.75b) for the SF phase, and with Eqs. (5.73a) and (5.73b) for the canted phase

As will be shown in the next section, in the spin-flop phase the frequency of one mode is zero and the other is $\omega_0 = \gamma(H_0^2 - H_c^2)^{1/2}$, as in Fig. 5.4b. Note that if the field is applied perpendicularly to the $c$-axis, the equilibrium positions of the sublattice magnetizations are at some angle with the field. In this case, the two modes are degenerate and there is no spin-flop transition.

## 5.2.2   Easy-Plane Antiferromagnets

For the easy-plane, or hard-axis, or biaxial, antiferromagnets, we use the energy expression (5.6), and with Eqs. (5.8) and (5.9) we obtain the equations of motion for the transverse components of the magnetizations

$$i\omega m_{1x} + \gamma m_{1y}(H_0 + H_E + H_{Az}) + \gamma H_E m_{2y} = 0, \qquad (5.16a)$$

$$-\gamma m_{1x}(H_0 + H_E + H_{Ax} + H_{Az}) + i\omega m_{1y} - \gamma H_E m_{2x} = 0, \qquad (5.16b)$$

$$-\gamma H_E m_{1y} + i\omega m_{2x} + \gamma m_{2y}(H_0 - H_E - H_{Az}) = 0, \qquad (5.16c)$$

$$\gamma H_E m_{1x} - \gamma m_{2x}(H_0 - H_E - H_{Ax} - H_{Az}) + i\omega m_{2y} = 0, \qquad (5.16d)$$

The AFMR frequencies are the eigenvalues of the resonance matrix [$A$], given by

$$[A] = \begin{bmatrix} 0 & 0 & -\gamma H_0 - (A - C) & -B \\ 0 & 0 & B & -\gamma H_0 + (A - C) \\ -\gamma H_0 - (A + C) & -B & 0 & 0 \\ B & -\gamma H_0 + (A + C) & 0 & 0 \end{bmatrix},$$

$$(5.17)$$

where the parameters are related to the effective fields by

$$A = \gamma[H_E + H_{Ax}/2 + H_{Az}], \tag{5.18a}$$

$$B = \gamma H_E, \tag{5.18b}$$

$$C = \gamma H_{Ax}/2. \tag{5.18c}$$

Equations (5.16a–5.16d) can be written as an eigenvalue equation in matrix form

$$[A][S] = [S][\omega], \tag{5.19}$$

where the columns of $[S]$ are eigenvectors of $[A]$ that represent the four normal modes with components $(m) = (m_{1x}, m_{2x}, im_{1y}, im_{2y})^T$, and $[\omega]$ is the diagonal matrix of the eigenvalues, which are given by the roots of

$$\det\{[A] - [\omega]\} = 0. \tag{5.20}$$

Solution of Eq. (5.20) for the case of an easy-axis AF, obtained by setting $H_{Ax} = 0$, that implies $C = 0$, gives for the eigenvalues $\omega_1 = -\omega_2 = \omega_\alpha$ and $\omega_3 = -\omega_4 = \omega_\beta$, where $\omega_\alpha$ and $\omega_\beta$ are the same AFMR frequencies as in Eqs. (5.14a) and (5.14b) with $H_A = H_{Az}$. Also, for $H_0 = 0$ it can be shown that Eq. (5.19) is satisfied by an eigenvector matrix that corresponds to the two sublattice magnetizations precessing circularly in the same sense, counterclockwise in mode $\alpha$ and clockwise in mode $\beta$, with the ratio between the precession amplitudes as in Eq. (5.15) [21].

For the case of an easy-plane AF, with $H_{Ax} > 0$, the eigenvalues of Eq. (5.19) given by the roots of $\det[A + i\omega] = 0$ are

$$\omega_{\alpha,\beta}^2 = (A^2 + \gamma^2 H_0^2) - (C^2 + B^2) \pm 2\sqrt{\gamma^2 H_0^2(A^2 - B^2) + B^2 C^2}. \tag{5.21}$$

Considering $H_E \gg H_{Ax} \gg H_{Az}$, substitution of the parameters defined by Eqs. (5.18a–5.18c) in Eq. (5.21) gives

$$\omega_{\alpha,\beta}^2 \approx \gamma[(2H_E H_{Az} + H_E H_{Ax} + H_0^2) \pm H_E H_{Ax} \pm 2H_0^2]. \tag{5.22}$$

With $H_{Ax} \gg H_{Az}$, appropriate for NiO, the two AFMR frequencies become simply

$$\omega_{\alpha 0}^2 \approx \gamma(2H_E H_{Ax} + 3H_0^2), \tag{5.23a}$$

$$\omega_{\beta 0}^2 \approx \gamma(2H_E H_{Az} - H_0^2). \tag{5.23b}$$

On the other hand, for $H_0 = 0$, regardless of the relative values of the anisotropy fields, Eq. (5.22) gives

$$\omega_{\alpha 0}^2 = \gamma^2 2H_E(H_{Az} + H_{Ax}), \quad \omega_{\beta 0}^2 = \gamma^2 2H_E H_{Az}. \tag{5.24}$$

These expressions show that, in contrast to the uniaxial AF, the two modes are not degenerate in the absence of an applied field. This feature gives NiO the property of transporting a magnonic spin current even with no applied field. Equations (5.23a) and (5.23b) show that as the field increases, the frequency of the $\alpha$-mode increases while the one for the $\beta$-mode decreases. The $\beta$-mode frequency goes to zero at a field $H_{SF} \approx (2H_E H_{Az})^{1/2}$, that represents the limit of stability of the AF phase. Using the values of the effective fields for NiO and $g = 2.18$, we find for the frequencies in zero-field 1.07 THz and 140 GHz. The temperature dependence of the spin-flop field in NiO has been measured by differential magnetization [17].

In order to obtain the eigenstates of the two uniform magnon modes in the AF phase of easy-plane antiferromagnets, we consider $H_0 = 0$ and use for the eigenvalues (frequencies) $\omega_1 = -\omega_2 = \omega_\alpha$ and $-\omega_3 = \omega_4 = \omega_\beta$, given by Eq. (5.21). Solution of the eigenvalue Eq. (5.19) gives for the eigenvector matrix [21]

$$[S] = \begin{pmatrix} -\varepsilon_\alpha & \varepsilon_\alpha & \varepsilon_\beta & -\varepsilon_\beta \\ \varepsilon_\alpha & -\varepsilon_\alpha & \varepsilon_\beta & -\varepsilon_\beta \\ 1 & 1 & -1 & -1 \\ 1 & 1 & 1 & 1 \end{pmatrix}, \tag{5.25}$$

where

$$\varepsilon_\alpha = \frac{\omega_\alpha}{A - (B - C)}, \quad \varepsilon_\beta = \frac{\omega_\beta}{A + (B + C)}, \tag{5.26}$$

are the ellipticities, defined as $\varepsilon = m_x/i\, m_y$. Using Eqs. (5.18a–5.18c) and (5.24), and considering $H_E \gg H_{Ax} \gg H_{Az}$, appropriate for NiO, Eqs. (5.26) give for the ellipticities

$$\varepsilon_\alpha \approx \left(\frac{2H_E}{H_{Ax}}\right)^{1/2}, \quad \varepsilon_\beta \approx \left(\frac{H_{Az}}{2H_E}\right)^{1/2}. \tag{5.27}$$

Using the parameters for NiO, Eqs. (5.27) give for the ellipticities $\varepsilon_\alpha \approx 55$ and $\varepsilon \approx 1/419$. From Eqs. (5.25) and (5.27) we see that in the $\alpha$-mode $\vec{M}_1$ precesses counterclockwise about the $z$-axis, while $\vec{M}_2$ precesses clockwise, with very elliptical trajectories, with the major axis oriented along the $x$-axis (hard anisotropy axis). The precessions in opposite directions give rise to an oscillating magnetization

**Fig. 5.5** Illustration of the precessions of the sublattice magnetizations of the two modes of antiferromagnetic resonance in an easy-plane (or biaxial) antiferromagnet

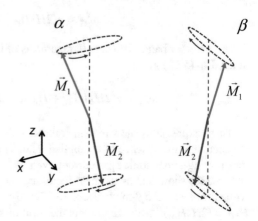

component in the $y$-direction, illustrated in Fig. 5.5. On the other hand, in the $\beta$-mode $\vec{M}_1$ precesses clockwise about the $z$-axis, while $\vec{M}_2$ precesses counterclockwise, with very elliptical trajectories, with the major axis oriented along the $y$-axis, giving rise to an oscillating magnetization component in the $x$-direction, as in Fig. 5.5. In Sect. 5.4 we shall return to the discussion of the eigenmodes in easy-plane AFs in connection with the spin waves in these materials.

## 5.3    Magnons in Easy-Axis Antiferromagnets

In this section we present a quantum formulation of spin waves in antiferromagnets with easy-axis anisotropy. We treat the quantized excitations of the magnetic system with the approach of Holstein–Primakoff (HP), used in Chap. 3 to study magnons in ferromagnets. We will treat separately magnons in the antiferromagnetic phase, in the spin-flop phase, and in the canted phase.

### 5.3.1    Magnons in Easy-Axis Antiferromagnets: AF Phase

Here we consider that the spins of the two sublattices point in the direction $\pm z$ of the symmetry axis, as in Fig. 5.1b, and write the Hamiltonian (5.1) in terms of raising and lowering spin operators

$$\mathrm{H} = -\sum_{i,j}\vec{S}_{i,j}\cdot\vec{H}_0 - \sum_{i\neq j}J_{ij}(S_i^+ S_j^- + S_i^- S_j^+ + 2S_i^z S_j^z) - D\sum_{i,j}(S_{i,j}^z)^2, \qquad (5.28)$$

where the subscripts $i$ and $j$ refer to the sites in sublattices 1 and 2, respectively. In the first HP transformation the components of the local spin operators are related to the creation and annihilation operators of spin deviations. For the up-spin sublattice, the lowering spin operator is related to the operator $a_i^\dagger$ that creates a spin deviation while the raising spin operator is related to the operator $a_i$ that destroys a spin deviation

$$S_{1i}^+ = (2S)^{1/2} \left(1 - \frac{a_i^\dagger a_i}{2S}\right)^{1/2} a_i, \tag{5.29a}$$

$$S_{1i}^- = (2S)^{1/2} a_i^\dagger \left(1 - \frac{a_i^\dagger a_i}{2S}\right)^{1/2}, \tag{5.29b}$$

$$S_{1i}^z = S - a_i^\dagger a_i. \tag{5.29c}$$

For the down-spin sublattice the raising spin operator is related to the operator that creates a spin deviation, and we have

$$S_{2j}^+ = (2S)^{1/2} b_j^\dagger \left(1 - \frac{b_j^\dagger b_j}{2S}\right)^{1/2}, \tag{5.30a}$$

$$S_{2j}^- = (2S)^{1/2} \left(1 - \frac{b_j^\dagger b_j}{2S}\right)^{1/2} b_j, \tag{5.30b}$$

$$S_{2j}^z = -S + b_j^\dagger b_j, \tag{5.30c}$$

where $a_i^\dagger$, $a_i$, and $b_j^\dagger$, $b_j$, are the creation, destruction operators for spin deviations at sites $i$, $j$ of sublattices 1 and 2, which satisfy the boson commutation rules $[a_i, a_{i'}^\dagger] = \delta_{ii'}, [a_i, a_{i'}] = 0, [b_j, b_{j'}^\dagger] = \delta_{jj'}$ and $[b_j, b_{j'}] = 0$. We shall consider only the inter-sublattice exchange interaction, represented by the (negative) parameter $J_2$, shown in Fig. 5.1a, which is the strongest one in $MnF_2$ and $FeF_2$. Substitution in Eq. (5.28) of Eqs. (5.29a–5.29c) and (5.30a–5.30c), with the binomial expansions of the square roots, gives for the quadratic part of the Hamiltonian

$$H^{(2)} = \sum_{i,j} [(g\mu_B H_0 + 2SD)a_i^\dagger a_i + (-g\mu_B H_0 + 2SD)b_j^\dagger b_j$$
$$+ 2S|J_2|(a_i b_j + a_i^\dagger b_j^\dagger + a_i^\dagger a_i + b_j^\dagger b_j)] \tag{5.31}$$

The next step consists of introducing a transformation from the localized field operators to collective boson operators that satisfy the commutation rules $[a_k, a_{k'}^\dagger] = \delta_{kk'}, [a_k, a_{k'}] = 0, [b_k, b_{k'}^\dagger] = \delta_{kk'}, [b_k, b_{k'}] = 0$,

$$a_i = N^{-1/2} \sum_k e^{i\vec{k}\cdot\vec{r}_i} a_k, \quad b_j = N^{-1/2} \sum_k e^{i\vec{k}\cdot\vec{r}_j} b_k, \tag{5.32}$$

where $N$ is the number of spins in each sublattice and $\vec{k}$ is a wave vector. Using Eq. (5.32) in Eq. (5.31), and with the orthonormality condition

$$N^{-1}\sum_i e^{i(\vec{k}-\vec{k}').\vec{r}_i} = \delta_{k,k'},\tag{5.33}$$

we obtain for the quadratic part of the Hamiltonian

$$\mathrm{H}^{(2)} = \gamma\hbar\sum_k (H_E + H_A + H_0)a_k^\dagger a_k + (H_E + H_A - H_0)b_k^\dagger b_k$$
$$+ \gamma_k H_E(a_k b_{-k} + a_k^\dagger b_{-k}^\dagger).\tag{5.34}$$

where the effective fields are related to the exchange and anisotropy parameters as in Eq. (5.3), and $\gamma_k$ is a structure factor defined by $\gamma_k = (1/z)\sum_\delta \exp(i\vec{k}.\vec{\delta})$, $\vec{\delta}$ are the vectors connecting nearest neighbors in opposite sublattices. The next step consists of performing canonical transformations from the collective boson operators $a_k^\dagger, a_k, b_k^\dagger, b_k$ into magnon creation and annihilation operators $\alpha_k^\dagger, \alpha_k, \beta_k^\dagger, \beta_k$, similarly to the procedure used in the diagonalization of Hamiltonians in Sects. 3.3 and 3.5. Here we use the canonical transformation

$$a_k = u_k\alpha_k - v_k\beta_{-k}^\dagger,\tag{5.35a}$$

$$b_{-k}^\dagger = -v_k\alpha_k + u_k\beta_{-k}^\dagger,\tag{5.35b}$$

in Eq. (5.34), and impose that the Hamiltonian be cast in the diagonal form

$$\mathrm{H}^{(2)} = \sum_k \hbar(\omega_{\alpha k}\alpha_k^\dagger\alpha_k + \omega_{\beta k}\beta_k^\dagger\beta_k),\tag{5.36}$$

where $\omega_{\alpha k}$ and $\omega_{\beta k}$ are the frequencies of the two magnon modes. One can then show that the frequencies of the two modes are given by

$$\omega_{\alpha_k} = \omega_k + \gamma H_0, \quad \omega_{\beta_k} = \omega_k - \gamma H_0,\tag{5.37a}$$

$$\omega_k = \gamma[2H_E H_A + H_A^2 + H_E^2(1 - \gamma_k^2)]^{1/2},\tag{5.37b}$$

and the coefficients of the transformations are

$$u_k = \left(\frac{\gamma H_E + \gamma H_A + \omega_k}{2\omega_k}\right)^{1/2},\tag{5.38a}$$

$$v_k = \left(\frac{\gamma H_E + \gamma H_A - \omega_k}{2\omega_k}\right)^{1/2}.\tag{5.38b}$$

Notice that the transformation coefficients satisfy the orthonormality condition $u_k^2 - v_k^2 = 1$. Using for the body-centered tetragonal structure of $MnF_2$ and $FeF_2$ the vectors connecting nearest neighbors $\vec{\delta} = \pm\hat{x}(a/2) \pm \hat{y}(a/2) \pm \hat{z}(c/2)$, one can show that the geometric structure factor is

$$\gamma_k = \cos\left(k_x a/2\right) \cos\left(k_y a/2\right) \cos\left(k_z c/2\right). \tag{5.39}$$

Note that for magnons with $k = 0$, $\gamma_k = 1$, and Eqs. (5.37a) and (5.37b) are the same as Eqs. (5.14a) and (5.14b) for the AFMR frequencies, except that here we have chosen the frequencies to be positive.

The magnon dispersion relations for $MnF_2$ and $FeF_2$, calculated with Eqs. (5.37a), (5.37b) and (5.39), are shown in Fig. 5.6. For $MnF_2$ we have used $H_E = 526$ kOe, $H_A = 8.2$ kOe and $g = 2.0$, while for $FeF_2$ the values are $H_E = 540$ kOe, $H_A = 200$ kOe and $g = 2.22$. For $H_0 = 0$, in $MnF_2$ the frequency of the $k = 0$ magnon is 260 GHz. As the wave number increases, the frequency increases due to the effect of the exchange energy and reaches 1.5 THz at the zone boundary. In $FeF_2$ the $k = 0$ magnon has frequency 1.57 THz, quite larger than in $MnF_2$ because of the large anisotropy. The value at the zone boundary is 2.3 THz.

Figure 5.6 also shows the dispersion relations calculated for non-zero fields. In $MnF_2$ a field of 70 kOe, produced routinely by superconducting magnets, is sufficient to separate well the dispersion curves of the two modes and bring the $k = 0$ $\beta$-mode magnon frequency down to the microwave region. However, in $FeF_2$ the field necessary to separate well the frequencies of the two modes is of hundreds of kOe, available only in high magnetic field facilities. As mentioned in Sect. 5.2, the $k = 0$ magnon has been studied in $MnF_2$ with submillimeter and millimeter waves techniques [19, 20] as well as with microwave absorption [12]. The full magnon dispersion over the whole Brillouin zone was measured with inelastic neutron scattering [9, 22]. In $FeF_2$ the $k \approx 0$ magnon was studied with far-infrared techniques [23, 24] and with Raman light scattering [25], while the full magnon dispersion was measured with inelastic neutron scattering [26].

In order to understand the nature of the magnon modes in the easy-axis AF we calculate their transverse spin components using the relations $S_i^x = (S_i^+ + S_i^-)/2$ and

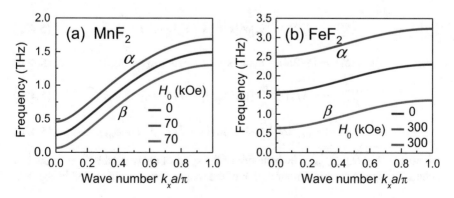

**Fig. 5.6** Calculated magnon dispersion relations for $MnF_2$ and $FeF_2$, with the magnetic field applied in the easy-anisotropy direction with the values indicated

$S_i^y = (S_i^+ - S_i^-)/2i$. Using the transformations (5.29a–5.29c) and (5.30a–5.30c) only with the terms linear in the boson operators, and with Eqs. (5.32), (5.35a) and (5.35b), we obtain the relations between the spin operators and the magnon operators

$$S_i^x = (S/2N)^{1/2} \sum_k (u_k \alpha_k - v_k \beta_k) e^{i \vec{k} \cdot \vec{r}_i} + H.c., \qquad (5.40a)$$

$$S_i^y = -i(S/2N)^{1/2} \sum_k (u_k \alpha_k + v_k \beta_k) e^{i \vec{k} \cdot \vec{r}_i} - H.c., \qquad (5.40b)$$

$$S_j^x = (S/2N)^{1/2} \sum_k (u_k \beta_k - v_k \alpha_k) e^{i \vec{k} \cdot \vec{r}_j} + H.c., \qquad (5.40c)$$

$$S_j^y = -i(S/2N)^{1/2} \sum_k (u_k \beta_k + v_k \alpha_k) e^{i \vec{k} \cdot \vec{r}_j} - H.c. \qquad (5.40d)$$

The eigenstates of the Hamiltonian (5.36) are the states $|n_{\alpha k}\rangle$ and $|n_{\beta k}\rangle$ that have well defined number of magnons. They are useful to study the thermodynamic and relaxation properties of antiferromagnets. However, as discussed in Sect. 3.3, they do not correspond to the classical picture of spin waves because they have vanishing expectation values for the transverse spin components. In analogy with ferromagnets, we define coherent AF magnon states $|\alpha_k\rangle$ and $|\beta_k\rangle$ as the eigenstates of the magnon annihilation operators

$$(\alpha_k)_{op} |\alpha_k\rangle = \alpha_k |\alpha_k\rangle, \quad (\beta_k)_{op} |\beta_k\rangle = \beta_k |\beta_k\rangle. \qquad (5.41)$$

These are the states that correspond to macroscopic spin waves in antiferromagnets. With Eqs. (5.40a–5.40d) and (5.41) one can readily obtain the expectation values of the transverse spin components in the coherent states. For a coherent state of the $\alpha$ mode we have

$$\langle S_i^x \rangle = (S/2N)^{1/2} u_k |\alpha_k| \cos (\vec{k} \cdot \vec{r}_i - \omega_{\alpha k} t + \phi_{\alpha k}), \qquad (5.42a)$$

$$\langle S_i^y \rangle = (S/2N)^{1/2} u_k |\alpha_k| \sin (\vec{k} \cdot \vec{r}_i - \omega_{\alpha k} t + \phi_{\alpha k}), \qquad (5.42b)$$

$$\langle S_j^x \rangle = -(S/2N)^{1/2} v_k |\alpha_k| \cos (\vec{k} \cdot \vec{r}_j - \omega_{\alpha k} t + \phi_{\alpha k}), \qquad (5.42c)$$

$$\langle S_j^y \rangle = -(S/2N)^{1/2} v_k |\alpha_k| \sin (\vec{k} \cdot \vec{r}_j - \omega_{\alpha k} t + \phi_{\alpha k}), \qquad (5.42d)$$

where $\alpha_k = |\alpha_k| \exp (i\phi_{\alpha k})$. The behavior of the spins of the two sublattices in the $|\alpha_k\rangle$ coherent magnon state is similar to that of the magnetizations, illustrated in Fig. 5.3.

The spins $\langle \vec{S}_i \rangle$ and $\langle \vec{S}_j \rangle$ are nearly opposite and precess clockwise in the same sense with frequency $\omega_{\alpha k}$. The ratio of the cone angles of the spins in the two sublattices is

$$\frac{\theta_i}{\theta_j} = \frac{u_k}{v_k} = \left[ \frac{\gamma(H_E + H_A) + \omega_k}{\gamma(H_E + H_A) - \omega_k} \right]^{1/2}. \tag{5.43}$$

For the $\beta$-mode the ratio of the cone angles is given by the inverse of Eq. (5.43), $\theta_i/\theta_j = v_k/u_k$. Equation (5.43) gives only the ratio of the cone angles. The actual values depend on the intensity of the driving. In order to see how the ratio (Eq. 5.43) varies with wave number, we show in Fig. 5.7 the variation of the transformation coefficients with wave number calculated with Eqs. (5.38a) and (5.38b) for $MnF_2$ and $FeF_2$.

For magnons with $k = 0$, $\gamma_k = 1$, we obtain with Eqs. (5.38a) and (5.38b)

$$\left( \frac{\theta_i}{\theta_j} \right)_\alpha = \left( \frac{\theta_j}{\theta_i} \right)_\beta = \frac{H_E + H_A - H_c}{H_E} = \frac{H_E}{H_E + H_A - H_c}, \tag{5.44}$$

which agrees with the result in Eq. (5.15). This shows that for $k = 0$ the precession angles are similar and the spins of the two sublattices are nearly in opposite directions. However, as $k$ increases, $v_k$ decreases faster than $u_k$ and the difference in the cone angles increases. For magnons at the Brillouin zone boundary, $\gamma_k = 0$, and from Eqs. (5.38a) and (5.38b) we have $u_k = 1$ and $v_k = 0$. Thus, with the relation (5.43) and the corresponding one for the $\beta$-mode, we have

$$\left( \frac{\theta_j}{\theta_i} \right)_\alpha = \left( \frac{\theta_i}{\theta_j} \right)_\beta = 0. \tag{5.45}$$

This shows that for magnons at the zone boundaries, the $\alpha$-mode involves the precession of the spins in sublattice $i$ (1) only, while the spins in sublattice $j$ (2) are

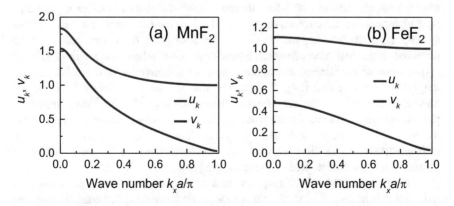

**Fig. 5.7** Variation of the transformation coefficients $u_k$ and $v_k$ with wave number in the [100] direction, calculated with Eqs. (5.38a) and (5.38b) for $MnF_2$ and $FeF_2$

fixed. On the other hand, in the $\beta$-mode only the spins in sublattice $j$ (2) precess, while the spins in sublattice $i$ (1) are fixed. This result could have been anticipated by setting $v_k = 0$ in Eqs. (5.40a–5.40d).

As mentioned earlier, the $k \approx 0$ magnon in FeF$_2$ was studied with Raman light scattering. The mechanism of the inelastic light scattering by magnons in antiferromagnets is the same as in ferromagnets, discussed in Sect. 3.6. The photons of the incident light interact with magnons by means of the magneto-optical coupling, generating scattered photons with shifted frequency, as illustrated in the diagram of Fig. 2.17. The difference from the experiments with ferromagnets is that in AFs the frequency of the scattering excitation is in the range of hundreds of GHz and few THz, not few GHz. Thus, the equipment used to analyze the scattered light is a grating spectrometer and not a Fabry–Perot interferometer. The technique is thus called Raman light scattering, instead of Brillouin light scattering.

Figure 5.8 shows the Raman light scattering spectra measured in FeF$_2$ by Fleury, Porto and co-workers [25] at various temperatures. Only the Stokes lines are observed because at low temperatures the intensity of the anti-Stokes peak is very small. The peaks at about 52 cm$^{-1}$ (1.56 THz) correspond to the scattering by magnons with $k \approx 10^5$ cm$^{-1}$, close to the Brillouin zone center, and thus with frequency close to the AFMR. The peaks at about 154 cm$^{-1}$ (4.62 THz) correspond to scattering by two magnons with wave vectors at the zone boundaries and frequency 2.31 THz. The strong exchange interaction and the large density of magnon states at the zone boundary favor the two-magnon scattering and result in the large peaks observed experimentally [27]. Since the zone boundary magnons have wave numbers two orders of magnitude larger than the photon wave numbers, the magnon pairs have wave vectors nearly opposite to each other, so as to conserve momentum in the scattering process. Notice that as the temperature increases, the peaks shift to lower frequencies, due to magnon energy renormalization. They also broaden due to the increase in the relaxation rate, decrease in amplitude and disappear above the Néel temperature.

Magnons near the zone center in FeF$_2$ were first observed by absorption experiments carried out with a far-infrared monochromator and radiation generated by a high-pressure mercury vapor lamp [23]. Since the frequency resolution of this system is relatively poor, the details of the absorption lineshapes could not be observed in the first measurements. Many years later, when the experiments were carried out with far-infrared lasers, the details of the interaction of the radiation with magnons were revealed [24]. It turns out that the direct coupling between the magnetization and the magnetic field of the radiation results in a hybrid magnon–photon excitation, the antiferromagnetic polariton. This has the same origin as the ferromagnetic polariton, and bears some similarities with the phonon polariton extensively studied in dielectrics and semiconductors, where the lattice vibration couples with the electric field of the radiation [28].

In order to understand the response of $k \approx 0$ magnons in FeF$_2$ to far-infrared radiation it is necessary to study the AF magnetic polaritons. For completeness we shall present the full derivation, repeating some steps presented in Sect. 2.5. The starting point is the response of the AF transverse components of the magnetizations,

**Fig. 5.8** Stokes peaks of the Raman light scattering by magnons in FeF$_2$ with no applied magnetic field at various temperatures. The lines at ~52 cm$^{-1}$ and ~154 cm$^{-1}$ are due to photons scattered by one and two magnons, respectively [25]. Reprinted with permission from P. A. Fleury et al., Phys. Rev. Lett. **17**, 84 (1966). Copyright (1966) by the American Physical Society

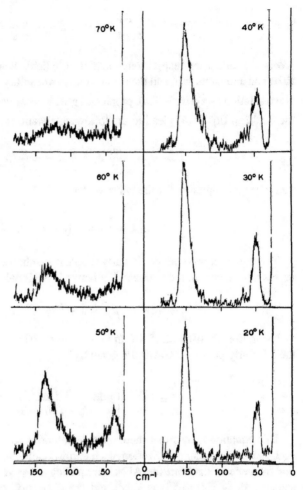

$m^{\pm} = m_1^{\pm} + m_2^{\pm}$, to the magnetic field of the radiation. This is expressed by the rf magnetic susceptibility, that can be calculated for antiferromagnets with a treatment similarly to the one for ferromagnets as in Sect. 1.2.2. It can be shown that for an easy-axis AF with the field $\vec{H}_0$ parallel to the $c$-axis, and the rotating rf magnetic field $\vec{h}^{\pm}$ applied perpendicularly to $\vec{H}_0$, the susceptibility is given by [29]

$$\chi^{\pm} = \frac{m^{\pm}}{h^{\pm}} = \frac{2H_A M_S}{H_c^2 - (\omega/\gamma \mp H_0)^2}, \tag{5.46}$$

where $M_S$ is the sublattice saturation magnetization and $H_c$ is defined in Eq. (5.14b). The propagation of radiation in the magnetic material is governed by Maxwell's equations (in CGS units)

$$\nabla \times \vec{h} = \frac{\varepsilon}{c} \frac{\partial \vec{e}}{\partial t}, \quad \nabla \times \vec{e} = -\frac{1}{c} \frac{\partial}{\partial t} (\bar{\mu} \cdot \vec{h}), \tag{5.47}$$

where $\vec{e}$, $\varepsilon$, and $\bar{\mu}$ are, respectively the rf electric field, dielectric constant (considered to be real and isotropic), and the rf magnetic permeability tensor. Considering plane-wave solutions, $\exp(ikz - i\omega t)$, propagating with wave vector $\vec{k}$ in the $z$-direction of the field $\vec{H}_0$, Eqs. (5.47) lead to the following equations

$$\vec{k} \times \vec{h} = -\frac{\omega \varepsilon}{c} \vec{e}, \quad \vec{k} \times \vec{e} = \frac{\omega}{c} (\bar{\mu} \cdot \vec{h}),$$

that give an expression for the wave vector

$$\vec{k} \times (\vec{k} \times \vec{h}) = -(\varepsilon \omega^2 / c^2) \bar{\mu} \cdot \vec{h}. \tag{5.48}$$

This equation becomes particularly simple if we choose rotating coordinates since in this representation the permeability tensor is diagonal

$$\mu_{xx} = 1 + 4\pi \chi^+, \quad \mu_{yy} = 1 + 4\pi \chi^-, \quad \mu_{xy} = \mu_{yx} = 0. \tag{5.49}$$

Using Eqs. (5.49) with (5.46) in (5.47) one can show that the wave numbers for the circularly polarized waves are given by

$$k^\pm = \frac{\omega \varepsilon^{1/2}}{c} \left[ 1 + \frac{8\pi H_A M_S}{H_c^2 - (\omega / \gamma \mp H_0)^2} \right]^{1/2}. \tag{5.50}$$

This equation shows that there are two solutions for the wave number, one for each sense of circular polarization, each corresponding to one of the AFMR modes. For frequencies far from the AFMR frequency $\omega_c = \gamma H_c$, the second term in the square root of Eq. (5.50) is small and the dispersion approaches $k^\pm = \omega \varepsilon^{1/2}/c$, characteristic of photons. If the frequency approaches $\omega_c$ from below, the wave number diverges, characterizing a flat dispersion, appropriate for magnons with negligible exchange. In the region near the crossing of the two dispersions the coupling of photons and magnons is strong and the excitation is a magnetic polariton. In the frequency range between $\omega_c$ and $\omega_c + \gamma(4\pi M_S H_A/H_c)$, the term in the square root is negative and there is no real solution for the wave number. This is a forbidden frequency band, or frequency gap, for which the radiation does not propagate in the AF material along the symmetry axis. If $H_0 = 0$ the two modes are degenerate and the two solutions coincide. Figure 5.9a shows the dispersion relation calculated with Eq. (5.50) for $FeF_2$ with no applied field.

Figure 5.9b shows the measured transmission of the 1.36 THz radiation of a $H_2O$ laser, through a 77 μm thick single crystal film of $FeF_2$, oriented with the $c$-axis perpendicular to the plane [24]. The measurements were made at $T = 4.2$ K, with a magnetic field applied in the direction of propagation, perpendicularly to the film.

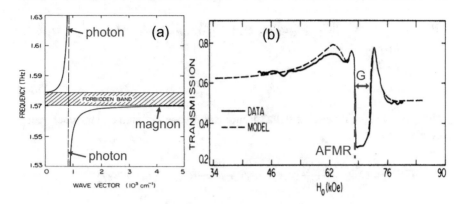

**Fig. 5.9** Magnetic polariton in $FeF_2$, adapted from Sanders et al. [24]. (**a**) Calculated magnetic polariton dispersion considering no magnetic damping and no applied static field. (**b**) Transmission of 1.36 THz radiation versus magnetic field $H_0$ in a 77 μm thick sample of in $FeF_2$. The solid line represents measurements at $T = 4.2$ K and the dashed line is the transmission calculated with $\varepsilon = 5.4$ and $\Delta H = 20$ Oe [24]. Reprinted with permission from R. W. Sanders et al., Phys. Rev. B **23**, 1190 (1981). Copyright (1981) by the American Physical Society

The field is necessary to tune the AFMR frequency of the $\beta$ mode to a range near the laser frequency. For small fields the transmission is only about 60%, due to the reflections at the surfaces. As the field increases and approaches the value for AFMR, 68.5 kOe, the magnetic permeability changes and so does the transmission. At the field for AFMR, the transmission falls abruptly to about half the value because the component with + circular polarization does not propagate through. Since the laser radiation is linearly polarized, the component with − circular polarization goes through and accounts for the observed transmission. The transmission curve shows clearly the gap G in field for which the + circularly polarized radiation does not propagate. Figure 5.9b also shows the dashed curve obtained with the calculated transmission using Eq. (5.50) including magnetic damping [24].

## 5.3.2   Magnon Energy Renormalization

In the transformation of spin operators into magnon operators in the Hamiltonian (Eq. 5.28), we retained in the binomial expansion of the square roots in Eqs. (5.29a–5.29c) and (5.30a–5.30c) only terms that lead to products of two magnon operators. If other terms in the binomial expansion are used, the Hamiltonian (Eq. 5.28) can expressed in the form $H = E_0 + H^{(2)} + H^{(4)} + H^{(6)}\ldots$, where each term contains an even number of magnon operators. We will consider here the 4-magnon interaction that is the most important one for the magnon energy renormalization and for magnon relaxation.

The 4-magnon Hamiltonian obtained from Eq. (5.28) contains terms for intramode and intermode scattering. The terms that conserve energy have the form [30]

$$H^{(4)} = \sum_{\substack{k_1, k_2, \\ k_3, k_4}} \Delta(\vec{k}) [C^{\alpha-\alpha}_{1234} \alpha^\dagger_{k_1} \alpha^\dagger_{k_2} \alpha_{k_3} \alpha_{k_4} + C^{\beta-\beta}_{1234} \beta^\dagger_{k_1} \beta^\dagger_{k_2} \beta_{k_3} \beta_{k_4} + C^{\alpha-\beta}_{1234} \alpha^\dagger_{k_1} \beta^\dagger_{k_2} \alpha_{k_3} \beta_{k_4}],$$

$$(5.51)$$

where $\Delta(\vec{k})$ is the Kronecker delta for momentum conservation and the coefficients are given by

$$C^{\alpha-\alpha}_{1234} = C^{\beta-\beta}_{1234} = \frac{z|J_2|}{2N} (u_1 v_2 v_3 v_4 \gamma_1 + u_1 u_2 u_3 v_4 \gamma_4 + u_1 v_2 u_3 v_4 \gamma_2$$

$$+ v_1 v_2 u_3 v_4 \gamma_3 - 4 u_1 v_2 u_3 v_4 \gamma_{2-4}) - \frac{D}{N} (u_1 u_2 u_3 u_4 + v_1 v_2 v_3 v_4),$$

$$(5.52a)$$

$$C^{\alpha-\beta}_{1234} = \frac{z|J_2|}{2N} (u_1 u_2 u_3 v_4 \gamma_2 + u_1 v_2 v_3 v_4 \gamma_3 + u_1 u_2 v_3 u_4 \gamma_1 + v_1 u_2 v_3 v_4 \gamma_4$$

$$+ v_1 v_2 v_3 u_4 \gamma_2 + v_1 u_2 u_3 u_4 \gamma_3 + v_1 v_2 u_3 v_4 \gamma_1 + u_1 v_2 u_3 u_4 \gamma_4$$

$$- 2 u_1 u_2 u_3 u_4 \gamma_{2-4} - 2 u_1 v_2 v_3 u_4 \gamma_{3-4} - 2 v_1 u_2 v_3 v_4 \gamma_{2+1} - 2 v_1 v_2 v_3 v_4 \gamma_{3-1})$$

$$- \frac{4D}{N} (u_1 v_2 u_3 v_4 + v_1 u_2 v_3 u_4),$$

$$(5.52b)$$

where we have simplified the notation of the subscripts by omitting the symbol of the wave vector.

The 4-magnon interaction, represented by the Hamiltonian (5.51), is the main origin of the magnon energy renormalization that accounts for the decrease in frequency with increasing temperature, observed experimentally in several antiferromagnets. The calculation of the energy renormalization is done in the same way as described in Sect. 3.7 for ferromagnets.

We use a random-phase approximation in the Hamiltonian (5.51) and replace one pair of magnon operators by its thermal average, such that the term is cast in the quadratic form (5.36). For example, $\alpha_{k1} \alpha^\dagger_{k2} \beta^\dagger_{k3} \beta_{k4} \Delta(\vec{k}_1 + \vec{k}_4 - \vec{k}_2 - \vec{k}_3)$ is replaced by $\langle \alpha_q \alpha^\dagger_q \rangle \beta^\dagger_k \beta_k + \langle \beta^\dagger_q \beta_q \rangle \alpha^\dagger_k \alpha_k = (\bar{n}_{\alpha q} + 1) \beta^\dagger_k \beta_k + \bar{n}_{\beta q} \alpha^\dagger_k \alpha_k$, where $\bar{n}_{\alpha,\beta q}$ is the thermal occupation number of the $\alpha$- or $\beta$-mode with wave vector $\vec{q}$, given by the Bose–Einstein distribution $\bar{n}_{\alpha,\beta q} = 1/(e^{\hbar \omega_{\alpha,\beta q}/k_B T} - 1)$. Applying this approximation to each 4-magnon term, the Hamiltonian reduces to a quadratic form

$$H^{(4)} \Rightarrow \hbar \sum_k \Delta \omega_{\alpha k} \alpha^\dagger_k \alpha_k + \Delta \omega_{\beta k} \beta^\dagger_k \beta_k, \quad (5.53)$$

where $\hbar \Delta \omega_{\alpha k}$ and $\hbar \Delta \omega_{\beta k}$ represent the renormalization of the energies of the two modes, so that the total magnon energies become temperature dependent,

$\hbar\omega_{\alpha,\,\beta k}(T) = \hbar\omega_{\alpha,\,\beta k}(0) + \hbar\Delta\omega_{\alpha,\,\beta k}(T)$. Hence, the total 4-magnon contribution to the $\alpha$-mode energy then becomes [31].

$$\hbar\Delta\omega_{\alpha k}(T) = 2z|J_2|[(u_k^2 + v_k^2 - 2u_k v_k \gamma_k)C_q + (u_k v_k \gamma_k - v_k^2)E_q + (u_k v_k \gamma_k - u_k^2)F_q]$$
$$- 4D(u_k^2 E_q + v_k^2 F_q)$$

(5.54)

where

$$C_q = \frac{1}{NS}\sum_q u_q v_q \gamma_q (\bar{n}_{\alpha q} + \bar{n}_{\beta q} + 1),$$

(5.55a)

$$E_q = \frac{1}{NS}\sum_q u_q^2 \bar{n}_{\alpha q} + v_q^2 (\bar{n}_{\beta q} + 1),$$

(5.55b)

$$F_q = \frac{1}{NS}\sum_q u_q^2 \bar{n}_{\beta q} + v_q^2 (\bar{n}_{\alpha q} + 1).$$

(5.55c)

The change in the frequency of the $\beta$-mode is given by the same expressions as Eqs. (5.54) and (5.55a–5.55c) with the interchange $\alpha \leftrightarrow \beta$. It is important to note that even at $T = 0$ there is a correction to the magnon energies due to the magnon interactions. This zero-point correction is a characteristic feature of antiferromagnets that was first noted by Oguchi several decades ago [32].

The symbols in Fig. 5.10 represent the data for the magnon dispersion relations for $MnF_2$, with no applied field, with the wave vector along the $c$-axis, measured by inelastic neutron scattering at three temperatures [22]. The curve for $T = 3$ K in Fig. 5.10 is the unrenormalized energy calculated with Eqs. (5.37a), (5.37b), and (5.39), using $H_A = 8.2$ and $H_E = 570$ kOe. The value of the exchange field was adjusted to fit the data at $T = 0$ K and is about 7% higher than the value given before, because Eqs. (5.37a) and (5.37b) does not include the effect of the intra-sublattice interaction $J_1$. This enters in the dispersion relation for the wave vector along the $c$-axis [9].

The temperature dependence of the magnon energy renormalization for $MnF_2$ due to 4-magnon interactions was calculated numerically with Eq. (5.54) replacing the sum over wave vectors by an integral over the Brillouin zone as in Sect. 3.7. For each temperature the integrals in Eqs. (5.55a–5.55c) were evaluated numerically by a discrete sum in $\vec{k}$ space. In the first cycle of the evaluation of $\Delta\omega_{\alpha k}$ and $\Delta\omega_{\beta k}$, the frequencies and the Bose factors are calculated for each point $k$ in the Brillouin zone, without renormalization. In the following cycles the new Bose factors are calculated with the magnon frequencies $\hbar\omega_{\alpha,\,\beta k}(T) = \hbar\omega_{\alpha,\,\beta k}(0) + \hbar\Delta\omega_{\alpha,\,\beta k}(T)$, using the renormalization from the previous cycle. The process is repeated until the change in frequency at all points is smaller than 0.1%. Figure 5.10 shows that the calculated dispersions for three temperatures in $MnF_2$ agree well with experimental data.

**Fig. 5.10** Magnon dispersion relations for $MnF_2$ with no applied field with the wave vector along the $c$-axis. Symbols represent inelastic neutron scattering data at three temperatures [22] and the solid lines represent the dispersions calculated for the same temperatures with Eqs. (5.37a), (5.37b) and (5.39) and with the magnon energy renormalization in Eqs. (5.54) and (5.55a–5.55c)

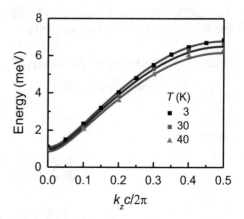

As shown in Sects. 3.4.3 and 3.7.2, the 4-magnon interaction also constitutes a mechanism for magnon damping. In antiferromagnets this is the main channel for the relaxation of magnons, since due the energy gap, the conservation of energy is inhibited in the 3-magnon interaction. Calculations of the 4-magnon relaxation rate for $MnF_2$ and $RbMnF_3$ done with Eq. (3.116) using the Hamiltonian (5.51) are in good agreement with experimental data [30, 31].

### 5.3.3   Magnons in Easy-Axis Antiferromagnets: SF Phase

If the applied field $H_0$ exceeds the critical value $H_c$, the energy of the $\beta$-mode at $k = 0$ becomes negative and the AF phase is no longer stable. Then the spins flip to the configuration illustrated in Fig. 5.1b, at an angle $\theta_1 = \theta_2 = \theta$ with the field, and in a plane determined by the small anisotropy that exists transverse to the symmetry axis. As shown in Sect. 5.1, the angle varies with the field and is given by $\cos\theta = [H_0/(2H_E - H_A)]$. We study the spin wave excitations in the SF phase using two different coordinate systems, one for each sublattice, as shown in Fig. 5.11. In each system the $z$-axis is chosen to point along the spin equilibrium direction in that sublattice, the

**Fig. 5.11** Coordinate systems used to represent the components of the spin operators for the two sublattices in the spin-flop phase

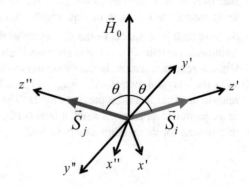

$y$-axes are in the direction of the $y$-axis of the crystal, as in Fig. 5.1, and the $x$-axes are determined by $\widehat{x} = \widehat{y} \times \widehat{z}$.

As in Sect. 5.3.1, we express the spin components in each sublattice in terms of spin deviation operators. Considering only the first order terms in the binomial expansions of the square roots in Eqs. (5.29a–5.29c) and (5.30a–5.30c) we have

$$S_i^+ = S_i^{x'} + iS_i^{y'} = \sqrt{2S}a_i, \quad S_j^+ = S_j^{x''} + iS_j^{y''} = \sqrt{2S}b_j, \tag{5.56a}$$

$$S_i^- = S_i^{x'} - iS_i^{y'} = \sqrt{2S}a_i^\dagger, \quad S_j^- = S_j^{x''} - iS_j^{y''} = \sqrt{2S}b_j^\dagger, \tag{5.56b}$$

$$S_i^{z'} = S - a_i^\dagger a_i, \quad S_j^{z''} = S - b_j^\dagger b_j. \tag{5.56c}$$

Using the transformation to collective boson operators in Eq. (5.32) and the relation (5.33), one can show that the Hamiltonian (5.28) leads to the following quadratic form

$$H^{(2)} = \hbar \sum_k A_k(a_k^\dagger a_k + b_k^\dagger b_k) + B_k(a_k b_{-k} + a_k^\dagger b_{-k}^\dagger)$$

$$+ \frac{1}{2} C_k(a_k a_{-k} + b_k b_{-k} + H.c.) + D_k(a_k b_k^\dagger + a_k^\dagger b_k), \tag{5.57}$$

where

$$A_k = \gamma[H_0\cos\theta - H_E\cos2\theta + H_A(\cos^2\theta - \frac{1}{2}\sin^2\theta)], \tag{5.58a}$$

$$B_k = \gamma\gamma_k H_E\sin^2\theta, \quad C_k = \frac{1}{2}\gamma H_A\sin^2\theta, \quad D_k = \gamma\gamma_k H_E\cos^2\theta. \tag{5.58b}$$

The next step consists of performing canonical transformations from the collective boson operators $a_k^\dagger, a_k, b_k^\dagger, b_k$ into magnon creation and annihilation operators $\alpha_k^\dagger, \alpha_k, \beta_k^\dagger, \beta_k$ such that the quadratic Hamiltonian is cast in the diagonal form

$$H^{(2)} = \sum_k \hbar(\omega_{\alpha k}\alpha_k^\dagger \alpha_k + \omega_{\beta k}\beta_k^\dagger \beta_k), \tag{5.59}$$

We follow the method of White, Sparks and Ortenburger [33] for generalizing the Bogoliubov transformation for diagonalizing Hamiltonians and write Eq. (5.57) in matrix form

$$H = \hbar \sum_{k>0} H_k, \quad H_k = (X)^\dagger[H](X), \tag{5.60}$$

where the matrices are

$$(X) = \begin{pmatrix} a_k \\ b_k \\ a_{-k}^\dagger \\ b_{-k}^\dagger \end{pmatrix}, \quad [H] = \begin{bmatrix} A_k & D_k & C_k & B_k \\ D_k & A_k & B_k & C_k \\ C_k & B_k & A_k & D_k \\ B_k & C_k & D_k & A_k \end{bmatrix}. \tag{5.61}$$

The next step consists of introducing a linear transformation to new operators

$$(X) = [Q](Z), \quad (Z) = \begin{pmatrix} \alpha_k \\ \beta_k \\ \alpha_{-k}^\dagger \\ \beta_{-k}^\dagger \end{pmatrix}, \tag{5.62}$$

such that the Hamiltonian in Eq. (5.60) can be written as

$$H = \hbar \sum_k (Z)^\dagger [\omega](Z), \tag{5.63}$$

where $[\omega]$ is a diagonal eigenvalue matrix. In order to find the transformation matrix $[Q]$ one needs a few relations. The first follows from the introduction of Eqs. (5.62) in (5.60) and comparison with Eq. (5.63). This leads to

$$[Q]^\dagger [H][Q] = \hbar[\omega]. \tag{5.64}$$

Another relation is obtained from the boson commutation rules. They can be written in matrix forms

$$[X, X^\dagger] = X(X)^\dagger - (X^* X^T)^T = [g], \tag{5.65a}$$

$$[Z, Z^\dagger] = Z(Z)^\dagger - (Z^* Z^T)^T = [g], \tag{5.65b}$$

where

$$[g] = \begin{bmatrix} 1 & 0 & 0 & 0 \\ 0 & 1 & 0 & 0 \\ 0 & 0 & -1 & 0 \\ 0 & 0 & 0 & -1 \end{bmatrix}. \tag{5.66}$$

Using Eqs. (5.61) and (5.62) in (5.65a) and in (5.65b), we obtain an orthonormality relation for the transformation matrix

$$[Q][g][Q]^\dagger = [g]. \tag{5.67}$$

With Eqs. (5.65a), (5.65b)–(5.67) one obtains an eigenvalue equation

$$[H][Q] = [g]^{-1}[Q][g]\hbar[\omega]. \tag{5.68}$$

Solution of Eq. (5.68) yields the elements of the transformation matrix $[Q]$ and the eigenfrequencies of the two magnon modes. In the present case, due to the symmetry of the $4\times4$ matrices, the equations involving them, can be reduced to relations between $2\times2$ matrices. This is done by first noticing that the Hamiltonian (5.61) can be written as

$$[H] = \begin{bmatrix} [H_1] & [H_2] \\ [H_2] & [H_1] \end{bmatrix}, \quad [H_1] = \begin{bmatrix} A_k & D_k \\ D_k & A_k \end{bmatrix}, \quad [H_2] = \begin{bmatrix} C_k & B_k \\ B_k & C_k \end{bmatrix}. \tag{5.69}$$

Also, note that the transformations of the operators $a_k$ and $b_k$ are the Hermitian conjugates of the ones for $a_{-k}^\dagger$ and $b_{-k}^\dagger$, so that the matrix $[Q]$ can also be written in terms of submatrices

$$[Q] = \begin{bmatrix} [Q_1] & [Q_2] \\ [Q_2^*] & [Q_1^*] \end{bmatrix}, \quad [Q_1] = \begin{bmatrix} Q_{11} & Q_{12} \\ Q_{21} & Q_{22} \end{bmatrix}, \quad [Q_2] = \begin{bmatrix} Q_{13} & Q_{14} \\ Q_{23} & Q_{24} \end{bmatrix}. \tag{5.70}$$

The matrix $[g]$ in Eq. (5.66) can also be written in the form

$$[g] = \begin{bmatrix} I & 0 \\ 0 & -I \end{bmatrix}, \tag{5.71}$$

where $I$ is a $2\times2$ diagonal unit matrix. Using Eqs. (5.69)–(5.71) in (5.68) we obtain the following relations between the $2\times2$ matrices

$$[H_1][Q_1] + [H_2][Q_2^*] = [Q_1][\omega], \tag{5.72a}$$

$$[H_1][Q_2] + [H_2][Q_1^*] = -[Q_2][\omega]. \tag{5.72b}$$

The solution of Eqs. (5.72a) and (5.72b) gives the Hamiltonian in Eq. (5.59), where the eigenfrequencies are

$$\omega_{\alpha k}^2 = (A_k - D_k)^2 - (B_k - C_k)^2, \tag{5.73a}$$

$$\omega_{\beta k}^2 = (A_k + D_k)^2 - (B_k + C_k)^2, \tag{5.73b}$$

and the elements of the transformation matrix are

$$Q_{11} = \left[\frac{(A_k - D_k) + \omega_{\alpha k}}{4\omega_{\alpha k}}\right]^{1/2}, \quad Q_{12} = -\left[\frac{(A_k + D_k) + \omega_{\beta k}}{4\omega_{\beta k}}\right]^{1/2}, \tag{5.74a}$$

$$Q_{13} = \left[\frac{(A_k - D_k) - \omega_{\alpha k}}{4\omega_{\alpha k}}\right]^{1/2}, \quad Q_{14} = \left[\frac{(A_k + D_k) - \omega_{\beta k}}{4\omega_{\beta k}}\right]^{1/2}, \tag{5.74b}$$

and $Q_{21} = -Q_{11}$, $Q_{22} = Q_{12}$, $Q_{23} = -Q_{13}$, $Q_{24} = Q_{14}$. These results greatly simplify if $H_0$, $H_A \ll H_E$, which is the case of $MnF_2$. Using for $ka \ll 1$, $1 - \gamma_k \approx (ka)^2/8$, the two magnon frequencies become

$$\omega_{ak} \approx \frac{1}{2}\gamma H_E ak, \tag{5.75a}$$

$$\omega_{\beta k} \approx \gamma \left(H_0^2 - H_c^2 + \frac{1}{4}H_E^2 a^2 k^2\right)^{1/2}. \tag{5.75b}$$

As we saw in Sect. 5.3.1, in the AF phase the frequency of the $\alpha_k$ mode increases linearly with magnetic field intensity for any wave vector. However, in the SF phase the frequency becomes independent of the applied field and, moreover, vanishes for the $k = 0$ magnon, as can be seeing in Eq. (5.75a). This behavior has been clearly observed experimentally in $MnF_2$ [20], as shown in Fig. 5.4b. Initially the frequency increases linearly with field, but as the field approaches $H_c$ it falls abruptly to a very small value. This is so because this mode corresponds to the rotation of the spins about the symmetry axis at no cost of energy. Actually, as mentioned earlier, in crystals there is always a small anisotropy in the plane perpendicular to the symmetry axis, that can be represented by an effective field $H_{AT}$. In this case, as will be shown in the next section, the frequency of the $k = 0$, $\alpha_k$ magnon is $\omega_{a0} = \gamma(2H_E H_{AT})^{1/2}$, which is still independent of the applied field, but not zero. In regard to the $\beta_k$ mode, in the AF phase its frequency decreases linearly with field and, for the $k = 0$ magnon, it goes to zero at $H_0 = H_c$. For $H_0 > H_c$, according to Eq. (5.75b), the frequency increases with field as $\omega_{\beta k} = \gamma(H_0^2 - H_c^2)^{1/2}$, in agreement with the experimental data shown in Fig. 5.4b.

Analogously to what we did in Sect. 5.3.1, we can find the behavior of the sublattice spins of the magnon modes in the SF phase, by calculating the expectation values of the spin components in coherent magnon states. For the $\alpha_k$ mode we obtain

$$\langle S_i^{x'} \rangle = (2S/N)^{1/2}(Q_{11} + Q_{13})|\alpha_k|\cos(\vec{k} \cdot \vec{r}_i - \omega_{ak} + \phi_{ak}), \tag{5.76a}$$

$$\langle S_i^{y'} \rangle = (2S/N)^{1/2}(Q_{11} - Q_{13})|\alpha_k|\sin(\vec{k} \cdot \vec{r}_i - \omega_{ak} + \phi_{ak}), \tag{5.76b}$$

$$\langle S_j^{x''} \rangle = -(2S/N)^{1/2}(Q_{11} + Q_{13})|\alpha_k|\cos(\vec{k} \cdot \vec{r}_j - \omega_{ak} + \phi_{ak}), \tag{5.76c}$$

$$\langle S_j^{y''} \rangle = -(2S/N)^{1/2}(Q_{11} - Q_{13})|\alpha_k|\sin(\vec{k} \cdot \vec{r}_j - \omega_{ak} + \phi_{ak}), \tag{5.76d}$$

where $\alpha_k = |\alpha_k|\exp(i\phi_{ak})$. For the $\beta_k$ mode we have

$$\langle S_i^{x'} \rangle = (2S/N)^{1/2}(Q_{14} + Q_{12})|\beta_k|\cos(\vec{k} \cdot \vec{r}_i - \omega_{\beta k} + \phi_{\beta k}), \tag{5.77a}$$

$$\langle S_i^{y'} \rangle = (2S/N)^{1/2}(Q_{14} - Q_{12})|\beta_k| \sin(\vec{k} \cdot \vec{r}_i - \omega_{\beta k} + \phi_{\beta k}), \qquad (5.77b)$$

$$\langle S_j^{x''} \rangle = (2S/N)^{1/2}(Q_{14} + Q_{12})|\beta_k| \cos(\vec{k} \cdot \vec{r}_j - \omega_{\beta k} + \phi_{\beta k}), \qquad (5.77c)$$

$$\langle S_j^{y''} \rangle = (2S/N)^{1/2}(Q_{14} - Q_{12})|\beta_k| \sin(\vec{k} \cdot \vec{r}_j - \omega_{\beta k} + \phi_{\beta k}), \qquad (5.77d)$$

where $\beta_k = |\beta_k| \exp(i\phi_{\beta k})$. With Eqs. (5.76a–5.76d) and (5.77a–5.77d) we can find the ellipticities of the spin precessions. Considering $H_E \gg H_0, H_A$ in the coefficients (5.74a) and (5.74b), we obtain with Eqs. (5.76a–5.76d) and (5.77a–5.77d) the ellipticity in both sublattices for the $\alpha_k$ mode

$$e_\alpha = \frac{\langle S_{i,j}^{x'} \rangle_{max}}{\langle S_{i,j}^{y'} \rangle_{max}} = \frac{Q_{11} + Q_{13}}{Q_{11} - Q_{13}} \approx \frac{(\gamma H_E + \omega_{\alpha k})^{1/2} + (\gamma H_E - \omega_{\alpha k})^{1/2}}{(\gamma H_E + \omega_{\alpha k})^{1/2} - (\gamma H_E - \omega_{\alpha k})^{1/2}}, \qquad (5.78)$$

and for the $\beta_k$ mode

$$e_\beta = \frac{\langle S_{i,j}^{x'} \rangle_{max}}{\langle S_{i,j}^{y'} \rangle_{max}} = \frac{Q_{14} + Q_{12}}{Q_{14} - Q_{12}} \approx \frac{(\gamma H_E + \omega_{\beta k})^{1/2} - (\gamma H_E - \omega_{\beta k})^{1/2}}{(\gamma H_E + \omega_{\beta k})^{1/2} + (\gamma H_E - \omega_{\beta k})^{1/2}}. \qquad (5.79)$$

These equations show that the ellipticity of the spin precession varies strongly with the wave number. Near the center of the Brillouin zone, Eqs. (5.75a) and (5.75b) give $\gamma H_E \gg \omega_{\alpha k}, \omega_{\beta k}$, so that the ellipticities become $e_\alpha = 8\gamma H_E/\omega_{\alpha k}$ and $e_\beta = \omega_{\beta k}/(8\gamma H_E)$. In this case, the spin precessions are highly elliptical, with the major axis much larger than the minor. Figure 5.12 illustrates the precessions of the spins in the two sublattices for both modes for $k = 0$. In the field-independent $\alpha_k$ mode, the spins process with the major axes along the x-axes of the two sublattices, with a net spin component along the external field. Thus, this mode can be driven by a rf field parallel to the static field. On the other hand, in the field-dependent $\beta_k$ mode, the major axes of the elliptical precessions are along the y-direction for each

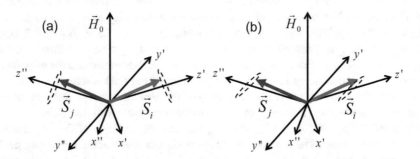

**Fig. 5.12** Illustration of the elliptical precessions of the sublattice spins in the $k = 0$ magnon modes in the SF phase. (**a**) $\alpha_k$ mode. (**b**) $\alpha_k$ mode

sublattice, with a net spin component along $y$, so that it can be driven by a rf field parallel to this direction.

### 5.3.4  Magnons in Easy-Axis Antiferromagnets: Canted Phase

We now consider the configuration where the external magnetic field is applied in a direction perpendicular to the easy axis. In this case, in the equilibrium configuration the two sublattice spins lie in the same plane in directions at an angle $\theta$ with the field, similar to the spin-flop phase, shown in Fig. 5.1b. Minimization of the energy gives $\cos\theta = H_0/(2H_E + H_A)$. In this phase, the Hamiltonian (5.28) expressed in terms of boson operators introduced by the transformations in Eqs. (5.56a–5.56c), with (5.32) and (5.33), becomes, with only the quadratic terms

$$H = \hbar\sum_k A_k(a_k^\dagger a_k + b_k^\dagger b_k) + B_k(a_k b_{-k} + a_k^\dagger b_{-k}^\dagger)$$
$$+ \frac{1}{2}C_k(a_k a_{-k} + b_k b_{-k} + H.c.) + D_k(a_k b_k^\dagger + a_k^\dagger b_k), \tag{5.80}$$

where

$$A_k = \gamma[H_0\cos\theta - H_E\cos2\theta + H_A(2\sin^2\theta - \cos^2\theta)/2], \tag{5.81a}$$

$$B_k = \gamma\gamma_k H_E\sin^2\theta, \tag{5.81b}$$

$$C_k = \gamma H_A(\cos^2\theta)/2, \tag{5.81c}$$

$$D_k = \gamma\gamma_k H_E(1 + \cos2\theta)/2, \tag{5.81d}$$

The Hamiltonian (5.80) has the same form as Eq. (5.57) for the spin-flop state, so that the frequencies and the transformation coefficients are given by the same expressions as (5.73a), (5.73b), (5.74a) and (5.74b), with the parameters defined in Eqs. (5.81a–5.81d).

Figure 5.13a shows the magnon dispersion relations for $MnF_2$ with a magnetic field of $H_0 = 200$ kOe applied perpendicularly to the easy axis, calculated with Eqs. (5.73a), (5.73b), and (5.81a–5.81d) using $H_E = 570$ kOe, $H_A = 8.2$ kOe, $g = 2.0$, and a geometric structure factor for a spherical Brillouin zone, $\gamma_k = \cos(\pi k/2k_{max})$. For zero-field the two modes are degenerate and the dispersion relations are the same as in Fig. 5.6. Application of the field results in a behavior similar to the spin-flop phase, the frequency of the $\alpha$-mode becomes independent of the field, while the frequency of the $\beta$-mode increases with the field. This is illustrated further in Fig. 5.13b, showing the field dependence of the frequencies for the $k = 0$ modes. The frequency of the $\beta$-mode increases continuously with field, as observed experimentally [20] and shown in Fig. 5.4b. Note that the frequency of the $\alpha$-mode is not shown in Fig. Fig. 5.4b because the experiments of Hagiwara et al. [20] were carried out with fixed frequency and scanning field, so that the $\alpha$-mode cannot be detected.

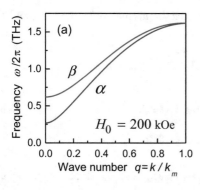
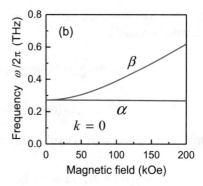

**Fig. 5.13** Magnon frequencies calculated with Eqs. (5.73a), (5.73b), and (5.81a–5.81d) for MnF$_2$, with the external field applied perpendicularly to the easy axis, in the spin canted configuration. (**a**) Dispersion relations for $H_0 = 200$ kOe. (**b**) Field dependence of the frequencies of the $k = 0$ modes [21]. Reproduced from S. M. Rezende et al., J. Appl. Phys. **126**, 151,101 (2019), with the permission of AIP Publishing

## 5.4  Magnons in Easy-Plane Antiferromagnets

In this section we shall calculate the magnon frequencies for easy-plane, or hard-axis, or biaxial, antiferromagnets in the AF phase, with no applied external field. We shall use the HP formalism as in the previous section. Initially we write the Hamiltonian (5.5), with $H_0 = 0$, in terms of spin raising and lowering operators

$$\mathrm{H} = -\gamma\hbar\sum_{i,j}J_{ij}(S_i^+ S_j^- + S_i^- S_j^+ + 2S_i^z S_j^z) + \sum_{i,j}D_x(S_{i,j}^x)^2 - D_z(S_{i,j}^z)^2. \qquad (5.82)$$

In the first HP transformation the components of the local spin operators are related to the creation and annihilation operators of spin deviations for the two opposite sublattices. Using in Eq. (5.82) the transformations (5.29a–5.29c) and (5.30a–5.30c), with only the first terms of the binomial expansions of the square roots, and with Eqs. (5.32) and (5.33), we obtain for the quadratic part of the Hamiltonian

$$\mathrm{H}^{(2)} = \gamma\hbar\Big[\sum_k (H_E + H_{Az})(a_k^\dagger a_k + b_k^\dagger b_k) + \gamma_k H_E(a_k b_{-k} + a_k^\dagger b_{-k}^\dagger)$$

$$+ \frac{H_{Ax}}{4}\big(a_k a_{-k} + a_k^\dagger a_{-k}^\dagger + a_k a_k^\dagger + a_k^\dagger a_k + b_k b_{-k} + b_k^\dagger b_{-k}^\dagger + b_k b_k^\dagger + b_k^\dagger b_k\big)\Big],$$

$$(5.83)$$

where $\gamma_k$ is a structure factor defined by $\gamma_k = (1/z)\sum_\delta \exp(i\vec{k}.\vec{\delta})$, $\vec{\delta}$ are the vectors connecting nearest neighbors, and we have considered only inter-sublattice exchange. The quadratic Hamiltonian written in normal order and without the constants becomes

$$H = \hbar \sum_k A_k(a_k^\dagger a_k + b_k^\dagger b_k) + B_k(a_k b_{-k} + a_k^\dagger b_{-k}^\dagger)$$

$$+ \frac{1}{2} C_k(a_k a_{-k} + b_k b_{-k} + H.c.), \tag{5.84}$$

where the parameters are related to the effective fields by

$$A_k = \gamma(H_E + H_{Ax}/2 + H_{Az}), \quad B_k = \gamma\gamma_k H_E, \quad C_k = \gamma\frac{H_{Ax}}{2}. \tag{5.85}$$

The next step consists of performing canonical transformations from the collective boson operators $a_k^\dagger, a_k, b_k^\dagger, b_k$ into magnon creation and annihilation operators $\alpha_k^\dagger, \alpha_k, \beta_k^\dagger, \beta_k$, such that the quadratic Hamiltonian is cast in the diagonal form

$$H^{(2)} = \sum_k \hbar(\omega_{\alpha k}\alpha_k^\dagger \alpha_k + \omega_{\beta k}\beta_k^\dagger \beta_k), \tag{5.86}$$

As shown in Sect. 5.3.3, the diagonalization procedure leads to the magnon frequencies of the two modes given by Eqs. (5.73a) and (5.73b), that with $D_k = 0$ lead to

$$\omega_{\alpha,\beta}^2 = A_k^2 - (B_k \mp C_k)^2. \tag{5.87}$$

Using the expressions for the parameters in Eqs. (5.85), one can write the frequencies of the two magnon modes in terms of the effective fields

$$\omega_{\alpha k}^2 = \gamma^2[H_{Az}^2 + 2H_E H_{Az} + H_{Ax}H_{Az} + H_E H_{Ax}(1 + \gamma_k) + H_E^2(1 - \gamma_k^2)], \tag{5.88a}$$

$$\omega_{\beta k}^2 = \gamma^2[H_{Az}^2 + 2H_E H_{Az} + H_{Ax}H_{Az} + H_E H_{Ax}(1 - \gamma_k) + H_E^2(1 - \gamma_k^2)]. \tag{5.88b}$$

For NiO, considering $H_E \gg H_{Ax}, H_{Az}$, the frequencies of the zone-center $k = 0$ ($\gamma_k = 1$) magnons are

$$\omega_{\alpha 0} \approx \gamma(2H_{Ax}H_E)^{1/2}, \quad \omega_{\beta 0} \approx \gamma(2H_{Az}H_E)^{1/2}, \tag{5.89}$$

which agree with Eqs. (5.22) for $H_0 = 0$. As shown in Machado et al. [17], the values of the parameters for NiO can be obtained by fitting Eqs. (5.88a) and (5.88b) to three sets of experimental data. Fitting to the neutron scattering measurements of Hutchings and Samuelsen [16] gives the value of the exchange field $H_E = 9\,684$ kOe considering for the $g$-factor $g = 2.18$. Since the neutron data do not have sufficient resolution to determine the frequencies of the zone-center magnons, one has to use the value $\omega_{\beta 0}/2\pi = 0.140$ THz measured by Brillouin light scattering [34] and $\omega_{\alpha 0}/2\pi = 1.07$ THz obtained from magnetization oscillations in the far infrared

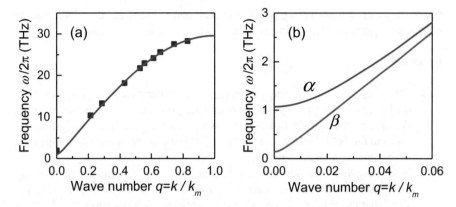

**Fig. 5.14**  Magnon dispersion curves in antiferromagnetic NiO at $T = 300$ K. (a) Solid curves show the magnon frequencies calculated with Eqs. (5.88a) and (5.88b). Symbols represent the neutron scattering data of Hutchings and Samuelsen [16]. (b) Blowup of the Brillouin zone center showing the separation of the frequencies of the $\alpha$ (upper blue curve) and $\beta$ (lower red curve) magnon modes [21]. Reproduced from S. M. Rezende et al., J. Appl. Phys. **126**, 151101 (2019), with the permission of AIP Publishing

[35]. With these values in Eq. (5.84) one can obtain the anisotropy fields, $H_{Ax} = 6.35$ and $H_{Az} = 0.11$ kOe [17]. Figure 5.14 shows the dispersion relations calculated with Eq. (5.83) assuming a spherical Brillouin zone and using for the structure factor $\gamma_k = \cos(\pi k / 2k_m)$, where $k_m = \pi/a_l$, $a_l$ being the lattice parameter and the data of Hutchings and Samuelsen [16]. The two magnon modes are nearly degenerate for large wave vectors, but the frequency separation is evident in the blowup near the Brillouin zone center, as shown in Fig. 5.14b. This frequency separation at small wave numbers lends to NiO important properties for antiferromagnetic spintronics applications [36].

To conclude this chapter, we note that similarly to the case of the spin-flop phase, studied in Sect. 5.3.3, we can calculate the expectation values of the spin components in coherent magnon states in the AF phase of the easy-plane antiferromagnets. In both modes the spin precessions in the two sublattices are characterized by an ellipticity $e = \langle S_{i,j}^x \rangle_{max} / \langle S_{i,j}^y \rangle_{max}$. Using the transformations in Eqs. (5.29a–5.29c) and (5.30a–5.30c), (5.32), and (5.62) one can express the spin components in terms of the coefficients in Eq. (5.70) and show that the ellipticities for modes $\alpha_k$ and $\beta_k$ are

$$e_\alpha = \frac{Q_{11} + Q_{13}}{Q_{11} - Q_{13}}, \quad e_\beta = \frac{Q_{14} - |Q_{12}|}{Q_{14} + |Q_{12}|}. \tag{5.90}$$

As in the spin-flop phase, the ellipticities of the spin precession vary strongly with the wave number. Near the center of the Brillouin zone, for $\gamma H_E \gg \omega_{ak}, \omega_{\beta k}$, Eqs. (5.90) give the same results as in Eqs. (5.27), showing that the spin precessions are highly elliptical, as in Fig. 5.5. However, for magnons near the Brillouin zone boundaries, $\omega_{\alpha k} \approx \omega_{\beta k} \approx \gamma H_E$, $Q_{13}, Q_{14} \rightarrow 0$, and one can see from Eq. (5.90) that the ellipticities approach unity, corresponding to circular spin precessions.

## Problems

5.1. Using Eq. (5.4) for the energy of an easy-axis antiferromagnet with an external field applied along the symmetry direction: (a) Calculate the polar angles of the sublattice magnetizations for the two possible equilibrium configurations. (b) Find the field above which the energy in the spin-flop phase is smaller than in the antiferromagnetic phase.

5.2. In an easy-axis two sublattice antiferromagnet with a Hamiltonian given by Eq. (5.28), use the transformations in Eqs. (5.29a–5.29c) and (5.30a–5.30c) to show that the Hamiltonian quadratic in spin-deviation operators has the form in Eq. (5.31).

5.3. Consider the quadratic Hamiltonian (5.34) with the boson operators for the easy-axis antiferromagnet, with a magnetic field applied along the symmetry axis. Using the transformations in Eq. (5.33), show the Hamiltonian becomes diagonal in the magnon operators if the magnon frequencies are given by Eq. (5.37a) and (5.37b) and the coefficients of the transformations by Eqs. (5.38a) and (5.38b).

5.4. Considering for $MnF_2$ the parameters $H_E = 526$ kOe, $H_A = 8.2$ kOe and $g = 2.0$, calculate the AFMR frequency for a field $H_0 = 200$ kOe applied along the $c$-axis, and compare with the experimental data in Fig. 5.4b.

5.5. In the body-centered tetragonal structure of $MnF_2$ and $FeF_2$, the eight vectors connecting nearest neighbors magnetic ions in different sublattices are given by: $\vec{\delta} = \pm \hat{x}(a/2) \pm \hat{y}(a/2) \pm \hat{z}(c/2)$. Show that the geometric structure factor is given by expression (5.39).

5.6. (a) Using the magnon dispersion relation Eqs. (5.37a) and (5.37b) for an easy-axis two-sublattice AF, calculate the magnon group velocity versus wave number, for $\vec{k} = \hat{x}k$. (b) Calculate the maximum value of the magnon group velocity in $MnF_2$ using the parameters $H_E = 526$ kOe, $H_A = 8.2$ kOe and $g = 2.0$.

5.7. Show that in an easy-axis two-sublattice AF subject to a static field $\vec{H}_0$ parallel to the symmetry axis, and a rotating rf magnetic field $\vec{h}^\pm$ with frequency $\omega$, applied perpendicularly to $\vec{H}_0$, the susceptibility is given by the expression (5.46).

5.8. Consider the following parameters for antiferromagnetic $FeF_2$: $H_E = 540$ kOe, $H_A = 200$ kOe, $g = 2.22$, and $\varepsilon = 5.4$. Calculate the AFMR frequency and the frequency for which the polariton wave number is $k = 0$, for the down-going mode, for a field $H_0 = 100$ kOe applied along the $c$-axis.

5.9. In a hard-axis two sublattice antiferromagnet, with a spin Hamiltonian given by Eq. (5.82), use the transformations to collective spin-deviation boson operators to show that the quadratic Hamiltonian has the form in Eq. (5.83).

5.10. Considering for NiO the parameters $H_E = 9\ 684$ kOe, $H_{Ax} = 6.35$ kOe and $H_{Az} = 0.11$ kOe, calculate the minimum value of the magnetic field applied in a $< 11\overline{2} >$ direction for which the AF phase becomes unstable.

## References

1. Néel, L.: Influence des fluctuations des champs moléculaires sur les propriétés magnétiques des corps. Ann. Phys. **18**, 5 (1932)
2. Baibich, M.N., Broto, J.M., Fert, A., Nguyen Van Dau, F., Petroff, F., Etienne, P., Creuzet, G., Friederich, A., Chazelas, J.: Giant magnetoresistance of (001)Fe/(001)Cr magnetic superlattices. Phys. Rev. Lett. **61**, 2472 (1988)
3. Binasch, G., Grünberg, P., Saurenbach, F., Zinn, W.: Enhanced magnetoresistance in layered magnetic structures with antiferromagnetic interlayer exchange. Phys. Rev. B. **39**, 4828 (1989)
4. Hoffmann, A., Bader, S.D.: Opportunities at the frontiers of spintronics. Phys. Rev. Appl. **4**, 047001 (2015)
5. Nogués, J., Schuller, I.K.: Exchange bias. J. Magn. Magn. Mater. **192**, 203 (1999)
6. Jungwirth, T., Marti, X., Wadley, P., Wunderlich, J.: Antiferromagnetic spintronics. Nat. Nano. **11**, 231 (2016)
7. Jungfleisch, M.B., Zhang, W., Hoffmann, A.: Perspectives of antiferromagnetic spintronics. Phys. Lett. A. **382**, 865 (2018)
8. Baltz, V., Manchon, A., Tsoi, M., Moriyama, T., Ono, T., Tserkovnyak, Y.: Antiferromagnetic spintronics. Rev. Mod. Phys. **90**, 015005 (2018)
9. Nikotin, O., Lindgard, P.A., Dietrich, O.W.: Magnon dispersion relation and exchange interactions in $MnF_2$. J. Phys. C. **2**, 1168 (1969)
10. Barak, J., Jaccarino, V., Rezende, S.M.: The magnetic anisotropy of $MnF_2$ at 0 K. J. Mag. Mag. Mat. **9**, 323 (1978)
11. Shapira, Y., Foner, S.: Magnetic phase diagram of $MnF_2$ from ultrasonic and differential magnetization measurements. Phys. Rev. B. **1**, 3083 (1970)
12. Rezende, S.M., King, A.R., White, R.M., Timbie, J.P.: Stability limit of the antiferromagnetic phase near the spin-flop boundary in $MnF_2$. Phys. Rev. B. **16**, 1126 (1977)
13. Wu, S.M., Zhang, W., Amit, K.C., Borisov, P., Pearson, J.E., Jiang, J.S., Lederman, D., Hoffmann, A., Bhattacharya, A.: Antiferromagnetic spin Seebeck effect. Phys. Rev. Lett. **116**, 097204 (2016)
14. Jaccarino, V., King, A.R., Motokawa, M., Sakakibara, T., Date, M.: Temperature dependence of $FeF_2$ spin flop field. J. Mag. Mag. Mat. **31–34**, 1117 (1983)
15. Roth, W.L.: Magnetic structures of MnO, FeO, CoO, and NiO. Phys. Rev. **110**, 1333 (1958)
16. Hutchings, M.T., Samuelsen, E.J.: Measurement of spin-wave dispersion in NiO by inelastic neutron scattering and its relation to magnetic properties. Phys. Rev. B. **6**, 3447 (1972)
17. Machado, F.L.A., Ribeiro, P.R.T., Holanda, J., Rodríguez-Suárez, R.L., Azevedo, A., Rezende, S.M.: Spin-flop transition in the easy-plane antiferromagnet niquel oxide. Phys. Rev. B. **95**, 104418 (2017)
18. Kittel, C.: Theory of antiferromagnetic resonance. Phys. Rev. **82**, 565 (1951)
19. Johnson, F.M., Nethercot Jr., A.H.: Antiferromagnetic Resonance in $MnF_2$. Phys. Rev. **114**, 705 (1959)
20. Hagiwara, M., Katsumata, K., Yamaguchi, H., Tokunaga, M., Yamada, I., Gross, M., Goy, P.: A complete frequency-field chart for the antiferromagnetic resonance in $MnF_2$. Int. J. Infrared Millimeter Waves. **20**, 617 (1999)
21. Rezende, S.M., Azevedo, A., Rodríguez-Suárez, R.L.: Introduction to antiferromagnetic magnons. J. Appl. Phys. **126**, 151101 (2019)
22. Bayrakci, S.P., Keller, T., Habicht, K., Keimer, B.: Spin-wave lifetimes throughout the brillouin zone. Science. **312**, 1926 (2006)

23. Ohlmann, R.C., Tinkham, M.: Antiferromagnetic resonance in $FeF_2$ at far-infrared frequencies. Phys. Rev. **123**, 425 (1961)
24. Sanders, R.W., Belanger, R.M., Motokawa, M., Jaccarino, V., Rezende, S.M.: Far-infrared laser study of magnetic polaritons in $FeF_2$ and Mn impurity mode in $FeF_2$.Mn. Phys. Rev. B. **23**, 1190 (1981)
25. Fleury, P.A., Porto, S.P.S., Cheesman, L.E., Guggenheim, H.J.: Light scattering by spin waves in $FeF_2$. Phys. Rev. Lett. **17**, 84 (1966)
26. Hutchings, M.T., Rainford, B.D., Guggenheim, H.J.: Spin waves in antiferromagnetic $FeF_2$. J. Phys. C: Solid St. Phys. **3**, 307 (1970)
27. Fleury, P.A., Loudon, R.: Scattering of light by one- and two-magnon excitations. Phys. Rev. **166**, 514 (1968)
28. Mills, D.L., Burnstein, E.: Polaritons: the electromagnetic modes of media. Rep. Prog. Phys. **37**, 817 (1974)
29. Keffer, F., Kittel, C.: Theory of antiferromagnetic resonance. Phys. Rev. **85**, 329 (1952)
30. Rezende, S.M., White, R.M.: Multimagnon theory of antiferromagnetic resonance relaxation. Phys. Rev. B. **14**, 2939 (1976)
31. Rezende, S.M., White, R.M.: Spin wave lifetimes in antiferromagnetic $RbMnF_3$. Phys. Rev. B. **18**, 2346 (1978)
32. Oguchi, T.: Theory of spin-wave interactions in ferro- and antiferromagnetism. Phys. Rev. **117**, 117 (1960)
33. White, R.M., Sparks, M., Ortenburger, I.: Diagonalization of the antiferromagnetic magnon-phonon interaction. Phys. Rev. **139**, A450 (1965)
34. Grimsditch, M., McNeil, L.E., Lockwood, D.J.: Unexpected behavior of the antiferromagnetic mode of NiO. Phys. Rev. B. **58**, 14462 (1998)
35. Kampfrath, T., et al.: Coherent terahertz control of antiferromagnetic spin waves. Nat. Photon. **5**, 31 (2010)
36. Rezende, S.M., Rodríguez-Suárez, R.L., Azevedo, A.: Diffusive magnonic spin transport in antiferromagnetic insulators. Phys. Rev. B. **93**, 054412 (2016)

## Further Reading

Akhiezer, A.I., Bar'yakhtar, V.G., Peletminskii, S.V.: Spin Waves. North-Holland, Amsterdam (1968)
de Jongh, L.J., Miedema, A.R.: Experiments on Simple Magnetic Model Systems. Barnes & Noble Books, New York (1974)
Gurevich, A.G., Melkov, G.A.: Magnetization Oscillations and Waves. CRC, Boca Raton (1994)
Keffer, F.: Spin Waves. In: Flugge, S. (ed.) Handbuch der Physik, vol. XVIII/B. Springer, Berlin (1966)
Kittel, C.: Introduction to Solid State Physics, 8th edn. Wiley, New York (2004)
Kittel, C.: Quantum Theory of Solids, 2nd edn. Wiley, New York (1987)
Lax e, B., Button, K.: Microwave Ferrites and Ferrimagnetics. McGraw-Hill, New York (1962)
Nagamiya, T., Yosida, K., Kubo, R.: Antiferromagnetism. Advances in Physics. Taylor & Francis, New York (1955)
White, R.M.: Quantum Theory of Magnetism, 3rd edn. Springer, Berlin (2007)

# Magnon Excitation and Nonlinear Dynamics

**6**

This chapter is devoted to the study of linear and nonlinear processes and techniques for the excitation of magnons. In the presentation of the linear excitation process we employ three different quantum approaches that are commonly used and show that they are all consistent. The last one serves to show that the magnons excited by microwave radiation are in quantum coherent states. The theory is used to interpret experimental measurements. A quantum approach is also used to study the parametric excitation of magnons by three different techniques, parallel pumping, perpendicular pumping in the first-order and in the second-order Suhl processes. The four-magnon interaction is discussed in the third section and shown to have three important effects: saturation of the parametric magnon population; coherence of the parametric magnon states; and nonlinear dynamics. The last section is devoted to study the nonlinear dynamics of parametric magnons that manifests in self-oscillations in the magnon populations and chaotic behavior.

## 6.1 Linear Excitation of Magnons

### 6.1.1 Theory

The techniques most widely used to study long wavelength spin waves in ferromagnets employ microwave radiation, as described in Sect. 2.6. One advantage of the microwave techniques is that the rf magnetic field of the radiation couples directly with the magnetic moments, so that the conversion from electromagnetic into spin wave energy is quite efficient. In the linear excitation mechanism, considered here, the driven spin wave has the same frequency and wave vector as the radiation. In this section we shall use a quantum approach to study the linear process to excite spin waves.

As described in previous chapters, spin waves can be generated in magnetic samples by means of a microwave magnetic field applied perpendicularly to the static magnetic field. Our previous treatment of this process used a semiclassical

© Springer Nature Switzerland AG 2020
S. M. Rezende, *Fundamentals of Magnonics*, Lecture Notes in Physics 969,
https://doi.org/10.1007/978-3-030-41317-0_6

approach based on the Landau–Lifshitz equation. As we have seen, the oscillating torque of the rf field sets the magnetization to precess around the equilibrium direction as long as the driving and spin wave frequencies are close enough. In the quantum formulation the first task is to obtain the perturbation Hamiltonian due the driving field. We consider a time-varying magnetic field $\hat{x}h(\vec{r}, t)$ applied at $t = 0$, perpendicularly to the static field $\hat{z}H_z$. The origin of the perturbation Hamiltonian lies in the Zeeman interaction of the spins

$$H_I(t) = -\gamma\hbar\sum_i \vec{S}_i \cdot \hat{x}h(\vec{r}, t). \tag{6.1}$$

Considering a microwave magnetic field with frequency $\omega$, introducing in Eq. (6.1) the transformations (3.22) and (3.30), the Hamiltonian for $t > 0$ can be written as

$$H_I(t) = -\frac{\gamma\hbar}{2}(2S/N)^{1/2}\sum_i\sum_k (e^{ikr_i}a_k + e^{-ikr_i}a_k^\dagger)h(\vec{r})\sin\omega t. \tag{6.2}$$

Using the transformation (3.53) to magnon operators and keeping the relevant terms, the interaction Hamiltonian becomes

$$H_I(t) = \theta(t)\sum_k [g_k c_k e^{i\omega t} + H.c.], \tag{6.3}$$

where $\theta(t)$ is the Heaviside function (0 for $t < 0$ and 1 for $t \geq 0$),

$$g_k = i\gamma\hbar(SN/8)^{1/2}(u_k - v_k^*)h(\vec{k}), \tag{6.4}$$

and

$$h(\vec{k}) = \sum_i e^{i\vec{k}\cdot\vec{r}}h(\vec{r}_i) \tag{6.5}$$

is the space Fourier transform of the driving field. Using time-dependent perturbation theory one can show that, for $t \gg 1/\eta_k$, where $\eta_k$ is the magnon relaxation rate, the transition probability per unit time for the system to evolve from the vacuum to a state with $\vec{k}$ magnons is

$$W_{0\to k} = \frac{2\pi}{\hbar^2}|H_{0k}'|^2\delta(\omega - \omega_k), \tag{6.6}$$

where the matrix element is

$$H_{0k}' = i\gamma\hbar(SN/8)^{1/2}(u_k - v_k)h(\vec{k}). \tag{6.7}$$

Using the property of the Dirac delta function

$$\delta(x) = \frac{1}{2\pi i} \lim_{\varepsilon \to 0^+} \left( \frac{1}{x - i\varepsilon} - \frac{1}{x + i\varepsilon} \right), \tag{6.8}$$

we write Eq. (6.6) as

$$W_{0 \to k} \approx \frac{2}{\hbar^2} |H'_{0k}|^2 \left[ \frac{\eta_k}{(\omega - \omega_k)^2 + \eta_k^2} \right], \tag{6.9}$$

where $\eta_k \ll \omega_k$. Using $M = NSg\mu_B/V$ and considering that magnons have energy $\hbar\omega_k$, we can write the power absorbed by the $k$-magnons from the microwave field as

$$P_k = \frac{V}{8\pi} \omega_k |(u_k - v_k)|^2 \chi''_{xx} |h(\vec{k})|^2, \tag{6.10}$$

where

$$\chi''_{xx} = \frac{\omega_M \eta_k / 2}{(\omega - \omega_k)^2 + \eta^2} \tag{6.11}$$

is the imaginary part of the *ac* susceptibility, as in Eq. (1.30). Equation (6.10) is consistent with Eq. (2.97), obtained with a semiclassical treatment for the $k = 0$ magnon. Here the factor $(u_k - v_k)^2$ accounts for the ellipticity introduced by the dipolar interaction, and the extra $4\pi$ is due to the relation $\omega_M = \gamma 4\pi M$ valid in the CGS system used here. As clearly expressed in Eqs. (6.11) and (6.12), in order to excite magnons with frequency $\omega_k = \omega$ and wave vector $\vec{k}$, the Fourier transform of the driving field must have a significant component with the same wave vector. The process described here, in which photons are linearly converted into magnons, is illustrated by the diagram in Fig. 6.1a. The power absorbed by magnons in this process is proportional to the imaginary part of the susceptibility. Thus, its frequency dependence, given by Eq. (6.11), has the shape of a Lorentzian line, shown in Fig. 6.1b. The full-width of the line at half-maximum is related to the relaxation rate by $\Delta\omega_F = 2\eta_k$.

It is instructive to present another quantum approach of the linear excitation of magnons by a microwave field. Consider the total Hamiltonian for the magnon system driven by the microwave field

**Fig. 6.1** (a) Diagramatic representation of the linear photon–magnon conversion in ferromagnets. (b) Magnon power absorption line as a function of the frequency of the microwave driving field

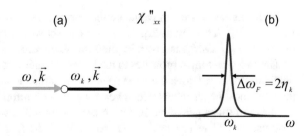

$$H(t) = H_0 + H_I(t), \tag{6.12}$$

where $H_0 = \sum_k \hbar \omega_k c_k^\dagger c_k$ is the unperturbed Hamiltonian and $H_I(t)$ is the time-dependent interaction Hamiltonian, given by Eqs. (6.3)–(6.5). In the Heisenberg picture, we can obtain the evolution of the magnon annihilation operator with the equation of motion

$$i\hbar \frac{d}{dt} c_k = [c_k, H(t)]. \tag{6.13}$$

Using the commutation relations (3.54) and the Hamiltonian (6.12), we obtain for $t \geq 0$

$$\frac{dc_k}{dt} = -i\omega_k c_k - \frac{i}{\hbar} g_k^* e^{-i\omega t}. \tag{6.14}$$

The damping can be introduced phenomenologically in this equation by means of the relaxation rate $\eta_k$, so that we have

$$\frac{dc_k}{dt} = -i\omega_k c_k - \eta_k c_k - \frac{i}{\hbar} g_k^* e^{-i\omega t}, \tag{6.15}$$

which gives for the expectation value of $c_k$ in the stationary steady state

$$\langle c_k \rangle = \frac{g_k^*}{\hbar[(\omega - \omega_k) + i\eta_k]} e^{-i\omega t}. \tag{6.16}$$

The equation of motion for the number of $k$-magnons is

$$\frac{dn_k}{dt} = \frac{dc_k^\dagger}{dt} c_k + c_k^\dagger \frac{dc_k}{dt}. \tag{6.17}$$

Using in this expression the result (6.16) and its Hermitian conjugate, one can show that in the steady state ($dn_k/dt = 0$) the number of $k$-magnons excited by the driving field is

$$\langle n_k \rangle = \frac{|g_k|^2/\hbar^2}{[(\omega - \omega_k)^2 + \eta_k^2]}. \tag{6.18}$$

With this result we can calculate the energy stored in the system of $k$-magnons $E_k = \hbar \omega_k \langle n_k \rangle$. Then, with Eq. (6.4), one can show that the power absorbed by the magnons, $P_k = 2\eta_k E_k$, is given by the same expression as in Eq. (6.10).

The two quantum approaches to the linear excitation of magnons by a microwave field, just presented, give the same result as the semiclassical one and are sufficient to interpret most experiments. However, they give no information about the nature of the magnon states generated in the process. We shall now present another quantum formulation of the process aimed explicitly in finding the nature of the driven quantum states. We consider the total Hamiltonian for the magnon system driven by the microwave field given by Eq. (6.12). In the Schrodinger picture, the state resulting from the driving process at an instant $t$ is related to the state at $t = 0$ through the time evolution operator

$$|t\rangle = U(t,0)|t = 0\rangle. \tag{6.19}$$

The evolution operator is conveniently calculated in the interaction picture, in which it is separated in two parts (see Cohen-Tannoudji 1977)

$$U(t,0) = U_0(t,0)U_I(t,0), \tag{6.20}$$

where the unperturbed part is

$$U_0(t,0) = e^{-iH_0 t/\hbar}. \tag{6.21}$$

and the intermediate evolution operator is governed by

$$i\hbar \frac{d}{dt} U_I = H_I(t)U_I. \tag{6.22}$$

Equation (6.22) with (6.3) and (6.4) can be solved analytically [1] to give for the evolution operator in Eq. (6.20)

$$U(t,0) = e^{-iH_0 t/\hbar + i\beta(t)} \exp\left[\sum_k (\alpha_k c_k^\dagger - \alpha_k^* c_k)\right], \tag{6.23}$$

where $\beta(t)$ is an unimportant phase and $\alpha_k$ is a c-number function given by

$$\alpha_k = \frac{1}{i\hbar} \int_0^t g_k^* e^{-i(\omega - \omega_k)t'} dt'. \tag{6.24}$$

Introducing the damping in the usual way, $\omega_k \to \omega_k - i\eta_k$, and performing the integration in Eq. (6.24), we have, for $t \gg 1/\eta_k$,

$$\alpha_k = -\frac{g_k^*/\hbar}{(\omega - \omega_k + i\eta_k)}. \tag{6.25}$$

The last exponential operator in Eq. (6.23) is the product of displacement operators for coherent states defined by Eq. (3.97). Therefore, if prior to $t = 0$ the state of the system is the vacuum, the driving field excites coherent magnon states

$$|t\rangle = U(t,0)|0\rangle = e^{-iH_0 t/\hbar + i\beta(t)} \prod_k |\alpha_k\rangle,$$

which can be written as

$$|t\rangle = e^{i\beta(t)} \prod_k |\alpha_k e^{-i\omega_k t}\rangle, \tag{6.26}$$

where, with Eqs. (6.4) and (6.25), the coherent state eigenvalue becomes

$$\alpha_k = i\gamma \, (SN/8)^{1/2} \frac{(u_k - v_k)}{(\omega - \omega_k + i\eta_k)} h(\vec{k}). \tag{6.27}$$

Equation (6.26) represents a product of the displacement operators defined in Eq. (3.97). This result shows that the transverse microwave driving field excites a superposition of coherent magnon states with eigenvalues given by Eq. (6.27), with wave vectors determined by the Fourier transform of the driving field. Thus, the driven states correspond to the classical view of the spin wave, with the spins precessing about the equilibrium direction with frequency $\omega_k = \omega$, and a phase that varies in space with wave vector $\vec{k}$. Note that the eigenvalue in Eq. (6.27) is proportional to the amplitude of the driving field, characterizing a linear excitation process. Using Eqs. (3.94) and (6.27), we obtain for the number of excited magnons in the coherent states

$$\langle n_k \rangle = |\alpha_k|^2 = \frac{\gamma^2 SN}{8} \frac{|(u_k - v_k)|^2}{(\omega - \omega_k)^2 + \eta_k^2} |h(\vec{k})|^2, \tag{6.28}$$

which is identical to the expression obtained by substituting Eq. (6.4) in (6.18).


## 6.1.2  Experiments for Linear Excitation of Magnons

In the experiments for driving spin waves by microwave radiation, described in Sects. 1.5 and 2.6, the main objective was to measure the field positions and the lineshapes of the power absorbed by the magnetic sample. In these experiments, the intensity of the microwave field is not a relevant parameter. One uses microwave radiation with sufficient power to produce good signals, but not high enough so as to avoid spurious effects, such as heating or nonlinear effects. Here we are interested in learning about the number of magnons excited linearly by a microwave radiation, thus we need to obtain a relation between the field intensity and the power.

We shall consider three structures to excite the ferromagnetic sample with microwave radiation. Figure 6.2a shows a hollow rectangular waveguide with a metallic wall at the end, that can be used in two ways: The first as a shorted waveguide, with an open input, to operate in a broad frequency range, determined by the band-pass of the waveguide; the second as a resonant microwave cavity, having a metallic wall also in the input side, with a small aperture to couple the external radiation to the cavity. In the latter case, one has to work at a single frequency, coinciding with the cavity resonance frequency, determined by the dimensions of the waveguide. Figure 6.2b shows another common structure, a microstrip line, made of a narrow metallic strip on a dielectric plate having a ground

**Fig. 6.2** Schematic illustration of structures used to excite the ferromagnetic resonance (FMR) with microwave radiation. (**a**) Hollow rectangular shorted waveguide. (**b**) Microstrip line used for broadband measurements

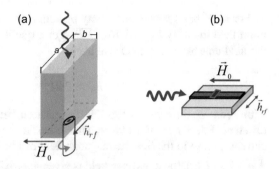

plane on the other side. In this case, the line supports the TEM mode that has no cutoff frequency, so that the operation can be done in a very wide frequency range.

First, we consider the structure with shorted rectangular waveguide operating in the $TE_{10}$ mode. In this mode the rf magnetic has maximum intensity in the middle of the transverse cross section, and direction along the long transverse lateral dimension, as indicated in Fig. 6.2a. Using the expressions for the electric and magnetic fields in this mode, one can show (see Pozar 2012) that in the middle of the short wall, the relation between the magnetic field intensity and the input power $P_i$ is, in SI units

$$h_{rf} = \frac{8[P_i(1 - \lambda_0^2/4a^2)^{1/2}]^{1/2}}{[(\mu_0/\varepsilon_0)^{1/2}ab]^{1/2}}, \tag{6.29}$$

where $a$ and $b$ are the long and short inner dimensions, shown in Fig. 6.2a, $\mu_0$ and $\varepsilon_0$ are the permeability and permittivity of free-space, and $\lambda_0 = c/f$ is the electromagnetic wavelength at the operating frequency $f$. Let us calculate the field in a shorted X-band waveguide, with standard dimensions $a = 2.3$ and $b = 1.1$ cm, that operates in the frequency range 8–12 GHz. For a frequency $f = 9.4$ GHz, the free space wavelength is $\lambda_0 = 3.19$ cm. With these numbers in Eq. (6.29), we obtain for the field intensity $h_{rf} \approx 20P_i^{1/2}$. Converting the field unit from A/m to Oe we have

$$h_{rf} \approx 0.25P_i^{1/2} \text{ Oe}, \tag{6.30}$$

where $P_i$ is in W. In the case of a rectangular cavity operating in the $TE_{102}$ mode, the magnetic field intensity is maximum at the short wall and also in the middle of the cavity. For a cavity with resonance frequency $f$ and loaded quality factor $Q_L$, one can show that the relation between the field intensity and the input power is

$$h_{rf} = \frac{[8P_iQ_L(1 - \lambda_0^2/4a^2)^{3/2}]^{1/2}}{[(\mu_0/\varepsilon_0)^{1/2}\pi ab]^{1/2}}. \tag{6.31}$$

For an X-band waveguide cavity operating at 9.4 GHz this gives for the maximum field intensity $h_{rf} \approx 3.16\sqrt{Q_L P_i}$. For a quality factor $Q_L = 3000$, converting the field unit from A/m to Oe we have

$$h_{rf} \approx 2.2 P_i^{1/2} \text{ Oe}, \tag{6.32}$$

with $P_i$ in W. In a microstrip line the relation between field and power is very different. For a line of width $w$, characteristic impedance $Z_0$, with a microwave current $i$, close to the line the magnetic field can be calculated with Ampère's law, $\oint \vec{h} \cdot \vec{dl} = i$. So, the transverse field is given approximately by $h_{rf} = i/2w$. Since the power is $P_i = Z_0 i^2$, the relation between field and power for a matched microstrip is

$$h_{rf} = \frac{1}{2w Z_0^{1/2}} P_i^{1/2}. \tag{6.33}$$

Note that the field does not depend on the frequency. For a microstrip of width $w$ in mm, characteristic impedance $Z_0 = 50\,\Omega$, Eq. (6.33) gives for the field very close to the line

$$h_{rf} = \frac{70.7}{w} P_i^{1/2}. \tag{6.34}$$

Considering a line of $w = 0.5$ mm and converting the field unit to Oe we have

$$h_{rf} \approx 1.77 P_i^{1/2}, \tag{6.35}$$

with $P_i$ in W. This value is similar to the one obtained with a cavity with $Q_L = 2000$ at 9.4 GHz.

To calculate the number of magnons linearly excited by a microwave radiation we consider a spherical YIG sample with diameter 0.5 mm, placed in the back wall of a shorted X-band waveguide, as in Fig. 6.2a. For a driving frequency of 9.4 GHz and input power of 100 mW, Eq. (6.30) gives $h_{rf} = 7.9 \times 10^{-2}$ Oe. Since the driving field is approximately uniform over the sample, we assume that only the uniform precession mode is excited. In Eq. (6.28) we use $h(k) \approx h(0) = h_{rf}$ and $(u_0 - v_0) = 1$, and consider in the volume of the YIG sample $N \approx 3.5 \times 10^{16}$ unit cells with lattice parameter $a = 1.23$ nm and spin $S = 20$. Assuming for YIG a FMR linewidth of 0.5 Oe and using the relation $\eta_0 = \gamma \Delta H$, Eq. (6.28) gives for the number of $k = 0$ magnons with $\omega_0 = \omega$ excited in the sample $\langle n_0 \rangle \approx 2 \times 10^{15}$. This very large number explains why magnonic quantum effects are not relevant in usual FMR experiments.

The experimental observation of quantum effects in magnon systems requires the use of elaborate methods to excite and detect magnons and work at very low temperatures [2–4]. We present here some results obtained with a spherical YIG sample with diameter 0.5 mm, cooled to a temperature of 10 mK, where very few thermally excited photons and magnons exist, reported in Lachance-Quirion et al. [4]. The authors employed the hybrid system illustrated in Fig. 6.3a to excite and detect magnons.

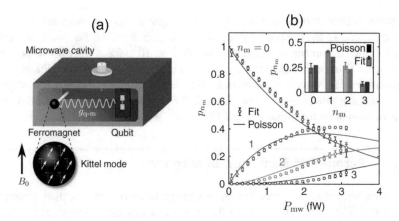

**Fig. 6.3** (a) Schematic illustration of a YIG sphere and a superconducting qubit inside a three-dimensional microwave cavity. A magnetic field $B_0$ is applied to the YIG sphere using permanent magnets and a coil. The magnetostatic mode in which spins uniformly precess in the ferromagnetic sphere, or the Kittel mode, couples to the magnetic field of the cavity modes. The qubit and the Kittel modes interact through virtual excitations in the cavity modes [4]. (b) Probability $p_{nm}$ of the first four magnon number states as a function of microwave power $P_{mw}$. Poisson distributions are shown as solid lines for $n_m > 0$. Inset: $p_{nm}$ for $P_{mw} = 3.1$ fW [4]. Reproduced and adapted from D. Lachance-Quirion et al., Sci. Adv. **3**, 1603150 (2017), with permission of the American Association for the Advancement of Science

The system consists of a YIG sphere subject to an external static magnetic field and a superconducting qubit, both placed in the same microwave cavity. The superconducting qubit is made of two Josephson junctions that behave as a two-level system, connected to two pads that act as an electric dipole antenna [3–5]. The YIG sphere is located in a position of maximum magnetic field of the $TE_{102}$ cavity mode, so that it couples to the cavity through the magnetic dipole of the $k = 0$ magnon, or Kittel mode, while the qubit couples to the same mode by its electric dipole. The coupling of the qubit and the magnon to the same cavity modes creates an effective interaction between them, mediated by virtual microwave photons [3]. In this way, the state of the qubit is sensitive to the number of magnons.

The magnon system is excited by the rf magnetic field of the cavity, fed with microwave frequency $\omega$ close to the $TE_{102}$ mode resonance. By tuning the applied field to a value such that $\omega_0 = \gamma H_0 \approx \omega$, magnons are excited in a state that can be probed by the qubit response, observed by changes in the cavity reflection measured at another microwave frequency. For each value of the input microwave power $P_{mw}$, the qubit spectrum exhibits peaks at frequencies higher than the zero-magnon peak. As shown in [4], these peaks correspond to different numbers of uniform precession magnons. By fitting an analytical model for the spectrum of a qubit dispersively coupled to a harmonic oscillator to the data, the authors obtain the probability $p_{nm}$ of having $n_m$ magnons in the excited mode. Figure 6.2b shows that the measured probability distributions for $n_m > 0$ are well represented by Poisson functions, as in Eq. (3.95), with a maximum that shifts to higher values as the number of magnons increases. This shows that the excited magnons are in coherent states, as demonstrated in the previous subsection. Note also that with Eq. (6.28) we calculated earlier a number of magnons $\langle n_0 \rangle \approx 2 \times 10^{15}$ for a microwave power of 100 mW. Thus, although our calculation was not done for the conditions of the experiments of Lachance-Quirion et al. [4], we expect that to excite a few magnons the necessary power is very low. Indeed, the microwave power level used in the experiments is on the order of a few fW ($10^{-15}$ W).

## 6.2  Parametric Excitation of Magnons

Magnons can also be generated by nonlinear processes, also called parametric processes. The origin of this name lies in the fact that the processes are based on the variation of a parameter of the medium, usually with a frequency different from the oscillation frequency. The nonlinear, or parametric, processes can also be used to amplify an existing wave, or oscillation. Nonlinear behavior in ferromagnets was first observed in microwave magnetic resonance experiments in the early 1950s. Bloembergen and Damon [6–8] observed the appearance of a subsidiary absorption and a premature saturation of the main resonance when the microwave power exceeded certain threshold levels. A few years later, Harry Suhl [9] produced a theory to explain those intriguing observations based on the parametric generation of spin waves. Later it was predicted theoretically by Morgenthaler [10] and Schlömann [11] that spin waves could also be parametrically pumped by a microwave field parallel to the static magnetic field. This was called parallel pumping to distinguish from the perpendicular pumping processes observed in the FMR configuration. In this section we shall study these three processes, starting with the parallel pumping because it is the simplest one.

### 6.2.1  Parallel Pumping Process

As pointed out in Sect. 2.2.2, the presence of the dipolar interaction in spin waves propagating at an angle with the magnetic field produces an ellipticity in the magnetization precession about the field. As a result, in a spin wave with frequency $\omega_k$, the $z$ component of the magnetization is not constant, it varies in time with frequency $2\omega_k$, as clearly seen in Eq. (2.40). Thus, if a driving microwave magnetic field is applied parallel to the static field, it can excite spin waves with half the driving frequency. As illustrated in Fig. 6.4, this is a process where one photon generates two magnons with half its frequency, so as to conserve energy. However, since this is a nonlinear process, magnons are excited in large numbers only if the rate at which they are created exceeds the relaxation rate. Thus, the process is characterized by a threshold pumping field, above which the magnon population grows exponentially, producing a spin wave instability.

The parallel pumping process can be studied with the semiclassical equations, or with a quantum approach. Here we will use the latter one because it is simpler for

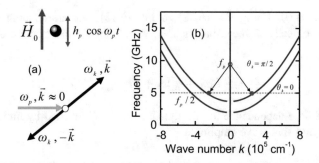

**Fig. 6.4** Illustration of the parallel pumping process. (**a**) Pumping field configuration and diagramatic representation of the excitation of two magnons by one microwave photon. (**b**) Magnon dispersion relations for YIG with an internal field of $H_z = 0.7$ kOe, showing the generation of two magnons with half the frequency of the microwave photon

introducing the magnon nonlinear interactions. We consider a microwave magnetic pumping field $\hat{z}h_p \cos \omega_p t$ applied parallel to the static field $\hat{z}H_z$. This creates a perturbation Hamiltonian that originates in the Zeeman interaction of the spins

$$H_I(t) = -\gamma\hbar\sum_i \vec{S}_i \cdot \hat{z}h_p\cos\omega_p t. \tag{6.36}$$

Introducing in Eq. (6.36) the transformations (3.22) and (3.30), the Hamiltonian can be written as

$$H_I(t) = \gamma\hbar\sum_k a_k^\dagger a_k h_p\cos\omega_p t. \tag{6.37}$$

Using the transformation (3.53) to magnon operators and keeping the relevant terms, the interaction Hamiltonian becomes

$$H_I(t) = -\frac{1}{2}\hbar\sum_k h_p\rho_k e^{-i\omega_p t}c_k^\dagger c_{-k}^\dagger + H.c., \tag{6.38}$$

where $\rho_k = u_k v_k$. Using Eqs. (3.51c), (3.71) and (3.73), one can show that the parameter representing the coupling of the driving field with the $\vec{k}, -\vec{k}$ magnon pair with frequency $\omega_k$ is

$$\rho_k = \frac{\gamma\omega_M}{4\omega_k}\sin^2\theta_k e^{-i2\varphi_k}, \tag{6.39}$$

where $\theta_k$ and $\varphi_k$ are the polar and azimuthal angles of the wave vector. Notice that the coupling is maximum for spin waves propagating perpendicularly to field, as expected, since they have the largest ellipticity. The evolution of the spin-wave amplitude can be derived from the Heisenberg equation with the total Hamiltonian

$H = H_0 + H_I(t)$. Using Eq. (6.38) in Eq. (6.13), and introducing dissipation phenomenologically in the usual manner, we obtain for the magnon annihilation operator

$$\frac{dc_k}{dt} = -(i\omega_k + \eta_k)c_k + i(h_p\rho_k)e^{-i\omega_p t}c_{-k}^{\dagger}. \tag{6.40}$$

The fast time dependence in this equation can be eliminated by introducing the slowly varying spin-wave amplitude

$$\tilde{c}_k = \langle c_k \rangle e^{i\omega_p t/2}, \tag{6.41}$$

for which Eq. (6.40) gives

$$\frac{d\tilde{c}_k}{dt} = -(i\Delta\omega_k + \eta_k)\tilde{c}_k + ih_p\rho_k\tilde{c}_{-k}^{\dagger}, \tag{6.42}$$

where $\Delta\omega_k = \omega_k - \omega_p/2$ is a detuning parameter. Equation (6.42) and its complex conjugate have solutions of the form $\tilde{c}_k, \tilde{c}_k^* \propto \exp(\gamma_k t)$, where

$$\gamma_k = -\eta_k + [h_p^2|\rho_k|^2 - \Delta\omega_k^2]^{1/2}. \tag{6.43}$$

Hence, the spin-wave amplitude grows exponentially in time ($\gamma_k > 0$) if the pumping field exceeds a critical value given by

$$h_{crit} = \frac{[\eta_k^2 + \Delta\omega_k^2]^{1/2}}{|\rho_k|}. \tag{6.44}$$

Using Eq. (6.39) and $\Delta\omega_k = 0$ in this equation, we obtain for the critical pumping field

$$h_{c\,min} = \frac{2\omega_p\eta_k}{\gamma\omega_M\sin^2\theta_k}\bigg|_{min}, \tag{6.45}$$

where the subscript min means the minimum value made possible by the maximum $\theta_k$ available for $\omega_k = \omega_p/2$. Figure 6.4b shows the magnon dispersion curves for yttrium iron garnet, calculated with Eq. (2.36), namely

$$\omega_k = \gamma\left[(H_z + Dk^2)(H_z + Dk^2 + 4\pi M\sin^2\theta_k)\right]^{1/2}, \tag{6.46}$$

assuming a static internal field of $H_z = 0.7$ kOe and using for YIG $\gamma/2\pi = 2.8$ GHz/kOe, $4\pi M = 1.76$ kG, and $D = 5.4 \times 10^{-12}$ kOe cm². Since the internal field is related to the applied field $H_0$ by $H_z = H_0 + H_A - N_z M$, for a spherical YIG sample the dispersion curves in Fig. 6.4b correspond to an applied field of $H_0 \approx 1.3$ kOe. The parallel pumping process is illustrated in the figure for a pump frequency $f_p = \omega_p/2\pi = 10$ GHz. The two circles represent the modes pumped at the minimum threshold, with frequency $f_p/2 = 5$ GHz and wave vectors with $k = 2.8 \times 10^5$ cm$^{-1}$ and $\theta_k = \pi/2$. Thus, by changing the applied static field and keeping fixed the pumping frequency, one can selectively excite modes with varying wave number. For this reason, the parallel pumping process is a very useful technique to measure the relaxation rate $\eta_k$ as a function of $k$ [12]. In YIG, at room temperature, $\eta_k/\gamma$ ~0.5 Oe, so that with a microwave pumping frequency $\omega_p \sim \omega_M$, the critical field in Eq. (6.45) is on the order of 1 Oe. As one can see in Eq. (6.32), this driving field can be obtained in typical X-band microwave cavities with $Q = 3000$ with an incident power on the order of 100 mW.

Figure 6.5a shows some results obtained in parallel pumping experiments aimed at studying nonlinear magnon dynamics [13]. The nonlinear dynamics will be described in the next section. Here we confine attention to measurements of the critical pumping field used to obtain the magnon relaxation rate. The experiments were carried out at room temperature, with a well-polished single crystal YIG sphere of diameter 1 mm. The sample was glued to an aluminum oxide rod and held at the center of a TE$_{102}$ rectangular microwave cavity with $Q = 3000$, with the rf magnetic field parallel to the applied static field. The power was provided by a pulsed microwave generator operating at the cavity resonance frequency of 9.4 GHz, amplified by a traveling-wave-tube amplifier, and directed to the cavity by a circulator, as in the FMR circuit shown in Fig. 1.7. The microwave signal reflected from the cavity is amplified and monitored by means of a diode detector connected to an oscilloscope. The experiment is done with discrete fixed values of the static field and varying microwave power. At low power levels the reflected signal is negligible. As the power is increased, an abrupt change in the signal occurs at the level for which the rf field reaches the critical value for driving the spin wave instability. In this way, one can measure the variation of the critical driving field, which is related to the power as in Eq. (6.31), as a function of the static field.

The measured critical field versus static field, shown in Fig. 6.5a, has a pattern that is called *butterfly curve*. The critical field on the left side of the curve

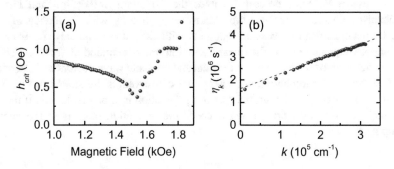

**Fig. 6.5** (a) Parallel pumping critical field as a function of the static field measured in a YIG sphere with diameter 1 mm pumped with frequency 9.4 GHz [13]. (b) Relaxation rate of magnons propagating perpendicularly to the field as a function of the wave number calculated from the data in (a) using Eqs. (6.45) and (6.47). (a) Reprinted with permission from F. M. de Aguiar and S. M. Rezende, Phys. Rev. Lett. **56**, 1070 (1986). Copyright (1986) by the American Physical Society

corresponds to the excitation of magnons with $\theta_k = \pi/2$. Thus, as the static field increases, the wave number decreases and so does the relaxation rate, resulting in a $h_{cmin}$ that decreases with the field. As the field is increased beyond the value for which the $\theta_k = \pi/2$ magnon has $k \approx 0$, $H_{zmin}$, the magnons that go unstable have $\theta_k < \pi/2$, so that $h_{crit}$ increases rapidly with increasing field as $\theta_k$ decreases. From the left side of the butterfly curve one can obtain the variation of relaxation rate, calculated with Eq. (6.45), with the wave number, for the $\theta_k = \pi/2$ magnon.

From Eq. (6.46) one can show that for a certain magnon frequency $\omega_k$, the wave number of the magnon with $\theta_k = \pi/2$ is given by

$$k = \left\{ \frac{-H_z - 2\pi M + [(2\pi M)^2 + \omega_k^2/\gamma^2]^{1/2}}{2H_z + 4\pi M} \right\}^{1/2}, \tag{6.47}$$

where $H_z = H_0 + H_A - N_z M$ and $\omega_k = \omega_p/2$. With this equation we can calculate the wave number $k$ as a function of $H_0$, for a given pumping frequency and $\sin\theta_k = 1$. From the data in Fig. 6.5a, using Eqs. (6.45) and (6.47) with $f_p = 9.4$ GHz, $H_A = 0.06$ kOe, and $N_z = 4\pi/3$, we obtain the wave number dependence of the relaxation rate shown in Fig. 6.5b. The linear variation of the relaxation rate with $k$ is very similar to the one reported in Kasuya and Le Craw [12], and, as discussed in Sect. 3.7.2, is explained by a three-magnon relaxation process.

## 6.2.2  Perpendicular Pumping: First-Order Suhl Process—Subsidiary Absorption

As mentioned in the introduction of this section, when the driving microwave field is perpendicular to the static field, two magnon instabilities processes are observed at high-power levels, that are called first- and second-order Suhl processes. Both processes require the existence of the magnon interactions presented in Sect. 3.4.

The first-order Suhl process involves the 3-magnon interaction and has the mechanism illustrated in Fig. 6.6. Microwave photons with frequency $\omega_p$ and $k \approx 0$ create uniform precession magnons with the same frequency, far from the resonance frequency $\omega_0$, by the linear excitation process studied in Sect. 6.1. These, in turn, decay into two magnons with equal and opposite wave vectors and frequency $\omega_p/2$, so that both momentum and energy are conserved, as illustrated in Fig. 6.6a. Since the process is nonlinear, the magnon population is appreciable only if the rate at which the magnons are created exceeds the relaxation rate, as in the parallel pumping process.

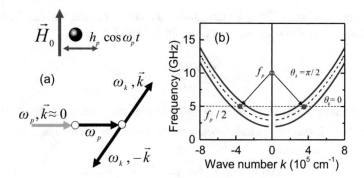

**Fig. 6.6** Illustration of the first-order Suhl instability pumping process. (**a**) Pumping field configuration and diagramatic representation of the excitation of two magnons by the $k = 0$ magnon generated by one microwave photon. (**b**) Magnon dispersion relations for YIG with an internal field of $H_z = 0.7$ kOe, showing the generation of two magnons with half the frequency of the microwave photon

We study this process using the equations for the magnon operators. First we use Eq. (6.15) to write the equation of motion for the annihilation operator of the $k = 0$ magnon, considering a uniform microwave driving field $h(0) = h_p$,

$$\frac{dc_0}{dt} = -i\omega_0 c_0 - \eta_0 c_0 - \gamma(SN/8)^{1/2} h_p e^{-i\omega_p t}, \tag{6.48}$$

where we have considered in Eq. (6.4) $u_0 - v_0 = 1$. In the stationary steady state we have $dc_0/dt = -i\omega_p c_0$, and this equation gives

$$c_0 = -\frac{i\gamma(SN/8)^{1/2}}{(\omega_p - \omega_0) + i\eta_0} h_p e^{-i\omega_p t}. \tag{6.49}$$

The uniform mode, or $k = 0$ magnon, creates a magnon pair $\vec{k}, -\vec{k}$, by means of the 3-magnon interaction arising from the dipolar coupling. From Eq. (3.103) we write the interaction Hamiltonian as

$$H^{(3)} = \frac{\hbar}{2}\sum_k V_{0k}^{(3)} c_0 c_k^\dagger c_{-k}^\dagger + H.c., \tag{6.50}$$

where

$$V_{0k}^{(3)} = -\frac{\omega_M}{\sqrt{2SN}}\sin 2\theta_k e^{i\varphi_k}. \tag{6.51}$$

If $\omega_p$ is very different from $\omega_0$, the uniform mode is driven far off-resonance, so that it behaves essentially like a virtual mode intermediate between the driving field and the magnon pairs. Thus, its amplitude acts like a classical variable. Assuming $\omega_p \approx 2\omega_0 \approx 2\omega_k$ and $\omega_p - \omega_0 \gg \eta_0$, we obtain from Eq. (6.49)

$$c_0 \approx -\frac{i\gamma(SN/2)^{1/2}}{\omega_p}h_p e^{-i\omega_p t}. \tag{6.52}$$

Substitution of this result in Eq. (6.50) gives for the interaction Hamiltonian for driving $\vec{k}, -\vec{k}$ magnon pairs by means of the 3-magnon interaction

$$\mathrm{H}_I(t) = -\frac{\hbar}{2}\sum_k h_p \rho_{\perp k} e^{-i\omega_p t} c_k^\dagger c_{-k}^\dagger + H.c., \tag{6.53}$$

where

$$\rho_{\perp k} = -i\frac{\gamma\omega_M}{2\omega_p}\sin 2\theta_k e^{i\varphi_k}. \tag{6.54}$$

Comparison of these equations with Eqs. (6.38) and (6.39) shows that the first-order Suhl instability is described by the same equation as in the parallel pumping. Thus, the critical driving field for the first-order Suhl process obtained with Eqs. (6.44) and (6.54), for $\omega_p = 2\omega_k$, is given by

$$h_{c\min} = \left.\frac{2\omega_p \eta_k}{\gamma\omega_M \sin 2\theta_k}\right|_{\min}. \tag{6.55}$$

Note that except for the angle factor, this expression is the same as the one for the parallel pumping process in Eq. (6.45). While in the parallel pumping process the modes with minimum threshold have $\theta_k = \pi/2$, here they have $\theta_k = \pi/4$. As indicated in Fig. 6.6b, in a YIG sample with an internal field $H_z = 0.7$ kOe, a pump field with frequency $f_p = 10$ GHz at the threshold excites magnon pairs with frequency 5 GHz and wave vectors with $k = 3.6 \times 10^5$ cm$^{-1}$ and $\theta_k = \pi/4$. Since the FMR frequency lies within the manifold band and the pumping frequency is much higher, the field value for which magnons are generated by the first-order Suhl process is quite smaller than the FMR field. The power absorption by the unstable magnons results in an extra peak in the FMR spectrum at high powers, at a field below the one for the main resonance. Figure 6.7a shows an FMR spectrum measured at 9.4 GHz in a single crystal sphere of nickel ferrite [14]. At low microwave power, only the main resonance is observed at a field of 2.6 kOe. At higher power, a peak at ~1.8 kOe shows up, that is called *subsidiary resonance*. Figure 6.7b shows the critical field for the subsidiary resonance versus static field intensity, measured in a YIG sphere with diameter 1 mm pumped by microwave radiation with frequency 8.7 GHz [15].

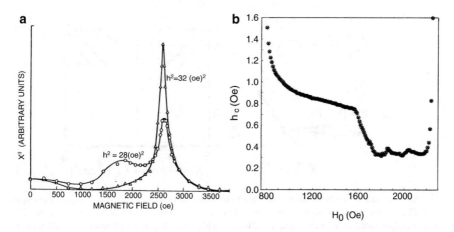

**Fig. 6.7** Illustration of the first-order Suhl instability. (**a**) FMR spectrum in a single crystal sphere of nickel ferrite showing the appearance of the subsidiary resonance at a field smaller than the field for FMR [14]. Reproduced with kind permission from Elsevier-Academic Press. (**b**) Critical field as a function of the static field measured in a YIG sphere with diameter 1 mm pumped with frequency 8.7 GHz [15]. Reprinted with permission from A. Azevedo and S. M. Rezende, Phys. Rev. Lett. **66**, 1342 (1991). Copyright (1991) by the American Physical Society

### 6.2.3   Second-Order Suhl Process: Premature Saturation of the FMR

The second-order Suhl process also manifests with the driving field perpendicular to the static field, but involves the 4-magnon interaction studied in Sect. 3.4.1. As illustrated in Fig. 6.8, microwave photons with frequency $\omega_p$ and $k \approx 0$ create uniform precession magnons on resonance, i.e., with the same frequency $\omega_0 = \omega_p$, by the linear excitation. Then, two magnons with $k = 0$ create a magnon pair with equal and opposite wave vectors and frequency $\omega_0/2$, so that both momentum and energy are conserved, as illustrated in Fig. 6.8a. Again, since the process is nonlinear, the magnon population is appreciable only if the rate at which the magnons are created exceeds the relaxation rate.

As in the previous sections, we study this process using the equations of motion for the magnon operators. First, we use Eq. (6.49) for the operator of the uniform mode, in the stationary steady state, considering $\omega_0 = \omega_p$

$$c_0 = -\frac{\gamma (SN/8)^{1/2}}{\eta_0} h_p e^{-i\omega_p t}. \tag{6.56}$$

Two $k = 0$ magnons driven on resonance create a $\vec{k}, -\vec{k}$ magnon pair by means of the 4-magnon interaction

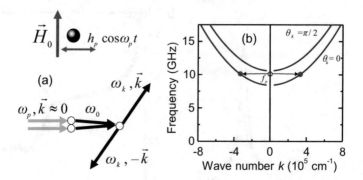

**Fig. 6.8** Illustration of the second-order Suhl instability process. (**a**) Pumping field configuration and diagramatic representation of the excitation of a magnon pair by two uniform precession magnons generated by two microwave photons. (**b**) Magnon dispersion relations for YIG with an internal field of $H_z = 3.0$ kOe, showing the generation of a $\vec{k}, -\vec{k}$ magnon pair by two $k = 0$ magnons with the same frequency of the microwave photons

$$H^{(4)} = \frac{\hbar}{2}\sum_k V_{0k}^{(4)} c_0 c_0 c_k^\dagger c_{-k}^\dagger + H.c.. \tag{6.57}$$

As we can see in Eq. (3.105), the 4-magnon interaction for $k_1 = k_2 = 0$ arising from the exchange energy is zero. Thus, only the dipolar interaction contributes to the vertex $V_{0,k}^{(4)}$ in Eq. (6.57), so that from Eq. (3.104) we have

$$V_{0k}^{(4)} = \frac{\omega_M}{4SN}\left(5\cos^2\theta_k - 5/3\right). \tag{6.58}$$

Using the Heisenberg equation (6.13) with the commutation relations (3.54), and introducing the relaxation phenomenologically in the usual manner, we obtain for the magnon annihilation operator of a $k \neq 0$ magnon

$$\frac{dc_k}{dt} = -(i\omega_k + \eta_k)c_k - iV_{0k}^{(4)} c_0 c_0 c_{-k}^\dagger. \tag{6.59}$$

The fast time dependence in Eqs. (6.56) and (6.59) can be eliminated by introducing the slowly varying spin-wave amplitude $\tilde{c}_k = \langle c_k \rangle e^{i\omega_p t/2}$ for all modes. Thus, substituting Eq. (6.56) in (6.59) with $\omega_k = \omega_0 = \omega_p$ we obtain

$$\frac{d\tilde{c}_k}{dt} = -\eta_k \tilde{c}_k - iV_{0k}^{(4)} \frac{\gamma^2(SN/8)}{\eta_0^2} h_p^2 \tilde{c}_{-k}^\dagger. \tag{6.60}$$

This equation and its complex conjugate have solutions of the form $\tilde{c}_k, \tilde{c}_k^* \propto \exp(\gamma_k t)$, where

$$\gamma_k = -\eta_k + \frac{\gamma^2 V_{0k}^{(4)} SN}{8\eta_0^2} h_p^2. \tag{6.61}$$

Hence, the $\vec{k}, -\vec{k}$ magnon-pair amplitude grows exponentially in time ($\gamma_k > 0$) if the pumping field exceeds a critical value given by

$$h_{crit} = \frac{\eta_0}{\gamma} \left( \frac{8\eta_k}{V_{0k}^{(4)} SN} \right)^{1/2}. \tag{6.62a}$$

Using Eq. (6.58) and $\cos\theta_k = 1$ in this equation, we obtain for the minimum critical pumping field

$$h_{c\,min} \approx \frac{2\eta_0}{\gamma} \left( \frac{2\eta_k}{\omega_M} \right)^{1/2}. \tag{6.62b}$$

Thus, in the second-order Suhl process the magnons that go unstable first have $\theta_k = 0$. This process is illustrated in Fig. 6.8b for a pump frequency $f_p = 10$ GHz. The two circles represent the modes pumped at the minimum threshold, with the same frequency and wave vectors with $k = 3.4 \times 10^5$ cm$^{-1}$ and $\theta_k = 0$. The fact that the uniform precession magnons transfer energy to $\vec{k}, -\vec{k}$ magnon pairs limits the FMR absorption above a certain microwave power, as shown in Fig. 6.7a. This phenomenon is called *premature saturation* of the FMR [6–9], because it occurs at power levels much smaller than in paramagnetic materials.

### 6.2.4 Three-Magnon Coincidence Process

In the subsidiary resonance process, the $k = 0$ magnon is driven far off-resonance, so its amplitude is quite small. Despite this, the threshold for driving spin waves unstable by the 3-magnon interaction is relatively low for materials with small damping, such as YIG. As pointed out by Suhl [9], the threshold can be much lower in the coincidence condition, where there are magnons with half the FMR frequency. This so because, in this case, $k = 0$ magnons are driven on-resonance with much larger amplitude than in the subsidiary resonance. This condition can be satisfied more easily in samples that have FMR frequency at the top of the magnon band under a relatively low static magnetic field. This is the case of magnetic films magnetized in the plane. We calculate the threshold for this process with the same steps used in the first-order Suhl process. First, we use for the $k = 0$ magnon amplitude the expression (6.56) for the excitation on-resonance. Substitution in Eq. (6.50a) gives for the interaction Hamiltonian

$$H_I(t) = -\frac{\hbar}{2}\sum_k h_p \rho_{\perp k} e^{-i\omega_p t} c_k^{\dagger} c_{-k}^{\dagger} + H.c., \qquad (6.63a)$$

where

$$\rho_{\perp kc} = \frac{\gamma\omega_M}{4\eta_0}\sin 2\theta_k e^{i\varphi_k}. \qquad (6.63b)$$

Using Eq. (6.44) with $\Delta\omega_k = 0$ we obtain the critical field for the coincidence process

$$h_{cmin} = \frac{4\eta_0\eta_k}{\gamma\omega_M\sin 2\theta_k}\bigg|_{min}. \qquad (6.64)$$

This result is smaller than the critical field for subsidiary resonance, Eq. (6.55), by a factor $\omega_p/2\eta_0 \sim 10^3$. The very low critical power for the three-magnon coincidence process has been observed experimentally in YIG films magnetized in the plane [16, 17].

We reproduce here some results reported in Cunha et al. [17]. The experiments were carried out with a 6 μm thick single-crystal YIG film grown by liquid-phase epitaxy onto a 0.5 mm thick [111]-oriented $Gd_3Ga_5O_{12}$ (GGG) substrate, cut in rectangular shape with dimensions of $5.0 \times 2.0$ mm$^2$. The microwave signal with amplitude adjusted by a variable attenuator feeds a microstrip line with characteristic impedance $Z_0 = 50\,\Omega$ made on a Duroid plate with a copper line of width 0.5 mm and a ground plane. The YIG film placed on top of the microstrip and separated by a 60 μm thick Mylar sheet is excited by the rf magnetic field $h_{rf}$ perpendicular to the static field $H_0$ applied in the film plane. With fixed frequency and sweeping static field, the variation of the transmitted power due to the FMR absorption is detected by a Schottky diode and measured with a lock-in amplifier (with the microwave signal amplitude modulated on-off with frequency 10 kHz).

Figure 6.9a shows the measured power absorption with scanning $H_0$ for several frequencies, keeping constant the incident power at the microstrip $P_i = 5\,\mu$W. The various lines at each frequency correspond to standing-wave magnetostatic spin-wave modes that have quantized in-plane wave numbers due to the boundary conditions at the edges of the film. The strongest line corresponds to the $k = 0$ magnon, or FMR mode. The symbols in Fig. 6.9b represent the measured FMR peak field $H_R$ versus frequency and the solid curve represents a fit with Eq. (4.43) for the FMR with parameters appropriate for YIG.

As the incident microwave power increases, the dependence of the spectra on the microwave frequency changes dramatically, as shown in Fig. 6.9d for a power of 5 mW. For frequencies below about 3 GHz, the FMR absorption peak decreases abruptly and almost disappears below 2 GHz. This behavior is due to three-magnon coincidence process that draws the energy from the FMR mode to magnon pairs, illustrated in Fig. 6.9c. For higher frequencies the condition for coincidence with the FMR frequency is not met, and the critical power is higher. The magnons with $k \neq 0$

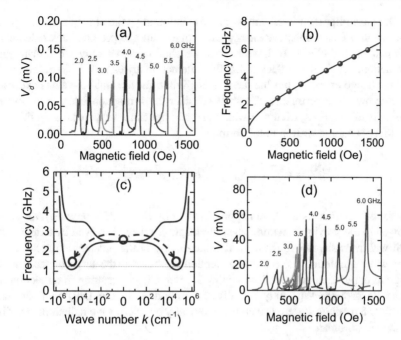

**Fig. 6.9** (a) FMR absorption in the YIG film expressed by the detector voltage versus magnetic field at the frequencies indicated and constant input power of 5 μW. (b) Symbols represent the measured microwave frequency versus the value of the field for the FMR peak. The solid line is a fit of the Kittel equation (4.43) for the FMR frequency. (c) Magnon dispersion curves for the modes with $\theta_k = 0, \pi/2$, calculated with Eq. (4.18) for a 6 μm thick YIG film under a field of 367 Oe. The symbols and arrows indicate the process by which the $k = 0$ magnon, driven by the microwave field, creates a pair of magnons with half frequency and opposite wave vectors. (d) FMR absorption versus magnetic field measured at the frequencies indicated and constant input power of 5 mW. Adapted from Figs. 1 and 2 of Cunha et al. [17]. Reproduced from R. O. Cunha et al., Appl. Phys. Lett. **106**, 192403 (2015), with the permission of AIP Publishing

excited by this process when the pumping frequency is below 3.2 GHz have been detected directly with Brillouin light scattering [16].

## 6.3 Magnon Interactions and Their Effects on Parametric Magnons

In the previous section we considered only a relatively simple role of the magnon interactions, namely a one-way conversion of $k = 0$ magnons into magnon pairs with $k \neq 0$ in the first- and second-order Suhl processes. Actually, the magnon interactions are responsible for a much richer class of phenomena, namely saturation, nonlinear dynamics, chaotic behavior, and Bose–Einstein condensation. In this section we shall study two important consequences of the interactions on parametric magnons, the saturation of their population and quantum coherence properties.

The most important interactions are those involving pairs of parametric magnons, because they easily conserve energy and momentum and provide back reactions of one pair on the others. In lowest order we shall consider only the four-magnon interaction. As shown in Sect. 3.4.1, this interaction arises mainly from the dipolar and exchange energies, but the latter is negligible for the small $k$-values excited in the microwave experiments. We shall consider the 4-magnon interaction between two pairs of magnons, each pair with equal and opposite wave vectors. In this case, Eq. (3.104) can be written in the form [18]

$$H^{(4)} = \hbar \sum_{k,k'} \left( \frac{1}{2} S_{kk'} c_k^\dagger c_{-k}^\dagger c_{k'} c_{-k'} + T_{kk'} c_k^\dagger c_{k'}^\dagger c_k c_{k'} \right), \qquad (6.65)$$

where the coefficients are given approximately by $S_{kk'} = 2T_{kk'} = 2\omega_M/NS$. As we showed in the previous section, when the magnon system is driven by a microwave field with intensity above the critical value, in any of the parametric processes, the spin wave amplitude increases exponentially. Actually, the large increase in the magnon population enhances the nonlinear interactions, resulting in a reaction that limits its growth. We study initially the saturation in the magnon population. Consider a magnon system driven parametrically by microwave radiation, described by the Hamiltonian

$$H = H_0 + H^{(4)} + H_I(t), \qquad (6.66)$$

where $H_0$ is the unperturbed Hamiltonian, $H^{(4)}$ represents the magnon interactions and $H_I(t)$ is the driving interaction Hamiltonian due either to the parallel pumping (6.38), or the first-order Suhl process (6.53). Considering that only one $k$-magnon mode is excited, using Eq. (6.66) in the Heisenberg equation (6.13) and introducing dissipation phenomenologically, we obtain the equation of motion for the magnon operator

$$\frac{dc_k}{dt} = -(i\omega_k + \eta_k)c_k + i(h_p\rho_k)e^{-i\omega_p t}c_{-k}^\dagger - i2(S_{kk}c_{-k}^\dagger c_k c_{-k} + 2T_{kk}c_k^\dagger c_k c_k), \quad (6.67)$$

where the factor 2 in the last term arises from the fact that $k'$ in (6.65) can be $k$ or $-k$. Since parametric magnons are driven in pairs, we assume $c_{-k} = c_k \exp(iq_k)$, where $q_k$ is a real phase. The fast time dependence in this equation can be eliminated by introducing the slowly varying spin-wave amplitude

$$\widetilde{c}_k = \langle c_k \rangle e^{iq_k/2} e^{i\omega_p t/2}. \qquad (6.68)$$

Considering $\widetilde{c}_k$ a classical variable, from Eq. (6.67) we obtain

$$\frac{d\widetilde{c}_k}{dt} = -(i\Delta\omega_k + \eta_k)\widetilde{c}_k + ih_p\rho_k\widetilde{c}_k^* e^{-iq_k/2} - iS_k\widetilde{c}_k^*\widetilde{c}_k\widetilde{c}_k. \qquad (6.69)$$

where $\Delta\omega_k = \omega_k - \omega_p/2$ and $S_k = S_{kk} + 2T_{kk}$. With Eq. (6.69) and its complex conjugate, one can show that in the steady-state, $d/dt = 0$, the number of magnons $\langle n_k \rangle = \widetilde{c}_k^* \widetilde{c}_k$ is given by

$$\langle n_k \rangle_{ss} = \frac{-\Delta\omega_k + [(h_p\rho_k)^2 - \eta_k^2]^{1/2}}{2S_k}.$$

Using Eq. (6.44) for the critical pumping field, $(h_{crit}\rho_k)^2 = \eta_k^2 + \Delta\omega_k^2$, this result can be written as

$$\langle n_k \rangle_{ss} = \frac{\eta_k[(h_p/h_{crit})^2 - 1]^{1/2}}{2S_{kk}}, \tag{6.70}$$

This shows that, for $h_p > h_{crit}$, the magnon interaction stabilizes the parametric mode population. Let us calculate the number of parametric magnons assuming a YIG film with thickness 5 μm and lateral dimensions $2 \times 2$ mm$^2$, considering a relaxation rate $\eta_k = 2 \times 10^6$ s$^{-1}$. Using the YIG lattice parameter $a = 1.23$ nm, one finds for the number of spins in the sample $N \approx 10^{16}$, and for the interaction parameter $S_k = 4\omega_M/NS \approx 6 \times 10^{-7}$ s$^{-1}$. Using these values in Eq. (6.70), we obtain the number of magnons as a function of the relative driving field $h_p/h_{crit}$ shown in Fig. 6.10a. Clearly, for $h_p < h_{crit}$ the magnon population is very small.

For $h_p > h_{crit}$ the steady-state magnon population increases rapidly with the pumping field and attains values comparable to those of the $k = 0$ magnon driven on resonance. The rapid growth of the population above threshold can be seen by calculating the number of magnons for a driving field just above the critical value. Using the parameters for the YIG film, we obtain with Eq. (6.70) for $h_p/h_{crit} = 1.00001$ the value $\langle n_k \rangle_{ss} = 6.3 \times 10^9$. Although the number of magnons shown in Fig. 6.10a is extremely large compared to the thermal magnon population

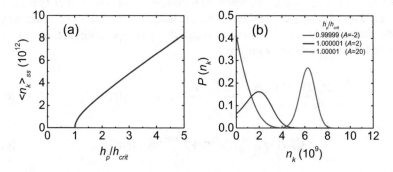

**Fig. 6.10** Parallel pumping in a YIG film. (**a**) Number of magnons in the steady-state as a function of the driving field intensity relative to the critical field. (**b**) Magnon distributions for three values of the driving field

($10^3$), it is much smaller than $NS$ ($\sim 10^{17}$ in the YIG film), so that the condition $\langle n_k \rangle \ll NS$ justifies comfortably the use of the spin-wave formalism.

The treatment just presented does not give information on the quantum nature of the magnons in the steady state above the threshold. This can be obtained with a formal treatment of the relaxation processes by using quantum-statistical techniques. We follow Araújo et al. [19] in the study presented here. The Hamiltonian for the system is the sum of Eq. (6.66) with two other terms

$$H_R + H_{RS} = \sum_p \omega_p R_p^\dagger R_p + \sum_{k,p} (\beta_{kp}^* R_p^\dagger c_k + \beta_{kp} R_p c_k^\dagger), \qquad (6.71)$$

where the first term represents a heat bath with large thermal capacity, containing a very large number of boson modes, and the second represents a linear interaction between the magnon system and the heat bath [20]. As done previously, since we are interested in studying the behavior near the threshold, we assume that only one magnon pair is present. Thus, the total Hamiltonian given by the sum of Eqs. (6.66) and (6.71) is

$$H = H_0 + H^{(4)} + H_R + H_{RS} + H_I(t), \qquad (6.72)$$

Using the Heisenberg equation we obtain the equation of motion for the magnon operator

$$\frac{dc_k}{dt} = -(i\omega_k + \eta_k)c_k + i(h_p\rho_k)e^{-i\omega_p t}c_{-k}^\dagger - i2S_k c_{-k}^\dagger c_{-k}c_k + F_k(t), \qquad (6.73)$$

where $\eta_k = \pi f(\omega_k)|\beta_{kp}|^2$ is the relaxation rate of the $k$-magnon, $f(\omega_k)$ is the density of heat-bath modes, and

$$F_k(t) = -i\sum_p \beta_{kp} R_p(t \to -\infty)e^{-i\omega_p t} \qquad (6.74)$$

represents a Langevin random force with correlators of Markoffian systems type [19–21]. Now it is very convenient to work in the representation of coherent magnon states, defined as in Eq. (3.92), $c_k|\alpha_k\rangle = \alpha_k|\alpha_k\rangle$. Thus, operating with (6.73) on the coherent state one can show that [19]

$$\frac{d\alpha_k(t)}{dt} - \left[\frac{|h_p\rho_k|^2 - \eta_k^2}{2\eta_k} - \frac{2S_k^2}{\eta_k}|\alpha_k(t)|^4\right]\alpha_k(t) = G_k(t), \qquad (6.75)$$

$$G_k(t) = \frac{1}{2}F_k(t)e^{i\omega_k t} + i\frac{h_p\rho_k}{2\eta_k}F_{-k}(t)e^{-i\omega_k t}, \qquad (6.76)$$

where we have considered $|\alpha_k| = |\alpha_{-k}|$ and $\Delta\omega_k = 0$. Equation (6.75) is a typical nonlinear Langevin equation which appears in Brownian motion studies and laser theory (see Nussenzweig 1973). It shows that the magnon-pair modes, that have amplitude $\alpha_k(t)$, are driven coherently by the pumping field and thermally by the heat-bath modes. The first term inside the brackets can be negative or positive, depending on whether the pumping field is below or above threshold. Therefore, it can produce either damping or amplification of the modes. The second term in the brackets represents a nonlinear damping effect which tends to saturate the rate of creation of the magnon-pair modes. Clearly the onset of the instability process is obtained when the first term is zero, in agreement with Eq. (6.44) with $\Delta\omega_k = 0$. Also, since $\langle n_k \rangle = |\alpha_k|^2$, in the steady-state condition, $d\alpha_k/dt = 0$, Eq. (6.75) gives for the number of magnons the same result as in Eq. (6.70).

In order to study the coherence properties of the magnon states generated in the parametric process, we need to find a linear Fokker–Planck equation that is stochastically equivalent to the Langevin equation (6.75). The standard procedure consists in calculating the density matrix $\rho$, and then its P representation, defined in Eq. (3.98), using $i(d\rho/dt) = [H, \rho]$, where H is the total Hamiltonian. Using $\alpha_k = re^{i\phi}$, one can show [19] that the equation of motion for $P$ is

$$\frac{dP}{dt_n} + \frac{1}{x}\frac{\partial}{\partial x}[(A - x^4)x^2 P] = \frac{1}{x}\frac{\partial}{\partial x}(x\partial P/\partial x) + \frac{1}{x^2}(\partial^2 P/\partial\phi^2), \tag{6.77}$$

where

$$t_n = \left(\frac{S_k^2}{8}\eta_k \bar{n}_k^2\right)^{1/3} t, \quad \text{and} \quad x = \left(\frac{8 S_k^2}{\eta_k^2 \bar{n}_k}\right)^{1/6} r, \tag{6.78}$$

represent, respectively, normalized dimensionless time and magnon amplitude, while the driving parameter $A$, also a dimensionless quantity, is given by

$$A = 4\left(\frac{S_k^2}{\eta_k^2 \bar{n}_k}\right)^{2/3}\frac{\left(|h_p \rho_k|^2 - \eta_k^2\right)}{(2 S_k)^2}, \tag{6.79}$$

which can be written in terms of the critical field in Eq. (6.44), $h_{crit} = \eta_k/|\rho_k|$, as

$$A = \left(\frac{\eta_k}{S_k \bar{n}_k}\right)^{2/3}(h_p^2/h_{crit}^2 - 1). \tag{6.80}$$

We are interested in the steady state solution of the $P$ representation, which is given by the solution of Eq. (6.77) with $\partial/\partial t_n = 0$. Straightforward integration of the result gives

$$P(x) = N\exp\left(\frac{1}{2}Ax^2 - \frac{1}{6}x^6\right). \tag{6.81}$$

Figure 6.10b shows the distribution functions calculated with Eq. (6.81) for three values of the relative pumping field, for the same YIG film considered previously. In order to obtain numerical quantities, we use the same parameters given before and from Eq. (6.80) we find for the pumping parameter $A \approx 10^6(h_p^2/h_{crit}^2 - 1)$. Also, with the same parameters, we obtain from Eq. (6.78) a relation between the number of magnons $n_k = r^2$ and the variable $x$

$$n_k = \left( \frac{\eta_k^2 \bar{n}_k}{8 S_k^2} \right)^{1/3} x^2, \tag{6.82}$$

that gives for the YIG film $n_k = 1.4 \times 10^9 x^2$. Thus, we can plot the distribution function in terms of the magnon number. As shown in Fig. 6.10b, for a pumping field slightly below threshold, $h_p/h_{crit} = 0.99999$, the pumping parameter is negative, $A = -2$, and $P(n_k)$ is a Gaussian function. For a pumping field slightly above threshold, $h_p/h_{crit} = 1.000001$, the pumping parameter is positive, $A = +2$, and $P(n_k)$ consists of two components, a coherent one convoluted with a smaller fluctuation with Gaussian distribution. Then, for larger pumping field, but still close to threshold, $h_p/h_{crit} = 1.00001$, $A = 20$, and $P(n_k)$ is clearly a Poisson distribution, characteristic of a coherent state, as in Fig. 3.4. Note that the peak of the function occurs at $n_k = 6.3 \times 10^9$, in agreement with the value for $\langle n_k \rangle_{ss}$ calculated earlier for the same pumping field. In fact, one can show that the maximum of the distribution function, calculated with $dP/dx = 0$ in Eq. (6.81), that is $x_m = A^{1/4}$, gives with (6.80) and (6.82), the same expression as in Eq. (6.70). In conclusion, the 4-magnon interaction act as to make the parametric magnon state a coherent one. This behavior is similar to that of photon states in lasers and it is the one we would expect from the well-known semiclassical analysis of the parallel pumping. Finally notice that, as shown in [1, 19], if the magnon interaction is not considered, the parametric magnons are not in a coherent state.

## 6.4   Nonlinear Dynamics of Magnons and Chaotic Behavior

The four-magnon interaction has another important consequence in magnonics, namely, the nonlinear dynamics of magnons that result in the so-called self-oscillations in the magnetization and chaotic behavior. Since the early days of studies of parametric excitation of magnons in the 1960s, it is known that when the microwave driving power is increased beyond threshold, the radiation reflected from the sample may exhibit an amplitude modulation, characterized by low-frequency oscillations, self-pulsations, or turbulent behavior. Suhl first wrote [9] that the spin-wave instability above threshold in perpendicular pumping experiments "bears a certain resemblance to the turbulent state in fluid dynamics." Later it was observed in parallel pumping experiments that above the threshold a variety of low-frequency wave forms could develop in the amplitude of the reflected microwave signal, which were generally called *relaxation oscillations* [22]. The explanation given at that time, based on the beating of several spin-wave modes allowed by the boundary conditions [23], was later questioned by several authors who showed that the spin wave behavior above threshold was governed by the nonlinear dynamics [18, 24, 25]. Several years later it was predicted theoretically that interacting magnon modes driven by the parallel pumping process should display chaotic dynamics similar to other physical systems [26]. The prediction was soon confirmed in experiments for driving magnons by the second-order Suhl instability [27], by parallel pumping [13] and by the first-order Suhl instability [15, 28].

## 6.4.1  Nonlinear Equations of Motion

The nonlinear differential equations that govern the dynamics of large $k$, plane spin-wave instabilities have been derived by many authors [26, 28, 29]. In all approaches the interacting spin-wave system is considered to be driven by a uniform microwave field in such a way that the pumping term of the Hamiltonian conserves momentum, i.e., magnons are generated in pairs $\vec{k}, -\vec{k}$. This is strictly correct only for an infinite medium, hence it cannot predict the observed sample size dependence for the self-oscillation frequency and threshold. In a finite sample it was shown [30] that in addition to the pumping of magnon pairs with equal and opposite wave vectors, it is possible to have excitation of pairs $+\vec{k}, -(\vec{k} + \Delta\vec{k})$. In this case, instead of Eq. (6.38), the driving Hamiltonian is [30],

$$H_I(t) = \frac{\hbar}{2}\sum_k h_p\rho_k e^{-i\omega_p t} c_k^\dagger c_{-k'}^\dagger (\delta_{k,k'} + \alpha_{\Delta k}\delta_{k,k\pm\Delta k} + \cdots) + H.c., \qquad (6.83)$$

where the coupling parameter $\rho_k$ is given either by Eq. (6.39), for the parallel pumping process, or by Eq. (6.54) for the first-order Suhl process, and $\alpha_{\Delta k}$ is a factor that depends on the wave vector mismatch $\Delta k_x$, $\Delta k_y$, $\Delta k_z$. For two neighboring standing waves along $x$ (or $y$ or $z$) with $\Delta k_x = \pi/L_x$, where $L_x$ is the dimension of the sample, $\alpha_{\Delta k}$ attains its maximum value of $2/\pi$ [30]. In an infinite medium $\alpha_{\Delta k} = 0$, so that Eq. (6.83) reduces to the expression (6.38) used before in the study of spin-wave parametric pumping. Note that due to the symmetry-breaking term in a finite sample, the microwave field can drive directly two neighboring standing-wave modes with wave numbers $k$ and $k' = k \pm \Delta k$, satisfying the energy-conservation relation $\omega_k + \omega_{k'} = \omega_p$.

The essential ingredient for the nonlinear dynamics is the coupling between two pairs of parametric magnons provided by the four-magnon interaction, given by the Hamiltonian (6.65). Thus, the total Hamiltonian for a magnon system pumped by a microwave field in parallel pumping or first-order Suhl process is

$$H = H_0 + H^{(4)} + H_I(t), \qquad (6.84)$$

With the Heisenberg equation we obtain the following equation of motion for the slowly varying spin-wave amplitude defined in Eq. (6.68) that characterizes the magnon-pair mode

$$\frac{d\widetilde{c}_k}{dt} = -(i\Delta\omega_k + \eta_k)\widetilde{c}_k - ih_p\rho_k(\widetilde{c}_k^* + \alpha_{\Delta k}\widetilde{c}_{k'}^* e^{i\beta_{kk'}/2}) - i\sum_{k'}(S_{kk'}\widetilde{c}_{k'}^2\widetilde{c}_k^* + 2T_{kk'}|\widetilde{c}_{k'}|^2\widetilde{c}_k),$$

(6.85)

where $\Delta\omega_k = \omega_k - \omega_p/2$ is the detuning and $\beta_{kk'} = q_k - q_{k'}$ is the phase-difference between modes $k$ and $k'$. Since the spin-wave variable has real and imaginary parts, Eq. (6.85) represents a set of $2M$ ($M$ is the number of nearly degenerate modes excited by the pump field) nonlinear differential coupled equations. Physically there is a very large number of modes involved as soon as the pumping field exceeds the threshold, so that the problem of solving the whole set of equations is in principle very difficult. However, it was shown [31] that the linearized equations have *collective*, normal mode solutions, formed by linear superpositions of the spin-wave variables, each with an associated pair of complex eigenvalues. Thus, in general, at not too intense driving, the dynamics is governed by essentially two modes, each resulting from the collective coupling of many spin-wave pair modes in a certain region of $k$-space. This explains the fact that numerical calculations with only two modes give quite good qualitative agreement with the experimental observations [29–31].

Our analysis of the magnon nonlinear dynamics is based on the equations for two modes obtained from Eq. (6.85)

$$\frac{dc_1}{dt} = -(i\Delta\omega_1 + \eta_1)c_1 - ih_p\rho_1(c_1^* + \alpha e^{i\beta/2}c_2^*) - i2(S_1c_1^2c_1^* + S_{12}c_1^*c_2^2 + 2T_{12}c_2^*c_2c_1),$$

(6.86a)

$$\frac{dc_2}{dt} = -(i\Delta\omega_2 + \eta_2)c_2 - ih_p\rho_2(c_2^* + \alpha e^{-i\beta/2}c_1^*) - i2(S_2c_2^2c_2^* + S_{12}c_2^*c_1^2 + 2T_{12}c_1^*c_1c_2),$$

(6.86b)

where the subscripts 1 and 2 denote the two collective modes, $S_k = S_{kk} + 2T_{kk}$, $\alpha = \alpha_{\Delta k}$, $\beta = \beta_{12}$, and the tilde in $\widetilde{c}_k$ was dropped to simplify the notation. Since $c_k$ is complex, the set (6.86a) and (6.86b) represents four coupled equations.

Although Eqs. (6.86a) and (6.86b) represent an enormous simplification of the original problem, they are still mathematically quite complex. They have 12 independent parameters so that a multitude of bifurcation phenomena can be encountered depending on the parameter values. In order to fully explore these phenomena it is necessary to make use of numerical methods and integrate the equations in a computer. In the numerical studies with an arbitrary set of parameters, one usually finds that the solutions are attracted to stable fixed points. However, in certain regions of the parameter space some fixed points become unstable and the solutions may exhibit a variety of dynamic behavior. Due to the complex nature of the modes involved in the dynamics, it is difficult to relate the parameters with the microscopic ones. However, it has been found that some sets of parameters yield results very similar to those observed experimentally. The numerical solutions of the equations can be obtained using Runge–Kutta subroutines in simple computer programs. In order to work with quantities of order 1, it is convenient to rescale the time, the magnon number $n_k = c_k^* c_k$, and the interaction parameters as $t \to t\eta_k$, $n_k \to Fn_k$, $S_{kk'} \to S_{kk'}/F$, and $T_{kk'} \to T_{kk'}/F$, where $F = \omega_M/(4SN\eta_k)$. In YIG, at room temperature, $\eta_k \sim 2 \times 10^6$ s$^{-1}$, $\omega_M \approx 3 \times 10^{10}$ s$^{-1}$, so that $F \sim 10^{17}$ in a sample with volume $\sim 1$ mm$^3$. Following [32], numerical solutions of the coupled two-mode equations (6.86a) and (6.86b) with the renormalized parameters will be presented in the next sections.

## 6.4.2 Self-Oscillations in the Parametric Magnon Populations

We consider initially the following set of parameters: $\Delta\omega_1 = \Delta\omega_2 = 0$, $\rho_1/\rho_2 = 0.7$, $\alpha = 0$, $\eta_2/\eta_1 = 5.0$, $S_1/\eta_1 = -1.0$, $S_2/\eta_1 = 0.5$, $S_{21}/\eta_1 = S_{12}/\eta_1 = 2.5$, and $T_{21}/\eta_1 = T_{12}/\eta_1 = 1.125$. Note that the parameters are quite different for the two modes, indicating that they are degenerate modes far in $k$-space. This justifies making $\alpha = 0$. Figure 6.11a shows the results of the numerical integration for the time evolution of the magnon populations $n_1$ and $n_2$ for a microwave driving applied at the instant $t = 0$. The amplitude of the pumping field is represented by the control parameter $R = h_p/h_{crit}$, so that $R = 1$ corresponds to the critical field for driving the magnon instability. For a driving field in the range $1.0 < R < 2.4$, only the *strong mode* 1 is excited, since it has a smaller damping. For $R > 2.4$ the *weak mode* 2 is also excited, even though its threshold field is not reached until $R = 5.0/0.7$. This happens because sufficient driving is supplied to mode 2 via the nonlinear coupling with mode 1.

For pumping intensities in the range $1.0 < R < 3.17$, initially the magnon populations have a transient behavior, shown in Fig. 6.11a for two values of $R$. Then, after some time $t\eta_1 \gg 1$, the magnon numbers stabilize at steady-state values that depend on the driving intensity. Figure 6.11b shows the projections on the $n_1 - n_2$ plane of the steady-state solutions for various microwave pumping intensities. For $R < 3.17$ the solutions of the nonlinear equations are said to be attracted to *fixed points* and the magnon numbers are constant. The heavy line in

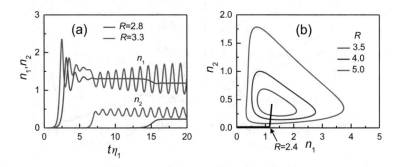

**Fig. 6.11** Numerical solutions of two-mode equations with parameters: $\Delta\omega_1 = \Delta\omega_2 = 0$, $\rho_1/\rho_2 = 0.7$, $\alpha = 0$, $\eta_2/\eta_1 = 5.0$, $S_1/\eta_1 = -1.0$, $S_2/\eta_1 = 0.5$, $S_{21}/\eta_1 = S_{12}/\eta_1 = 2.5$, and $T_{21}/\eta_1 = T_{12}/\eta_1 = 1.125$. (a) Transient behavior of the magnon numbers for two pumping levels $R = h_p/h_{crit}$ as indicated. (b) Lines of fixed points for $R < 3.17$ and limit cycles for three pumping fields calculated in the steady state

Fig. 6.11b shows the path on the $n_1 - n_2$ plane of the stable fixed points with varying $R$. As the driving increases beyond the critical value $R'_c = 3.17$, the fixed point becomes unstable and a *limit cycle* develops around it. In other words, there is an oscillation of the magnon numbers $n_1$ and $n_2$ produced by a dynamic energy shuffling between the two modes, that is a consequence of the nonlinear nature of the system. The amplitude of the limit cycle increases with increasing pumping, as shown in Fig. 6.11b. Experimentally this effect is observed in a self-oscillation in the microwave signal reflected by the sample.

Figure 6.12 shows experimental data [30] obtained with parallel pumping at a frequency 9.5 GHz, in a 1 mm YIG sphere, with the static field $H_0$ applied in the [111] direction. The lower butterfly curve in Fig. 6.12a represents the threshold for the magnon instability, while the upper curve corresponds the critical field above which the microwave signal reflected from the cavity develops a low-frequency self-oscillation. This change in behavior at the second critical field is associated with a *bifurcation* in the response of the nonlinear magnon system. A bifurcation is the analog of a phase transition in equilibrium thermodynamics and corresponds to a qualitative change in the state of the system as some relevant parameter is varied.

Figure 6.12b shows the measured self-oscillation frequency as a function of the static field. The data are explained qualitatively by the two-mode model, that shows that the limit cycle trajectory is described at a frequency $f_0 \sim \eta_1$. In the left-hand side of the curve in Fig. 6.12b, the relaxation rate of the parametric magnons decreases with increasing field, and so does the oscillation frequency. As shown in [32], the amplitude of the oscillation of $n_1$ increases proportionally to $(R - R'_c)^{1/2}$. This is characteristic of a Hopf bifurcation, that was observed experimentally in the parametric excitation of magnons in some field ranges of the subsidiary resonance in a YIG sphere [28, 30]. In other field ranges, different types of bifurcations are observed experimentally, as illustrated in Fig. 6.13 [28]. All types of dynamic

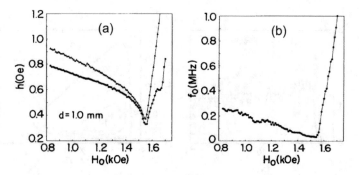

**Fig. 6.12** Experimental data for self-oscillations obtained with 9.5 GHz parallel pumping in a 1 mm [111] YIG sphere. (**a**) Critical fields for magnons instability and for self-oscillations. (**b**) Self-oscillation frequency versus static field [30]. Reprinted with permission from S. M. Rezende and A. Azevedo, Phys. Rev. B **45**, 10,387 (1992). Copyright (1992) by the American Physical Society

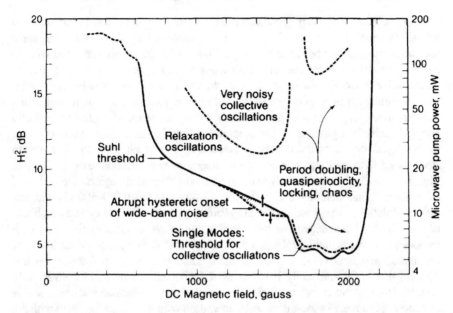

**Fig. 6.13** Regions and types of dynamic behavior experimentally observed in the perpendicular-pumping first-order Suhl instability in a YIG sphere as a function of static magnetic field and microwave driving power. After Bryant et al. [28]. Reprinted with permission from P. Bryant, C. Jeffries, and K. Nakamura, Phys. Rev. Lett. **60**, 1185 (1988). Copyright (1988) by the American Physical Society

behavior observed experimentally can be reproduced with the two-mode model with appropriate sets of parameters.

### 6.4.3   Period Doubling, Strange Attractors, and Chaos

As mentioned earlier, chaotic behavior of interacting parametric magnons was first predicted theoretically for the parallel pumping process [26]. However, the first experimental observation of chaotic magnon dynamics was achieved with the perpendicular-pumping second-order Suhl process [27]. The experiments were carried out with a Ga–YIG sphere, excited by a microwave field with frequency 1.3 GHz, with the oscillations in the magnetization detected by a pickup coil. A period-doubling cascade in the oscillation was observed with increasing microwave power, leading to chaotic behavior. The experiments were explained by a two-mode model involving the uniform FMR mode and a degenerate $k \neq 0$ magnon [29].

Period-doubling cascade in the oscillation and chaotic behavior were also observed in the configurations of parallel pumping and first-order Suhl process (subsidiary resonance) [13, 15, 29, 30, 32, 33]. The experiments of de Aguiar et al. [33] were performed with a spherical YIG sample with diameter 1.0 mm, held in the center of a $TE_{102}$ cavity ($Q = 3000$), operating at 9.4 GHz, with the microwave magnetic field perpendicular to the static field, in the configuration of the subsidiary resonance process. The power was provided by a 2-W traveling-wave tube amplifier fed by a backward-wave oscillator, with frequency stabilized with an external crystal oscillator and manually adjusted to the center of the cavity resonance. Microwave signals reflected from the cavity were monitored either with a spectrum analyzer or with a crystal detector to have its time variation recorded in a storage oscilloscope. At low power levels the signal reflected from the critically coupled cavity is negligible. As the pumping field $h_p$ is increased, an abrupt change in the signal occurs at the threshold value $h_{crit}$, for which a pump photon with frequency $f_p/2$ excites a pair of parametric magnons with wave vectors $k$, $-k$, as in Fig. 6.6. As $h_p$ is increased further in the post-threshold region, the reflected microwave signal increases until it develops a low-frequency (50–500 kHz) amplitude modulation, corresponding to a Hopf bifurcation. The frequency and amplitude of this self-oscillation depend on the pumping configuration, crystal orientation with respect to the static field, and the values of $H_0$ and $R = h_p/h_{crit}$. At an arbitrary orientation and field value, except for the gradual increase in self-oscillation frequency, nothing dramatically happens as $R$ increases. However, at conveniently selected configurations the modulation shows period multiplication and chaos as $R$ is increased. Figure 6.14 shows the spectra measured with the sample oriented with the [100] direction along the field, and $H_0 = 1950$ Oe, in which a self-oscillation appears

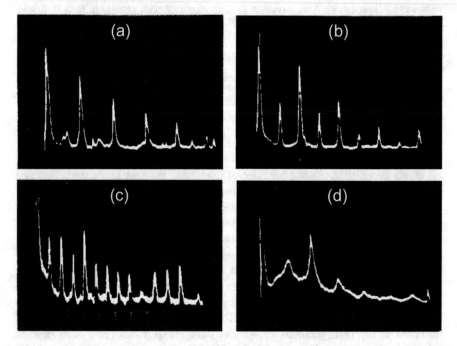

**Fig. 6.14**  Power spectra of the self-oscillation signal measured in perpendicular-pumping subsidiary resonance experiments, showing (**a**) period-1, (**b**) period-2, (**c**) period-4, and (**d**) chaotic behavior corresponding, respectively, to $R = h_p/h_{crit} = 2.12$, 2.19, 2.29, and 2.54 [33]. Reprinted with permission from F. M. de Aguiar, A. Azevedo, and S. M. Rezende, Phys. Rev. B **39**, 9448 (1989). Copyright (1989) by the American Physical Society

initially with frequency 450 kHz at $R = 2.12$. The frequency increases with increasing pumping intensity, and at $R = 2.19$ a period doubling (2T) is observed. Then successive bifurcation periods 4T, 8T, and 16T are observed with increasing control parameter $R$ before the signal becomes chaotic at $R = 2.54$. This is a typical Feigenbaum route to chaos, also observed in other nonlinear physical systems.

All experimental results can be explained qualitatively by the two-mode nonlinear equations (6.86a) and (6.86b) with appropriate parameters [32]. Let us consider here the same parameters used before but at higher pumping levels. The results are presented in Fig. 6.15. In (a)–(c) we show the trajectories in phase-space $n_2 \times n_1$ for several values of the control parameter $R$. For $R = 6.0$ we have a stable limit cycle characterizing a self-oscillation with frequency $f_0 \sim \eta_1$. Period doubling is observed at $R = 7.5$. The parameter $R$ was varied up to the fifth decimal place in the range 7.50–7.76, but no traces of periods 4 or 8 were seen. At $R = 7.76$ the trajectory jumps to a chaotic pattern, just like experimental results obtained with parallel pumping [13]. The strange attractor characteristic of chaos is shown in Fig. 6.15c for $R = 8.0$. Figure 6.15d shows a $n_1 \times R$ bifurcation diagram, made of the points at which the trajectories intercept the Poincaré section shown by the black line in (c). Note that by

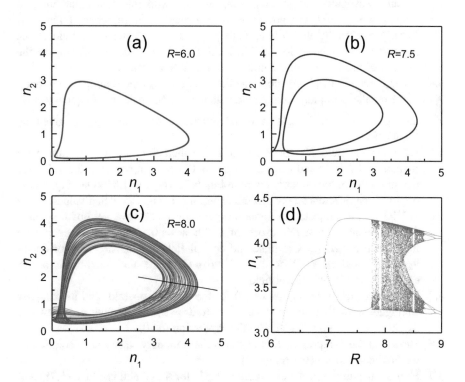

**Fig. 6.15** (**a**), (**b**) and (**c**) Phase-space trajectories $n_2 \times n_1$, calculated numerically with the two-mode model with the same parameters as in Fig. 6.11. Period 1 at $R = 6.0$, period 2 at $R = 7.5$, and strange attractor for fully developed chaos at $R = 8.0$. (**d**) Bifurcation diagram calculated at the Poincaré section represented by the black line in (**c**)

increasing $R$ further in the chaotic region several windows appear, such as the period-4 windows at about $R = 7.9$ and $8.5$. Interestingly, an inverted period-doubling cascade is observed in the range $8.7$–$9.0$. By varying the mode parameters one can obtain a variety of oscillation shapes, trajectories, and bifurcation diagrams, as observed experimentally by varying the sample shape, crystal orientation, and intensity of the applied field [32].

## Problems

6.1. Using expression (6.16) for the expectation value of the magnon operator in a linear excitation process and considering for the energy stored in the $k$-magnons $E_k = \hbar\omega_k n_k = \hbar\omega_k |\langle c_k \rangle|^2$, show that the power absorbed by the magnons is given by Eq. (6.10).

6.2. Use the equation of motion (6.17) for the $k$-magnon operator in a linear excitation process to show that the number of magnons generated is, in the steady state $(dn_k/dt = 0)$, given by Eq. (6.18).

6.3. A spherical YIG sample with diameter 1 mm is placed at the end of a shorted X-band waveguide, as in Fig. 6.2a, and fed with microwave radiation of frequency 9.4 GHz and power 50 mW. Calculate the number of magnons linearly excited by the driving field perpendicular to the static field at the resonance, considering for YIG a FMR linewidth of 0.5 Oe. Compare the number of magnons with the number of spins in the sample.

6.4. Using the relations (3.53) show that the parameter $\rho_k$ representing the coupling of the parallel pumping driving field with the magnon pair with frequency $\omega_p/2$, having $\vec{k}, -\vec{k}$ with polar and azimuthal angles $\theta_k$ and $\varphi_k$, is given by Eq. (6.39).

6.5. Calculate the field $H_0$ applied to a YIG sphere for which a magnon with frequency 4.7 GHz and $\theta_k = \pi/2$ has $k = 0$, and compare with the value for the minimum of the butterfly curve in Fig. 6.5a, $H_{0min} = 1.54$ kOe. Explain the origin of the difference between the experimental and calculated values.

6.6. A YIG film is subjected to a parallel pumping microwave field of frequency 6 GHz, created by a microstrip of width $w = 0.5$ mm and characteristic impedance 50 $\Omega$. (a) Calculate the critical field for driving magnons with $\theta_k = \pi/2$ and $\eta_k = 4 \times 10^6$ s$^{-1}$. (b) Calculate the microwave power corresponding to the critical field.

6.7. From the equation of motion (6.69) for the magnon operator and its complex conjugate show that, in the steady-state $(d/dt = 0)$, the number of parametric magnons $\langle n_k \rangle = \tilde{c}_k^* \tilde{c}_k$ is given by the expression (6.70).

6.8. Show that the $P$-representation (6.81) is the stationary steady-state solution of the Fokker–Planck equation (6.77).

6.9. Find the maximum of the distribution (6.81) for $A > 0$ and use Eqs. (6.78) and (6.80) to show that it corresponds to the number of magnons in the steady-state given by Eq. (6.70).

# References

1. Zagury, N., Rezende, S.M.: Theory of macroscopic excitations of magnons. Phys. Rev. B. **4**, 201 (1971)
2. Huebl, H., Zollitsch, C.W., Lotze, J., Hocke, F., Greifenstein, M., Marx, A., Gross, R., Goennenwein, S.T.B.: High cooperativity in coupled microwave resonator ferrimagnetic insulator hybrids. Phys. Rev. Lett. **111**, 127003 (2013)
3. Tabuchi, Y., Ishino, S., Noguchi, A., Ishikawa, T., Yamazaki, R., Usami, K., Nakamura, Y.: Coherent coupling between a ferromagnetic magnon and a superconducting qubit. Science. **349**, 405 (2015)
4. Lachance-Quirion, D., Tabuchi, Y., Ishino, S., Noguchi, A., Ishikawa, T., Yamazaki, R., Nakamura, Y.: Resolving quanta of collective spin excitations in a millimeter-sized ferromagnet. Sci. Adv. **3**(1603150), e1603150 (2017)
5. Koch, J., Yu, T.M., Gambetta, J., Houck, A.A., Schuster, D.I., Majer, J., Blais, A., Devoret, M. H., Girvin, S.M., Schoelkopf, R.J.: Charge-insensitive qubit design derived from the Cooper pair box. Phys. Rev. A. **76**, 042319 (2007)
6. Bloembergen, N., Damon, R.W.: Relaxation effects in ferromagnetic resonance. Phys. Rev. **85**, 699 (1952)
7. Damon, R.W.: Relaxation effects in the ferromagnetic resonance. Rev. Mod. Phys. **25**, 239 (1953)
8. Bloembergen, N., Wang, S.: Relaxation effects in para-and ferromagnetic resonance. Phys. Rev. **93**, 72 (1954)
9. Suhl, H.: The theory of ferromagnetic resonance at high signal powers. J. Phys. Chem. Solids. **1**, 209 (1957)
10. Morgenthaler, F.R.: Survey of ferromagnetic resonance in small ferrimagnetic ellipsoids. J. Appl. Phys. **31**, 95 S (1960)
11. Schlömann, E., Green, J.J., Milano, U.: Recent developments in ferromagnetic resonance at high power levels. J. Appl. Phys. **31**, 386 S (1960)
12. Kasuya, T., Le Craw, R.C.: Relaxation mechanisms in ferromagnetic resonance. Phys. Rev. Lett. **6**, 223 (1961)
13. de Aguiar, F.M., Rezende, S.M.: Observation of subharmonic routes to chaos in parallel-pumped spin waves in yttrium iron garnet. Phys. Rev. Lett. **56**, 1070 (1986)
14. Damon, R.W.: Ferromagnetic resonance at high power. In: Rado, G.T., Suhl, H. (eds.) Magnetism, vol. I, p. 551. Academic, New York (1963)
15. Azevedo, A., Rezende, S.M.: Controlling chaos in spin-wave instabilities. Phys. Rev. Lett. **66**, 1342 (1991)
16. Kurebayashi, H., Dzyapko, O., Demidov, V.E., Fang, D., Ferguson, A.J., Demokritov, S.O.: Controlled enhancement of spin-current emission by three-magnon splitting. Nat. Mat. **10**, 660 (2011)
17. Cunha, R.O., Holanda, J., Vilela-Leão, L.H., Azevedo, A., Rodríguez-Suárez, R.L., Rezende, S. M.: Nonlinear dynamics of three-magnon process driven by ferromagnetic resonance in yttrium iron garnet. Appl. Phys. Lett. **106**, 192403 (2015)
18. Zakharov, V.E., L'vov, V.S., Starobinets, S.S.: Spin-wave turbulence beyond the parametric excitation threshold. Usp. Fiz. Nauk. **114**, 609 (1974). [Sov. Phys. Usp. 17, 896 (1975)]
19. de Araújo, C.B.: Quantum-statistical theory of the nonlinear excitation of magnons in parallel pumping experiments. Phys. Rev. B. **10**, 3961 (1974)
20. Balucani, U., Barocchi, F., Tognetti, V.: Green's-function theory of a damped Boson system (Application to an interacting Magnon system). Phys. Rev. A. **5**, 442 (1972)
21. ter Haar, D.: Theory and applications of the density matrix. Rep. Prog. Phys. **24**, 304 (1961)
22. Hartwick, T.S., Peressini, E.R., Weiss, M.T.: Subsidiary resonance in YIG. J. Appl. Phys. **32**, 223 S (1961)
23. Wang, S., Thomas, G., Hsu, T.-l.: Standing-spin-wave modulation of the reflected microwave power in YIG. J. Appl. Phys. **39**, 2719 (1968)

24. L'vov, V.S., Musher, S.L., Starobinets, S.S.: Theory of magnetization self-oscillations on parametric excitation of spin waves. Zh. Eksp. Teor. Fiz. **64**, 1074 (1973) [Sov. Phys.-JETP **37**, 546 (1973)]
25. Ozhogin, V.I., Yakubovskii, A.Y.: Parametric pairs in antiferromagnet with easy plane anisotropy. Z. Eksp. Teor. Fiz. **67**, 287 (1974). [Sov. Phys. JETP 40, 144 (1975)]
26. Nakamura, K., Ohta, S., Kawasaki, K.: Chaotic states of ferromagnets in strong parallel pumping fields. J. Phys. C: D. **15**, L143 (1982)
27. Gibson, G., Jeffries, C.: Observation of period doubling and chaos in spin-wave instabilities in yttrium iron garnet. Phys. Rev. A. **29**, 811 (1984)
28. Bryant, P., Jeffries, C., Nakamura, K.: Spin-wave nonlinear dynamics in an yttrium iron garnet sphere. Phys. Rev. Lett. **60**, 1185 (1988)
29. Rezende, S.M., de Aguiar, F.M.: Spin-wave instabilities, auto-oscillations, and chaos in yttrium-iron-garnet. Proc. IEEE. **78**, 893 (1990)
30. Rezende, S.M., Azevedo, A.: Self-oscillations in spin-wave instabilities. Phys. Rev. B. **45**(10), 387 (1992)
31. Suhl, H., Zhang, X.Y.: Theory of auto-oscillations in high-power ferromagnetic resonance. Phys. Rev. B. **38**, 4893 (1988)
32. Rezende, S.M., Azevedo, A., de Aguiar, F.M.: Spin-wave instabilities, auto-oscillations, chaos, and control of chaos in YIG spheres. In: Cottam, M.G. (ed.) Linear and Nonlinear Spin Waves in Magnetic Films and Superlattices. World Scientific, Singapore (1994)
33. de Aguiar, F.M., Azevedo, A., Rezende, S.M.: Characterization of strange attractors in spin-wave chaos. Phys. Rev. B. **39**(9), 448 (1989)

## Further Reading

Akhiezer, A.I., Bar'yakhtar, V.G., Peletminskii, S.V.: Spin Waves. North-Holland, Amsterdam (1968)
Cohen-Tannoudji, C., Diu, B., Laloë, F.: Quantum Mechanics. Wiley, New York (1977)
Cottam, M.G. (ed.): Linear and nonlinear spin waves in magnetic films and superlattices. World Scientific, Singapore (1994)
Gurevich, A.G., Melkov, G.A.: Magnetization Oscillations and Waves. CRC, Boca Raton (1994)
Kabos, P., Stalmachov, V.S.: Magnetostatic Waves and Their Applications. Chapman and Hall, London (1994)
Keffer, F.: Spin Waves. In: Flugge, S. (ed.) Handbuch der Physik, vol. XVIII/B. Springer, Berlin (1966)
Lax, B., Button, K.J.: Microwave Ferrites and Ferrimagnetics. McGraw-Hill, New York (1962)
L'vov, V.S.: Turbulence Under Parametric Excitation, Applications to Magnets. Springer, Berlin (1994)
Nussenzweig, H.M.: Topics in Quantum Optics. Gordon and Breach, New York (1973)
Pozar, D.M.: Microwave Engineering, 4th edn. Wiley, New York (2012)
Sparks, M.: Ferromagnetic Relaxation. McGraw-Hill, New York (1964)
Stancil, D.D., Prabhakar, A.: Spin Waves: Theory and Applications. Springer Science, New York (2009)
White, R.M.: Quantum Theory of Magnetism, 3rd edn. Springer, Berlin (2007)
Wigen, P.E. (ed.): Nonlinear phenomena and chaos in magnetic materials. World Scientific, Singapore (1994)

# Bose–Einstein Condensation of Magnons

**7**

In this chapter we study the phenomenon of quasi-equilibrium Bose–Einstein condensation (BEC) of magnons at room temperature, experimentally observed in films of yttrium iron garnet (YIG) parametrically excited by microwave radiation. In the first section we briefly present the concept of BEC, as observed in systems of boson particles in ultralow temperatures, and describe some of the experimental results obtained with magnons excited by microwave parallel pumping in YIG films at room temperature. Then, in three sections, we present a theoretical model, based on the formalism used in previous chapters, for the dynamics of the magnon gas driven by a microwave field far out of equilibrium, that provides rigorous support for the formation of a BEC of magnons as observed experimentally. The theory demonstrates that if the microwave-driving power exceeds a threshold value, the nonlinear magnetic interactions create cooperative mechanisms for the onset of a phase transition, leading to spontaneous quantum coherence. The theory also leads to an equation for the wave function of the condensed magnons that is similar to the one established for other BEC systems.

## 7.1 Experimental Observation of Magnon Condensation

### 7.1.1 Bose–Einstein Condensation

The phenomenon of Bose–Einstein condensation (BEC) was predicted by Albert Einstein in 1924 on the basis of ideas of Satyendra Bose concerning photons [1–3]. The BEC consists in the formation of a collective state in which a macroscopic number of particles occupy the same quantum state, or a small number of closely spaced states, such that the system is governed by a single wave function. Under appropriate conditions this phenomenon occurs in systems of particles, or quasiparticles, obeying Bose–Einstein statistics and whose total number is conserved. The formation of a BEC is characterized by a phase transition that falls

© Springer Nature Switzerland AG 2020
S. M. Rezende, *Fundamentals of Magnonics*, Lecture Notes in Physics 969,
https://doi.org/10.1007/978-3-030-41317-0_7

in the same category of other transitions to condensed quantum phases, like super-fluidity and superconductivity [4].

It took several decades after the theoretical prediction for the experimental realization of a BEC. Only in 1995 two different groups achieved experimental observation of BEC, using dilute atomic alkali gases trapped by a magnetic field and subject to very low temperatures [5, 6]. The observation of BEC triggered intense theoretical and experimental research in the field and today it is recognized that there are three types of BEC [7]. The first corresponds to the original proposal, realized by the atomic alkali gases, in which the system consists of boson particles in equilibrium at very low temperatures, which undergo condensation when their de Broglie thermal wave length becomes larger than their mean separation distance.

A second type of BEC is the one of boson-like quasiparticles, that is, those associated to elementary excitations in solids (e.g., phonons, excitons, magnons, hybrid excitations, etc.), when in equilibrium at extremely low temperatures. A well-studied case is that of an exciton-polariton system confined in microcavities (a near two-dimensional sheet), exhibiting the classic features of a BEC [8, 9]. This type of BEC has also been observed in spin systems of certain quantum magnets with axial symmetry at ultralow temperatures [10].

Finally, a third type of BEC is the case of boson-like quasiparticles associated to elementary excitations in solids that are driven out of equilibrium by some external perturbation so as to increase the particle density above a critical value. As predicted by Fröhlich [11], a system of bosons in quasi-equilibrium can undergo Bose–Einstein condensation even at relatively high temperatures, if the flow rate of energy pumped into the system exceeds a critical value. This type of BEC was proposed to occur in certain biosystems [12], but was actually observed at room temperature only in a gas of magnons generated by parallel pumping in a film of yttrium iron garnet [13]. This is the system that will be studied in this chapter.

A basic property of the Bose–Einstein condensation phenomenon can be understood by considering the case of noninteracting bosons in thermal equilibrium, at a temperature $T$, and described by the standard grand canonical ensemble. The mean number of particles in the single-particle energy eigenstate $i$ with energy $\varepsilon_i$ has the standard Bose–Einstein distribution form

$$n_{BE}(\varepsilon_i, \mu, T) = \frac{1}{e^{(\varepsilon_i - \mu)/k_B T} - 1},  \tag{7.1}$$

where $\mu$ is the chemical potential. Considering that the system has a fixed number of particles $N$, we have

$$N = \sum_i n_{BE}(\varepsilon_i, \mu, T).  \tag{7.2a}$$

Consider a system with a distribution of energy levels such that in some region it has the shape of a well, with minimum energy $\varepsilon_0$. Equations (7.1) and (7.2a) show that for a given number $N$, the value of the chemical potential is determined by the

value of the temperature. At higher temperatures the chemical potential lies below
$\varepsilon_0$. Then, clearly, since $N$ is fixed, as $T$ decreases the value of $\mu$ increases and
approaches $\varepsilon_0$. At the critical temperature $T_c$ for which $\mu \rightarrow \varepsilon_0$ the occupation of
the lowest single-particle state (energy $\varepsilon_0$) is of order $N$, while the other $n_i$ are still
generally of order unity or less, so that the system undergoes Bose-Einstein condensation. This simple view is valid to all types of BEC. However, in the third type of
BEC described before, the temperature is kept fixed and what makes $\mu \rightarrow \varepsilon_0$ is the
increase in the number of quasiparticles by means of some external pumping. This is
the case of the magnon BEC that will be presented in the next section.

### 7.1.2   Experiments with BEC of Parametrically Driven Magnons

The phenomenon of room-temperature nonequilibrium Bose–Einstein condensation
of parametrically driven magnons was proposed theoretically in a series of articles
[14–16] based on quite simple thermodynamic arguments. Perhaps, for this reason,
the proposal was not immediately tackled by experimentalists. It took almost two
decades for the experimental realization of the BEC of magnons and the publication
of very exciting results by several authors [13, 17–20].

All reported experiments have been carried out with single-crystal YIG films
magnetized in the plane, in which, as studied in Chap. 4, magnons have a well-
shaped dispersion relation, with minimum energy for the wave vector along the
applied magnetic field. Since the energy well corresponds to magnons with $k > 0$,
the density of states has a peak at the minimum energy, favoring the formation of the
BEC. Another important property of YIG for the realization of the nonequilibrium
BEC is that it has a very long spin–lattice relaxation time, above 1 µs, that is much
longer than the magnon–magnon thermalization time due to two- and four-magnon
relaxation mechanisms, which are as low as 100–200 ns. The basic idea of the
experiments is to inject magnons in the system by microwave parallel pumping, so as
to raise the chemical potential. This is done with pulsed pumping, such that in a short
interval of time the parametrically driven magnons are redistributed by magnon–
magnon scattering processes to form a condensate in states around the minimum
energy. The evolution in time of the magnon population in phase space is monitored
by time-resolved Brillouin light scattering (BLS).

Figure 7.1 shows a combination of two figures of the pioneering paper by
Demokritov and co-workers [13] that reported the first observation of the magnon
BEC. Figure 7.1a shows a schematic illustration of the experimental setup used to
pump magnons parametrically and to detect the evolution of the magnon population.
The experiments were performed with optically transparent YIG films of 2–10 µm
thickness and lateral sizes of 2 mm $\times$ 20 mm, grown on GGG. A microstrip
resonator underneath the YIG film creates a pulsed microwave pumping field with
frequency in the range of 6–9 GHz and duration of several µs. The film was placed in
a uniform static magnetic field, $H$, up to 1 kOe. The pumping process is illustrated in
the inset of Fig. 7.1a. As studied in Chap. 4, the low-frequency part of the magnon
dispersion with wave vectors parallel to the field has a minimum at $k_m = 5 \times 10^4$,

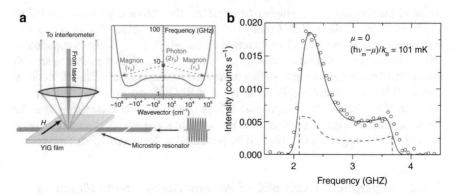

**Fig. 7.1** (a) Illustration of the setup for magnon excitation and detection [13]. The resonator attached to the bottom of the YIG film is fed by microwave pulses. The laser beam is focused onto the resonator, and the scattered light is directed to the Fabry–Perot interferometer. The inset shows the process of creation of two magnons by a microwave photon. The low-frequency part of the magnon dispersion for the field applied in the film plane is shown by the solid line. It has a minimum at $k_m = 5 \times 10^4$ cm$^{-1}$. The wave vector interval indicated by the red hatching corresponds to the interval of the wave vectors accessible for Brillouin light scattering (BLS). (b) BLS spectrum of thermal magnons recorded without pumping. The reduced density of states, shown by the dashed line, is obtained from the fit of the experimental data (solid line) to a calculation using the distribution in Eq. (7.1) with zero chemical potential, $\mu = 0$, where $\nu_m$ is the minimum magnon frequency and $h$ is Planck's constant [13]. Reproduced from S.O. Demokritov et al., Nature **443**, 430 (2006), with permission of Springer-Nature

shown by the solid line in the log–scale plot. A microwave photon with a certain frequency $\nu_p$ creates two primary excited magnons with frequency $\nu_p/2$ and opposite wave vectors. These primary magnons relax fast and create a quasi-equilibrium distribution of thermalized magnons, forming a magnon gas. As the chemical potential of the gas increases with pumping power, a possible BEC transition can take place near the minimum in the spectrum, as it corresponds to the state with the absolute minimum in magnon energy.

The distribution of the magnons over the spectrum was monitored with BLS spectroscopy. As shown in Fig. 7.1a, the incident laser beam is focused onto the resonator. The beam passes through the YIG film, is reflected by the resonator, and passes through the film again, in such a way that the reflected beam acts as the incident laser in the diagram of Fig. 2.17. The wave vector of the magnon scatterer, given by the difference between the wave vectors of the incident and scattered lights, is determined by the angle of the scattered beam [21]. The scattered light is collected by a wide-aperture objective lens and sent to the interferometer for frequency analysis of light photons inelastically scattered by the magnons. This approach allows a simultaneous detection of the magnons in a wide interval of in-plane wave vectors, estimated as $\pm 2 \times 10^5$ cm$^{-1}$, which exceeds $k_m$, as indicated by the red hatching in the inset of Fig. 7.1a. Thus, the BLS setup is able to detect all magnons at and close to the frequency minimum, where the condensation takes place. The BLS experiments are performed with time resolution, wherein the start of

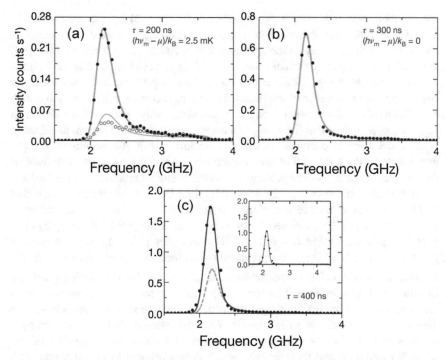

**Fig. 7.2** BLS spectra from pumped magnons at different delay times: (**a**) 200 ns; (**b**) 300 ns; (**c**) 400 ns. In (**a**) the open circles represent the data recorded at $P = 4$ W. All other spectra (solid circles) were measured $P = 5.9$ W. The solid lines represent fits of the spectra based on the equation for Bose–Einstein distribution with the chemical potential as a fitting parameter [13]. Reproduced from S.O. Demokritov et al., Nature **443**, 430 (2006), with permission of Springer-Nature

the pumping pulse plays the role of the reference stroboscopic clock, so that it is possible to monitor the time evolution of the frequency-dependent magnon density. Figure 7.1b shows the BLS spectrum of thermal magnons recorded without pumping. The reduced density of states, shown by the dashed line, is obtained from the fit of the experimental data (solid line) with a calculation using the distribution in Eq. (7.1) with zero chemical potential, $\mu = 0$, where $\nu_m$ is the minimum frequency of magnons, and $h$ is Planck's constant [13].

Figure 7.2 shows the BLS spectra of the pumped magnons recorded at different delay times, $\tau$, after the start of the pumping pulse that has duration of 1 µs and repetition period 20 µs [13]. One can see in Fig. 7.2a that with a pumping microwave power of $P = 4$ W the spectrum is not much different from the one in Fig. 7.1b, indicating that the system is essentially in thermal equilibrium. However, for $P = 5.9$ W, the magnon population at the bottom of the spectrum is already appreciable in a time $\tau = 200$ ns, and it continues to increase with time and concentrate around the minimum frequency, as shown in Fig. 7.2b, c. The solid lines in the figures represent the fits of a calculation based on the distribution of Eq. (7.1), with the chemical potential used as a fitting parameter. Thus, all data for the pumping power $P = 5.9$ W are described very well by the Bose–Einstein statistics at room temperature and a non-zero chemical potential. As shown in Fig. 7.2a, the fit of theory to the spectrum at $\tau = 200$ ns yields a distance between the minimum energy and the chemical potential, in units of temperature, of 2.5 mK. The other panels in Fig. 7.2 show that the magnon population shifts toward lower frequencies as the number of the pumped magnons increases with time, and that at $\tau = 200$ ns the redistribution process is still under way. However, as shown in Fig. 7.2b, c, for $\tau > 300$ ns the fitting of the spectra with a Bose–Einstein distribution is obtained with a chemical potential coinciding with the minimum energy, demonstrating the formation of the magnon BEC. At longer times the amplitude of the spectrum eventually decays away due to the magnetic relaxation to the lattice.

Additional confirmation of the non-equilibrium BEC of magnons was provided by measurements of the kinetics of the magnon population in phase space, reported in another article [18] published by authors of the same group of Ref. [13]. The experiments were carried out with a YIG film under pulsed microwave parallel pumping, similarly to the setup previously described, but using resolution in the in-plane wave vector. The resolution in $\vec{k}_m$ is achieved with the scattering geometry illustrated in Fig. 7.3a. By using a movable pinhole aperture it is possible to select to the wave vector of the scattered light and hence also of the magnons participating in the inelastic scattering, as shown in Fig. 7.3b. The selected light is analyzed by a six-pass tandem Fabry–Perot interferometer to determine the frequency of the scattering magnons and their density, which is proportional to the BLS intensity. The setup allows the detection of magnons with wave vectors up to approximately $4.2 \times 10^4$ cm$^{-1}$, with a resolution of $3 \times 10^3$ cm$^{-1}$, determined by the diameter of the pinhole.

The panels in Fig. 7.3c–h show graphs combining the information about the magnon dispersion spectrum and the color-coded representation of the measured BLS intensity, reflecting the distributions of magnon density in wave vector space $k_{//}$, $k_{\perp}$, representing the components parallel and perpendicular to the field. Each panel corresponds to a different delay after the start of the pumping pulse, with duration 30 ns. The measurements were made with $H = 1.0$ kOe, $f_p = 8.24$ GHz, and $P = 8$ W. Lines in the graphs are the constant frequency contours, whereas the cross indicates the point of minimum frequency of magnons, the bottom of the dispersion spectrum, 2.91 GHz. Analysis of the panels in Fig. 7.3 reveals the dynamic redistribution of the parametrically pumped magnons in phase space.

Panel (c) in Fig. 7.3 shows that, already at 20 ns after the start of the pumping pulse, the primary parametric magnons appear. They have frequency 4.1 GHz, corresponding to half the pumping frequency, and wave vectors with components parallel and perpendicular to the field. The pumping pulse finishes at a time 30 ns, but the redistribution of the magnon population over the phase space evolves continuously in time, as shown in the other panels. At 40 ns panel (d) shows that the magnon population has already shifted toward the point corresponding to the bottom of the magnon spectrum. At 200 ns [panel (g)] the magnons mainly occupy states in the proximity of the bottom of the spectrum, but the width of their distribution is still rather large, $\Delta k > 10^4$ cm$^{-1}$. Further modification of this distribution can be described as a gradual narrowing with a rate that is significantly

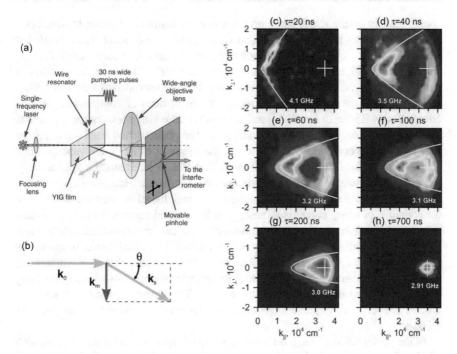

**Fig. 7.3** (**a**) Illustration of the experimental setup for magnon excitation with microwave pumping and detection by Brillouin light scattering with wave vector resolution [18]. (**b**) Diagram indicating the wave vector $k_m$ of the probed magnon. (**c–h**) Graphs of the magnon spectrum in the form of constant frequency contours in the two-dimensional magnon wave vector space, with color-coded representation of normalized BLS intensity. Each panel corresponds to a different delay after the start of the microwave pumping pulse, as indicated. The cross indicates the position of the bottom of the magnon spectrum. Measurements made with $H = 1.0$ kOe, $f_p = 8.24$ GHz, and $P = 8$ W [18]. Reprinted with permission from V.E. Demidov et al., Phys. Rev. Lett. **101**, 257201 (2008). Copyright (2008) by the American Physical Society

smaller than that of the first stage. At the end of this stage, at $\tau \approx 700$ ns, a very narrow distribution of magnons is formed having the form of a peak centered at the position $k_{//}^{min}, k_{\perp}^{min}$, corresponding to the absolute minimum of the magnon energy. This peak-like distribution provides a clear evidence of the formation of the Bose–Einstein magnon condensate. Further experimental results will be presented in the following sections to illustrate the application of the theoretical interpretation of the phenomenon.

## 7.2    Theory for the Dynamics of the Microwave-Driven Magnon Gas in $k$-Space

In the original concept of the phenomenon of Bose–Einstein condensation, the basic idea was that by lowering the temperature of a gas of integer spin particles, or bosons, it would be possible to cause a significant fraction of the particles to spontaneously enter a single quantum state, so as to form a condensate, in which they act collectively as a coherent, classical wave. Later, the concept was extended to quasiparticles, or elementary excitations in solids, and it became clear that the condensate could be formed by pumping the system to increase the number of particles instead of lowering the temperature. In the experimental search for BEC, several systems were found that exhibited some features of condensates, but not others. Today it is recognized [22] that three conditions are necessary for a system to exhibit a true Bose–Einstein condensation: (1) The phenomenon must show a phase transition to the condensate state through the behavior of some relevant quantity; (2) The condensate must have coherence, that has to be attained spontaneously. Experiments showing interference between two condensates are generally regarded as more convincing evidence of coherence, and therefore condensation, than observations of many particles seemingly in the ground state; and (3) The condensate must be characterized by a wave function, that can be measured in many ways. In this and the following sections we will present a theoretical model based on the quantum formulation of spin waves, studied in Chaps. 3 and 6, that demonstrates that the magnon system in a YIG film under parametric pumping satisfies all necessary conditions for the Bose–Einstein condensation. The model is based on a series of articles [23–26] that explained the experimental results presented in the previous section.

In the presentation of the model we shall use the approximate analytic dispersion relation for magnons with wave vector in the plane of a thin magnetic film, given by Eq. (4.40)

$$\omega_k^2 = \gamma^2 [H_0 + Dk^2 + 4\pi M (1 - F_k) \sin^2 \theta_k] [H_0 + Dk^2 + 4\pi M F_k], \qquad (7.2)$$

where $\theta_k$ is the angle between the wave vector $\vec{k}$ and the applied field $\vec{H}_0$ in the plane and the factor $F_k$ is

$$F_k = (1 - e^{-kt})/kt, \qquad (7.3)$$

where $t$ is the film thickness. Figure 7.4 shows the magnon dispersion relations calculated with Eqs. (7.2) and (7.3) for a 5 μm thick YIG film under a field of 1.0 kOe applied in the plane, for several angles of the in-plane wave vector. For angles $\theta_k < 50°$ the dispersion relations exhibit a minimum due to the competition between the dipolar and exchange energies, and the absolute minimum frequency of $f_{min} \approx 2.9$ GHz occurs for $\theta_k = 0$ and $k \approx 5.4 \times 10^4$ cm$^{-1}$.

In the experiments of [13, 17–19], briefly described in the previous section, magnon pairs are parametrically driven in a YIG film at room temperature by parallel pumping with short microwave pulses with frequency $f_p$ and studied by Brillouin light scattering. As studied in Chap. 6, if the microwave power exceeds a threshold value, there is a large increase in the population of the magnons with frequencies in a narrow range around $f_p/2$. Then the energy of these primary magnons quickly redistributes through modes with lower frequencies down to the minimum frequency $f_{min}$ as a result of magnon scattering. During a time interval of about 1 μs after the

**Fig. 7.4** Magnon dispersion relations calculated with Eqs. (7.2) and (7.3) for a 5 μm thick YIG film magnetized in the plane by a magnetic field of 1.0 kOe

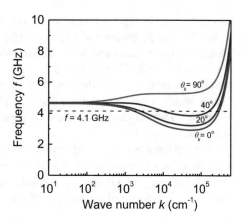

start of the driving, the number of additional magnons pumped into the system remains approximately the same. This number, which determines the chemical potential, can be calculated with the results obtained in the study of the parallel pumping process. For completeness we repeat the main results here. The driving Hamiltonian for the parallel pumping process follows from the Zeeman interaction of the microwave pumping field $\hat{z}h_p \cos(\omega_p t)$ with the magnetic system. For a magnetic film one can express the Zeeman interaction in terms of the magnon operators using the transformations in Eqs. (3.22), (3.30), and (4.36). Keeping only terms that conserve energy one can show that the driving Hamiltonian for a ferromagnetic film becomes

$$H_I(t) = -\frac{1}{2}\hbar\sum_k h_p\rho_k e^{-i\omega_p t}c_k^\dagger c_{-k}^\dagger + H.c.,\qquad(7.4)$$

where

$$\rho_k = \gamma u_k v_k = \gamma\omega_M\left[(1 - F_k)\sin^2\theta_k - F_k\right]/4\omega_k\qquad(7.5)$$

represents the coupling of the pumping field with the $\vec{k}, -\vec{k}$ magnon pair with frequency $\omega_k$ equal or close to $\omega_p/2$. Note that for a thick film, or a large wave vector, or a combination of both such that $kt >> 1$, Eq. (7.3) gives $F_k < < 1$. In this case the coupling coefficient approaches the value for bulk samples given in Eq. (6.39), $\rho_k = \gamma\omega_M\sin^2\theta_k/4\omega_k$. This is maximum for waves propagating perpendicularly to the field, since they have the largest ellipticity, and vanishes for waves with $\vec{k}$ along the field. However, in films with $kt$ on the order of 1 or less, $F_k$ is finite and the parallel pumping field can drive waves with any value of $\theta_k$. As seen in Fig. 7.4, in a YIG film with $t = 5$ μm, in a field $H = 1.0$ kOe, magnons with frequency 4.1 GHz and $\theta_k = 0$ can have two values for $k$, approximately, $2 \times 10^3$ cm$^{-1}$ and $5 \times 10^5$ cm$^{-1}$. The first value corresponds to $kt \cong 1$ and $F_k \cong 0.6$, while the second corresponds to $kt \cong 250$ and $F_k{\sim}0$. This means that magnons with frequency 4.1 GHz and $k \sim 2 \times 10^3$ cm$^{-1}$ with $\theta_k = 0$ have a finite ellipticity and can be parallel pumped. In fact, as can be seen in Fig. 7.4, for $H = 1.0$ kOe only waves with $\theta_k$ in the range of 0 to about 50° can be pumped at $f_p/2 = 4.1$ GHz. It turns out that as $\theta_k$ increases with fixed frequency, the wave vector$k$ increases so that $F_k$ decreases. In a film with thickness $t = 5$ μm this approximately compensates the increase in the $\sin^2\theta_k$ term so that the factor $u_k v_k$ which determines the parallel pumping coupling remains about 0.2 in the whole range of $\theta_k$ from 0 to 50°. The wide range of wave vector angles of the pumped primary magnons is clearly seen in the experiments of [18], as shown in Fig. 7.3c.

The Heisenberg equation of motion for the operators $c_k$ and $c_k^\dagger$ with the Hamiltonian $H = H_0 + H_I(t)$ given by Eqs. (4.37) and (7.4) can be easily solved assuming that the pumping field is applied at $t = 0$. From them we obtain for the evolution of the expectation value of the number of magnons

$$\langle n_k(t)\rangle = \langle n_k(0)\rangle e^{2\lambda_k t}, \tag{7.6a}$$

where

$$\lambda_k = [(\hbar\rho_k)^2 - \Delta\omega_k^2]^{1/2} - \eta_k, \tag{7.6b}$$

where $\langle n_k(0)\rangle$ is assumed to be the thermal number of magnons, $\Delta\omega_k = \omega_k - \omega_p/2$ is the detuning from the frequency of maximum pumping strength, and $\eta_k$ is the magnon relaxation rate which was introduced phenomenologically in the equations of motion. Equations (7.6a, b) express the well-known effect of the parallel pumping excitation. Magnon pairs with frequency $\omega_k$ equal or close to $\omega_p/2$ and wave vectors $\vec{k}, -\vec{k}$ determined by the dispersion relation are driven parametrically and their population grow exponentially when the field amplitude exceeds a critical value $h_{crit}$, given by the condition $\lambda_k = 0$ in (7.6b), as in Eq. (6.44)

$$h_{crit} = (\eta_k^2 + \Delta\omega_k^2)^{1/2}/\rho_k. \tag{7.7}$$

As studied in Chap. 6, the large increase in the magnon population enhances the nonlinear interactions causing a reaction that limits its growth. Due to energy and momentum conservation the important mechanism in this process is the four-magnon interaction [27], which can be represented by the Hamiltonian in Eq. (6.65)

$$H^{(4)} = \hbar \sum_{k,k'} \left( \frac{1}{2} S_{kk'} c_k^\dagger c_{-k}^\dagger c_{k'} c_{-k'} + T_{kk'} c_k^\dagger c_{k'}^\dagger c_k c_{k'} \right), \tag{7.8}$$

where the interaction coefficients are determined mainly by the dipolar and exchange energies. For the $k$-values relevant in the experiments the contribution from the exchange energy is negligible compared to the dipolar. From the full expression for the dipolar four-magnon interaction, that led to Eq. (3.104), using $F_k << 1$, $u_k \sim 1$ and $v_k << 1$, valid for the conditions of the experiments, one can show that the coefficients in the Eq. (7.8) are given approximately by $S_{kk'} = 2T_{kk'} = 2\omega_M/NS$.

Using the total Hamiltonian with the interaction term (7.8), one can write the Heisenberg equations for the operators $c_k$ and $c_k^\dagger$, from which several quantities of interest can be obtained. One of them is the correlation function $\sigma_k$ defined by [27]

$$\sigma_k = <c_k c_{-k}> = n_k e^{i\phi_k} e^{-i2\omega_k t}, \tag{7.9}$$

where $n_k$ is the magnon number operator and $\phi_k$ the phase between the states of the pair. From the equation of motion for $\sigma_k$ it can be shown that for $h_p > h_{crit}$, in steady state, the number of pumped magnons with wave vector $\vec{k}$ is

$$\langle n_k \rangle_{ss} = \frac{[(h_p \rho_k)^2 - \eta_k^2]^{1/2} - |\Delta\omega_k|}{2 V_{(4)}}, \tag{7.10}$$

where

$$V_{(4)} = S_{kk} + 2T_{kk} = 4\omega_M / NS. \tag{7.11}$$

It can also be shown that the phase $\phi_k$ varies from $-\pi/2$ to $\pi$ as $h_p$ increases from $h_{crit}$ to infinity. In the range of pumping power of the experiments $\phi_k \sim -\pi/2$. In the reported experiments the minimum $h_{crit}$ corresponds to a critical power $p_{c1}$ determined by the experimental geometry and the magnon relaxation rate in YIG. Using the fact that the driving microwave power $p$ is proportional to $h_p^2$, we can write from (7.10) an expression for the steady-state number of parametric magnons with frequency $\omega_k = \omega_p/2$ as a function of power,

$$\langle n_k \rangle_{ss} = \frac{[(p - p_{c1})/p_{c1}]^{1/2}}{2 V_{(4)}/\eta_m}, \tag{7.12}$$

where $p_{c1}$ is the critical power for driving parametrically magnons with relaxation rate $\eta_m$. Using numbers appropriate for the experiments [13], $\eta_m = 5 \times 10^7 \, \text{s}^{-1}$, $p_{c1} = 62.5 \, \text{mW}$, $V_{(4)} NS = 4\omega_M = 1.24 \times 10^{11} \, \text{s}^{-1}$, for a driving power $p = 4 \, \text{W}$, Eq. (7.12) gives for the normalized number of parametric magnons $\langle n_k \rangle_{ss}/NS = 1.6 \times 10^{-3}$. The number of magnons pumped by the microwave field is actually larger than this because many modes with frequency in the vicinity of $\omega_p/2$ are also driven. From (7.12) one can write an approximate equation for the total number of magnons pumped into the system as

$$N_p = r_p n_H [(p - p_{c1})/p_{c1}]^{1/2}, \tag{7.13a}$$

where

$$n_H \equiv \eta_m / 2V_{(4)} = \eta_m NS / 8\omega_M \tag{7.13b}$$

and $r_p$ is a factor that represents the number of pumped modes weighted by a factor relative to the number of magnons of the mode with maximum coupling.

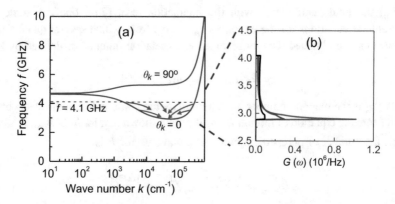

**Fig. 7.5** Illustration of the formation of the magnon BEC in a 5 μm thick YIG film with an in-plane field $H = 1.0$ kOe. (**a**) Dispersion relations for magnons with wave vector at an angle $\theta_k$ with the field showing the magnon redistribution. (**b**) Spectral density $G(\omega)$ calculated numerically with Eq. (7.18) for several values of the chemical potential $\mu$ in frequency units ($f_{min} = 2.9$ GHz and $h$ is Planck constant): black line- $\mu = 0$; blue line- $\mu/h = 2.75$ GHz; red line- $\mu/h = 2.898$ GHz

As described earlier, the primary magnons driven parametrically are quickly redistributed over a broad frequency range down to the minimum frequency $f_{min}$ by four-magnon scattering events, as illustrated in Fig. 7.5a. Since the spin-lattice relaxation time in YIG is much longer than the magnon–magnon decay time, the hot magnon gas remains practically decoupled from the lattice for several hundred ns with an essentially constant number of magnons. In this situation the occupation number of the system is given by the Bose–Einstein distribution as in Eq. (7.1). Without external driving the magnon system is in thermal equilibrium with the lattice and has uncertain number so that the chemical potential is $\mu = 0$. If a microwave driving is applied and the power exceeds the threshold for parallel pumping, the total number of particles in the magnon gas increases and can be expressed as

$$N_{\text{tot}} = \int D(\omega) n_{BE}(\omega, \mu, T) d\omega, \qquad (7.14)$$

where $D(\omega)$ is the magnon density of states and the integral in (7.14) is carried out over the whole range of magnon frequencies. Clearly, as the microwave power is raised the total number of magnons increases so that the effective temperature of the magnon gas and the chemical potential increase. Using (7.14) and the similar equation for the energy of the system it is possible to determine the values of $\mu$ and $T$ for a given $N_{tot}$. In the experiments with long pulses [13] the BLS spectra could be fitted with the spectral density function $D(\omega) n_{BE}(\omega, \mu, T)$, allowing the determination of $\mu$ and $T$ for each power value. At a high enough power, the chemical potential reaches the energy corresponding to $f_{min}$ resulting in an overpopulation of magnons with that frequency relative to the theoretical fit. It was then necessary to add a singularity at $f_{min}$ to fit the spectrum [13]. This was interpreted as a signature of the Bose–Einstein condensation of magnons, namely: when the number of magnons reaches a critical value defined by the condition $\mu_c = hf_{min}$ the gas is spontaneously divided in two parts, one with the magnons distributed according to $n_{BE}$ and another one with the magnons accumulated in the state of minimum energy.

The experiments with short microwave pulses (30 ns) [18] allowed the observation of the dynamics of the redistribution of energy from the primary magnons to the modes in the broader energy range and the formation of the sole BLS peak in a narrow range in phase space centered at the wave vector $k_0$ along the field and frequency $\omega_{k_0} = 2\pi f_{min}$. The behavior of the peak intensity and of the relaxation to the lattice with increasing microwave pumping power revealed that above a critical power level the magnons accumulated at the bottom of the spectrum develop a spontaneous emergence of coherence. The theory presented in this and the following sections show that the cooperative action of the magnon gas through the four-magnon interaction explains the observed spontaneous emergence of quantum coherence in the BEC.

We assume that after the hot magnon reservoir is formed by the redistribution of the primary magnons, the correlation between the phases of the magnon pairs lasts for a time that can be as large as $4/\eta_m$, which is about 100 ns in the experiments. This is a sufficient time for the four-magnon interaction to come into play for establishing a cooperative phenomenon to drive a specific $k$ mode. The effective driving Hamiltonian for this process is obtained from Eq. (7.8) by taking averages of pairs of destruction operators for reservoir magnons to form correlation functions as defined in (7.9), and is given by

$$H'(t) = \hbar \sum_{k_R} \frac{1}{2} S_{kk_R} n_{k_R} e^{i\phi_{k_R}} e^{-i2\omega_{k_R}t} c_k^\dagger c_{-k}^\dagger + H.c.. \tag{7.15}$$

This expression has the same form as the Hamiltonian (7.4) for parallel pumping, revealing that, under appropriate conditions, magnon pairs can be pumped out of equilibrium in the gas by other magnon pairs. This is the basic mechanism of the spontaneous emergence of the condensate. Since this is a nonlinear mechanism, we have to find the conditions for the critical number of magnons necessary to drive the process. Consider that the population of the primary magnons is distributed among the $N_R$ modes $k_R$ in the magnon reservoir, so that with Eq. (7.13a) we can write an expression for the average population of the modes as a function of the microwave power $p$

$$n_R = r n_H [(p - p_{c1})/p_{c1}]^{1/2}, \tag{7.16}$$

where $r = r_p/N_R$. If all the reservoir states had the same magnon number, the sum in $k_R$ in (7.15) would reproduce the density of states $D(\omega)$. Actually the number of magnons in each state $k_R$ depends on its energy and can be written approximately as $n_{k_R} = f_{BE}(\omega_{kR}) n_R$, where $f_{BE}(\omega_{kR})$ is a normalized Bose–Einstein distribution function such that its average over the frequency range of the reservoir modes is equal to unity, given by

$$f_{BE}(\omega) = n_{BE}(\omega)/C_{BE}, \tag{7.17a}$$

$$C_{BE} = \frac{1}{\Delta\omega_R} \int n_{BE} d\omega, \tag{7.17b}$$

where $\Delta\omega_R = \omega_p/2 - \omega_{k0}$ is the frequency range of the reservoir modes. Thus, the relevant quantity for determining the frequency dependence of the coefficient in the Hamiltonian (7.15) is the density of states weighted by the normalized Bose–Einstein distribution, given by

$$G(\omega) = D(\omega) f_{BE}(\omega). \tag{7.18}$$

Note that $f_{BE}(\omega)$ and $G(\omega)$ also vary with $\mu$ and $T$ but we omit them in the functions to simplify the notation. Figure 7.5b shows plots of (7.18) for several values of $\mu$ and the corresponding $T$ for a 5 μm thick YIG film. The density of states was calculated numerically using the dispersion relation (7.2a, 7.2b) by counting the number of states with $k_x = \pm n_x 2\pi/L_x$, $k_z = \pm n_z 2\pi/L_z$ having frequencies in discrete intervals $\delta\omega = 2\pi \times 1.0$ MHz in the range $0 - \omega_p/2$. The values of $\mu$ were chosen so that their differences to $\hbar\omega_{k0}$ are the same as the ones used in Ref. [13] to fit the measured BLS spectra with varying microwave power. The corresponding values of $T$ were estimated by the fits to the BLS spectra in [13]. The dimensions used to calculate the density of states are $L_x = L_z = 2$ mm. As expected, $G(\omega)$ has a peak at the minimum frequency that becomes sharper as the chemical potential rises and approaches the minimum energy. The consequence of this is that, as the microwave pumping power increases and $(\hbar\omega_{k0} - \mu)/k_B T$ becomes very small, the peak in $G(\omega)$ dominates the coefficient in the Hamiltonian (7.15), revealing that it is possible to establish a cooperative action of the modes with frequency $\omega_{k_R}$ close to $\omega_{k0}$ so as to drive magnon pairs nonlinearly as in the parallel pumping process. Considering that the pumping is effective for frequencies $\omega_{k_R}$ in the range $\omega_{k0} \pm \eta_m$, the sum over $k_R$ in (7.15) can be replaced by $D(\omega_{k0})\eta_m$, so that one can write a Hamiltonian for driving $\vec{k}_0, -\vec{k}_0$ magnon pairs as

$$H'_{eff}(t) \cong \hbar(h\rho)_{eff} e^{-i2\omega_{k0}t} c^\dagger_{k_0} c^\dagger_{-k_0} + H.c., \tag{7.19a}$$

where

$$(h\rho)_{eff} = -iG(\omega_{k0})\eta_m V_{(4)} n_R/2 \tag{7.19b}$$

represents an effective field proportional to the average number of magnons $n_R$ in the reservoir. Note that the factor $-i$ in (7.19b) arises from the phase between pairs that is approximately $-\pi/2$ in the range of power of interest. Similarly to the parallel pumping process, one can see that there is a critical number of reservoir modes above which they act cooperatively to pump the $\vec{k}_0, -\vec{k}_0$ magnons parametrically.

The condition $|(h\rho)_{eff}| = \eta_m$ gives for the critical average number of reservoir magnons

$$n_c = 2/[V_{(4)}G(\omega_{k0})]. \tag{7.20}$$

Since the Hamiltonian (7.19a) has the same form as (7.4), the population of the $k_0$ mode driven by the effective field and saturated by the effect of the four-magnon interaction is calculated in the same manner as done for the direct parallel pumping process. Thus, from (7.10) with $\Delta\omega_k = 0$, we have

$$n_{k0} = \frac{[|(h\rho)_{eff}|^2 - \eta_m^2]^{1/2}}{2V_{(4)}}. \tag{7.21}$$

Using (7.13b), (7.19b) and (7.20) in Eq. (7.21) one can write the population of the $k_0$ mode in terms of the average reservoir number $n_R$

$$n_{k0} = \frac{n_H}{n_c}(n_R^2 - n_c^2)^{1/2}. \tag{7.22}$$

Alternatively, $n_{k0}$ can be written in terms of the pumping microwave power using (7.16) and (7.20) in (7.22)

$$n_{k0} = n_H \left[\frac{p - p_{c2}}{p_{c2} - p_{c1}}\right]^{1/2}, \tag{7.23}$$

where $n_H$ is given by (7.13b) and

$$p_{c2} = p_{c1}\left\{1 + \frac{16}{[r\eta_m G(\omega_{k0})]^2}\right\} \tag{7.24}$$

is another threshold power level. Equation (7.23) reveals that for $p \geq p_{c2}$ the $k_0$ magnons are pumped-up out of equilibrium as a result of a spontaneous cooperative action of the reservoir modes and, as will be shown in next section, they are in coherent magnon states. This means that when the average reservoir magnon number reaches the critical value (7.20), the magnon gas separates in two parts, one in thermal equilibrium with the reservoir, having frequencies in a wide range, and one with a higher magnon number in a narrow range around the minimum frequency. This is one of the characteristic features of a Bose–Einstein condensate.

For future use, note that with Eqs. (7.16) and (7.24), for $p \geq p_{c2}$ the effective driving field (7.19b) can be expressed in terms of power as

$$(h\rho)_{eff} = -i\eta_m \left[\frac{p - p_{c2}}{p_{c2} - p_{c1}}\right]^{1/2}. \tag{7.25}$$

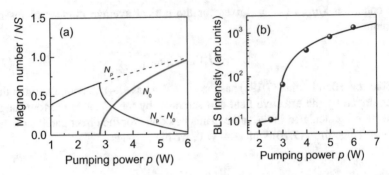

**Fig. 7.6** (a) Variation with microwave pumping power of the normalized number of primary pumped magnons $N_p$, of the BEC population $N_0$ and of the number of uncondensed magnons $N_p$—$N_0$. (b) Fit of the theory [23] for the number of BEC magnons, given by Eq. (7.23), to the BLS intensity versus microwave pumping power measured in Ref [17]. Reprinted with permission from S.M. Rezende, Phys. Rev. B **79**, 174411 (2009). Copyright (2009) by the American Physical Society

The calculations presented so far are valid for magnon pairs with frequencies and wave vectors in the vicinity of $\omega_{k0}$ and $k_0$, $-k_0$. The dynamics for several modes can be treated approximately assuming that the condensate consists of $p_{k0}$ modes, all governed by the single-mode equations. Thus, we can write the number of magnons in the condensate as

$$N_0 = p_{k0} n_{k0}. \tag{7.26}$$

Figure 7.6a shows the variation with microwave pumping power of the total number of magnons pumped into the system $N_p$, given by Eq. (7.13a), and the population of the BEC magnons $N_0$ calculated with Eqs. (7.23) using parameters obtained from fit of theory to data, $p_{k0} = 4.4 \times 10^3$ and $r_p = 5.0 \times 10^2$ [23]. Also shown is the number of uncondensed magnons $N_p - N_0$ as a function of $p$. Clearly, as the power increases above the critical value $p_{c2}$, the number of condensed magnons approaches the total number in the reservoir, while the number of uncondensed magnons vanishes. This is also a typical feature of a BEC. Figure 7.6b shows a fit of Eq. (7.23) to the experimental data for the BLS intensity versus microwave pumping power [17], evidencing the phase transition at $p_{c2} = 2.8W$, characterizing the onset of the Bose–Einstein condensation.

## 7.3    Quantum Coherence of the Bose–Einstein Magnon Condensate

For the study of the coherence properties of the $k_0$ mode pumped above the BEC threshold, we follow here the same procedure used to study the direct parallel pumping process in Sect. 6.3. The first step is to represent the magnon reservoir and its interactions with a specific $k$ mode by a Hamiltonian that allows a full description of the thermal and driving processes for the interacting magnon system. Similarly to Eq. (6.72), we write for the total Hamiltonian

$$H = H_0 + H^{(4)} + H_R + H_{RS} + H'_{eff}(t), \qquad (7.27)$$

where the four-magnon and the driving terms are given, respectively, by (7.8) and (7.19a, 7.19b), and

$$H_R = \hbar \sum_{k_R} \omega_{kR} c^{\dagger}_{kR} c_{kR} \qquad (7.28)$$

is the Hamiltonian for the magnon reservoir, assumed to be a system with large thermal capacity and in thermal equilibrium, and

$$H_{RS} = \hbar \sum_{k,k_R} \beta^*_{k,kR} c^{\dagger}_{kR} c_k + \beta_{k,kR} c_{kR} c^{\dagger}_k \qquad (7.29)$$

represents a linear interaction between the magnons $k$ and the heat reservoir. With the Heisenberg equation for the magnon operators for a mode $k$ in the vicinity of $k_0$, with the total Hamiltonian (7.27), and assuming that $n_k = n_{-k}$, we obtain

$$\frac{dc_k}{dt} = -(i\omega_k + \eta_m + i2V_{(4)}n_k)c_k - i(\hbar\rho)_{eff} e^{-i2\omega_0 t} c^{\dagger}_{-k} + F_k(t), \qquad (7.30)$$

where

$$\eta_m = \pi D(\omega_k) |\beta_{k,kR}|^2, \qquad (7.31a)$$

$$F_k(t) = -i \sum_{kR} \beta_{k,kR} c_{kR} e^{-i\omega_{kR} t}, \qquad (7.31b)$$

represent, respectively, the magnetic relaxation rate expressed in terms of the interaction between magnon $k$ and the heat reservoir, and a Langevin random force with correlators of Markoffian systems type [28]. Using Eq. (7.30) and the corresponding one for the operator $c^{\dagger}_{-k}$, transforming them to the representation of coherent magnon states $|\alpha_k\rangle$ and working with variables in a rotating frame $c_k |\alpha_k\rangle = \alpha_k(t) e^{-i\omega_k t} |\alpha_k\rangle$, we obtain an equation of motion for the coherent state eigenvalue with $k \approx k_0$,

$$\frac{d\alpha_k(t)}{dt} - \frac{2V^2_{(4)}}{\eta_m} \left( \frac{|(\hbar\rho)_{eff}|^2 - \eta_m^2}{4V^2_{(4)}} - |\alpha_k(t)|^4 \right) \alpha_k(t) = S_k(t), \qquad (7.32)$$

where

$$S_k(t) = F_k(t) e^{i\omega_k t} - i \frac{(\hbar\rho)_{eff}}{\eta_m} F^*_{-k}(t) e^{-i\omega_k t}. \qquad (7.33)$$

Equations (7.32) and (7.33) contain all the information carried by the equations of motion for the magnon operators. They show that the magnon modes with amplitude $\alpha_k$ are driven thermally by the hot magnon reservoir and also by an effective driving field. The solutions of (7.32) confirm the previous analysis. For negative values of the driving term $[|(h\rho)_{eff}|^2 - \eta_m^2]$ the magnon amplitudes are essentially the ones of the thermal reservoir. For positive values they grow exponentially and are limited by the effect of the four-magnon interactions. Above the threshold condition, the steady-state solution of (7.32) gives for the number of magnons $n_k = |\alpha_k|^2$ an expression identical to (7.21). The final step to obtain information about the coherence of the excited mode is to find an equation for the probability density $P(\alpha_k)$, defined in Sect. 3.3.3, that is stochastically equivalent to the Langevin equation. Using $\alpha_k = a_k \exp(i\varphi_k)$ we obtain a Fokker–Plank equation in the form [28].

$$\frac{\partial P}{\partial t'} + \frac{1}{x}\left(\frac{\partial}{\partial x}\right)[(A - x^4)x^2 P] = \frac{1}{x}\frac{\partial}{\partial x}\left(x\frac{\partial P}{\partial x}\right) + \frac{1}{x^2}\left(\frac{\partial^2 P}{\partial \varphi_k^2}\right),  \tag{7.34}$$

where

$$t' = (\bar{n}_{k0}^2 \eta_m^3/n_H^2)^{1/3} t \quad \text{and} \quad x = (2/n_H^2 \bar{n}_{k0})^{1/6} a_k  \tag{7.35a}$$

represent normalized time and magnon amplitude, and the parameter $A$ is given by

$$A = \left(\frac{2}{n_H^2 \bar{n}_{k0}}\right)^{2/3} \frac{[|(h\rho)_{eff}|^2 - \eta_m^2]^{1/2}}{2V_{(4)}} \equiv \left(\frac{2}{n_H^2 \bar{n}_{k0}}\right)^{2/3} n_{k0}^2.  \tag{7.35b}$$

Using Eqs. (7.22) and (7.23), the parameter $A$ can also be written in terms of the number of reservoir magnons, or in terms of the microwave pumping power $p$

$$A = \left[\frac{2n_H}{f_{BE}(\omega_{k0})n_R}\right]^{2/3}\left[\left(\frac{n_R}{n_c}\right)^2 - 1\right],  \tag{7.35c}$$

$$A = \left[\frac{2}{r f_{BE}(\omega_{k0})}\right]^{2/3}\left[\frac{p_{c1}}{p - p_{c1}}\right]^{1/3}\frac{(p - p_{c2})}{(p_{c2} - p_{c1})}.  \tag{7.35d}$$

Application of Eq. (7.34) to describe the full dynamics of the pulsed experiments must consider that the factors relating $t'$ to $t$ and $x$ to $a_k$, as well as the parameter $A$, are all time dependent. However, for typical numbers appropriate for the experiments $t' \sim t \times 2 \times 10^6$ s$^{-1}$, so that the dynamics of the pulsed experiments is relatively slow in the renormalized time scale. Thus, in a first approximation, we assume that all parameters are constant and obtain the stationary solution of (7.34) independent of $\varphi_k$ in the form,

$$P(x) = C\exp\left(\frac{1}{2}Ax^2 - \frac{1}{6}x^6\right),  \tag{7.36}$$

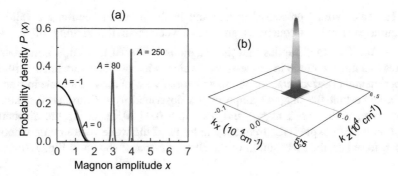

**Fig. 7.7** (a) Probability density of the magnon distribution for a microwave driven interacting magnon gas, calculated with Eq. (7.36) for several values of the parameter $A$ that represents the microwave pumping power [23]. Negative values correspond to $p < p_{c2}$; $A = 0$ corresponds to $p = p_{c2}$; $A = 80$ and 250 correspond, respectively, to $p/p_{c2} = 1.015$ and $1.047$. (b) Pictorial view of the condensate represented by the distribution in $k$-space of the magnon amplitude. Reprinted with permission from S.M. Rezende, Phys. Rev. B **79**, 174411 (2009). Copyright (2009) by the American Physical Society

where $C$ is a normalization constant such that the integral of $P(x)$ in the range of $x$ from zero to infinity is equal to unity. Note that for obtaining (7.36) all integration constants were set to zero to satisfy this condition. Figure 7.7a shows plots of $P(x)$ for four values of the parameter $A$, $-1$, 0, 80 and 250. In choosing the positive values we have considered parameters which enter in (7.35a) and (7.35d) appropriate for the experiments [23]: $p_{c1} = 0.0625$ W and $p_{c2} = 2.8$ W; $rf_{BE}(\omega_{k0}) \sim 7 \times 10^{-7}$, determined from Eqs. (7.18) and (7.24), and using $D(\omega_{k0}) \approx 10^5/\text{MHz}$, calculated numerically as described earlier, and $\eta_m/2\pi = 8$ MHz. With these numbers we obtain $A = 250$ for $p/p_{c2} = 1.047$, that is, just above the critical power for the BEC.

Equation (7.35d) shows that for a pumping power below the critical value, $p < p_{c2}$, the parameter $A$ is negative. In this case, the function $P(x)$ in (7.36) behaves as a Gaussian distribution, characteristic of systems in thermal equilibrium and described by incoherent magnon states. On the other hand, for $p > p_{c2}$, $A > 0$, and the stationary state consists of two components, a coherent one convoluted with a much smaller fluctuation with Gaussian distribution. Since the variance of $P(x)$ is proportional do $A^{-1}$, for $A >> 1$ the function $P(x)$ becomes a delta-like distribution, characteristic of a coherent magnon state [29]. Figure 7.7a shows that in the conditions of the experiments $P(x)$ becomes a delta-like function at power levels just above the critical value. Note that only in the presence of the four-magnon interaction the magnon state driven collectively by the reservoir modes is a coherent state [29]. Note also that $P(x)$ has a peak at $x_0 = A^{1/4}$, so that it represents a coherent state with an average number of magnons given by $x_0^2 = A^{1/2}$. From Eqs. (7.35a) and (7.35b) we see that this corresponds to a magnon number $a_0^2$ which is precisely the value $n_{k_0}$ given by Eq. (7.22). Thus, the magnon $\omega_{k0}$ driven cooperatively by the reservoir modes is in a quantum coherent state, demonstrating that the model satisfies an essential condition for characterizing the Bose–Einstein condensation.

The calculations presented in this and in the previous section are valid for magnon-pairs with frequencies and wave vectors in the vicinity of $\omega_{k0}$ and $\vec{k}_0, -\vec{k}_0$. This fact implies that the magnons in the BEC occupy a number of states in a narrow range in phase-space, which is in complete agreement with the experimental observations. To obtain a pictorial view of the condensate in $k$-space we assume that the magnon amplitude is described by a Gaussian function in frequency, with a peak at $f_{min}$ and linewidth 0.07 MHz. Using the dispersion relation we can express the frequency in terms of the wave vector components $k_x$ and $k_z$ to obtain the distribution of the BEC in $k$-space, shown in Fig. 7.7b.

## 7.4    Wave Function of the Magnon Condensate

Another important result of the theoretical model for the dynamics of the magnon system in $k$-space is the equation for the BEC wave function in configuration space. It is known that in BEC systems the condensate wave function is proportional to the order parameter. In the case of magnons, the order parameter is the small-signal magnetization, which, as we saw in Sect. 3.3.3, is proportional to the coherent state amplitude. Thus, we define the eigenfunction for the BEC of magnons as

$$\psi(\vec{r}, t) = 1/V^{1/2} \sum_k e^{i\vec{k}.\vec{r}} \alpha_k. \tag{7.37}$$

where $\alpha_k$ are the coherent state eigenvalues for $k$-states around $k_0$ and $V$ is the volume of the sample. It is important to note that the eigenfunction in (7.37) is a function of time and space, and since $|\alpha_k| = \langle n_k \rangle^{1/2}$ it implies that

$$\int d^3r \psi^* \psi = \sum_k n_k = N_0, \tag{7.38}$$

where $N_0$ is the number of magnons in the condensate given by Eq. (7.26). Consider that the wave function (7.37) corresponds to a narrow packet in wave vector space with a central wave number $k_0$ and with frequency around the minimum $\omega_{k0}$, so that the frequency of the states involved can be written as

$$\omega_k \cong \omega_{k0} + \lambda_x k_x^2 + \lambda_z k_z^2, \tag{7.39}$$

where

$$\lambda_x = \frac{1}{2} \frac{\partial^2 \omega_k}{\partial k_x^2}\bigg|_{k0}, \quad \lambda_z = \frac{1}{2} \frac{\partial^2 \omega_k}{\partial k_z^2}\bigg|_{k0}, \tag{7.40}$$

$x$ and $z$ being the coordinates in the film plane. Since the central mode corresponds to a pair $\vec{k}_0, -\vec{k}_0$, one can introduce a slowly varying envelope $\psi_0(\vec{r}, t)$ in a frame rotating with frequency $\omega_{k0}$ such that

$$\psi(\vec{r}, t) = 2 \cos(\vec{k}_0 \cdot \vec{r}) \psi_0(\vec{r}, t) e^{-i\omega_{k0} t}, \qquad (7.41)$$

where

$$\psi_0(\vec{r}, t) = \frac{e^{i\omega_{k0} t}}{V^{1/2}} \sum_{\delta k} e^{i\delta \vec{k} \cdot \vec{r}} \alpha_k \qquad (7.42)$$

and $\vec{k} = \vec{k}_0 + \delta \vec{k}$. Using the Hamiltonian for the system with a general four-magnon interaction and Eq. (7.19a) for the driving term one can write the Heisenberg equation for the magnon operator $c_k$ as

$$i \frac{dc_k}{dt} = (\omega_k - i\eta_m) c_k - |h\rho|_{eff} e^{-i2\omega_{k0} t} c_{-k}^\dagger +$$

$$V_{(4)} \sum_{1,2,3,4} (c_1^\dagger c_3 c_4 + c_2^\dagger c_3 c_4 + c_3^\dagger c_1 c_2 + c_4^\dagger c_1 c_3) \Delta(\vec{k}), \qquad (7.43)$$

where $V_{(4)} = 4 T_{1234} = 4 \omega_M / NS$, the summation runs over wave vectors $\vec{k}, -\vec{k}$, $\Delta(\vec{k})$ is the appropriate Kronecker delta for momentum conservation, and the relaxation rate $\eta_m$, assumed to be the same for the wave vectors involved, was introduced phenomenologically. If ones assumes that only one pair mode $\vec{k}_0, -\vec{k}_0$ is driven by the collective action, Eq. (7.43) and the corresponding one for the operator $c_{-k0}^\dagger$ can be solved in steady-state to give the population of the $k_0$-mode driven by the effective field and saturated by the four-magnon interaction. Considering that the four-magnon interaction process must conserve energy, one can do calculations similar to those for spin-wave solitons [30, 31] using Eqs. (7.39)–(7.42) and show that Eq. (7.43) leads to

$$i \frac{\partial \psi_0}{\partial t} = -i\eta_m \psi_0 - \lambda_x \frac{\partial^2 \psi_0}{\partial x^2} - \lambda_z \frac{\partial^2 \psi_0}{\partial z^2} + 2V_{(4)} V |\psi_0|^2 \psi_0 - |h\rho|_{eff} \psi_0^*. \qquad (7.44)$$

This equation has the form of the Gross–Pitaevskii equation (GPE) used to describe the wave function of several BEC systems [See Leggett, or Pitaevskii and S. Stringari]. Without the last term it is also the nonlinear Schrödinger equation used to study solitons [30, 31]. Note that the usual GPE has a term describing the non-uniform external potential, such as the trapping potential in atomic gas systems [4]. This is not present in Eq. (7.44) which has, instead, a nonuniform driving term $|h\rho|_{eff} \psi_0^*$ due to the spatial variation of the pumping microwave field. In order to check the consistency of the current approach with the results obtained from the analysis of the dynamics in wave vector space, let us obtain the solution of Eq. (7.44) in steady-state, $\partial/\partial t = 0$, considering a uniform driving field, $\partial/\partial x = \partial/\partial z = 0$. With these conditions we obtain from Eq. (7.44)

$$(2V_{(4)})^2 V|\psi_0|^2|\psi_0|^2 = |h\rho|^2_{eff} - \eta^2_m, \tag{7.45}$$

which, of course, is valid only for $|h\rho|^2_{eff} - \eta^2_m \geq 0$, or $p \geq p_{c2}$. Integrating Eq. (7.45) in the volume $V$ and using Eqs. (7.25) and (7.38), one can see that the number of magnons in the condensate is the same as in Eq. (7.26), as long as the factor $p_{k0}$ obeys the relation

$$\int d^3 r|\psi_0|^2|\psi_0|^2 = \frac{N_0^2}{p_{k0}^2 V}. \tag{7.46}$$

Note that Eq. (7.46) provides a formal definition of the factor $p_{k0}$ which was introduced earlier in an ad-hoc manner. We now assume that the pumping is uniform in the $x$ direction to reduce the problem to one dimension. This is the situation in the experiments of Ref. [32], where a thin wire or tape placed on top and close to the YIG film is used to generate the driving magnetic field. One can write the wave function in the same form as in other BEC systems

$$\psi_0 = (N_0)^{1/2}\chi(z), \tag{7.47}$$

where $\chi(z)$ is a normalized wave function independent of the number of magnons in the condensate. To find $\chi(z)$ we use Eq. (7.44) without the driving and relaxation terms, and consider that their effect is expressed in the number of magnons in the condensate given by (7.23) and (7.26). With $\partial/\partial t = \partial/\partial x = 0$ in Eq. (7.44) we obtain

$$\lambda_z \frac{d^2\chi}{dz^2} - b|\chi|^2\chi = 0, \tag{7.48}$$

where the coefficient $b$ obtained with Eqs. (7.23) and (7.26), and considering $p_{c2} >> p_{c1}$, is given by

$$b = \eta_m (p/p_{c2} - 1)^{1/2}. \tag{7.49}$$

Notice that there is a characteristic length associated with the Gross–Pitaevskii Eq. (7.48) given by

$$\xi = (\lambda_z/b)^{1/2}, \tag{7.50}$$

which is identified as the healing length [4]. With parameter values appropriate for the YIG film used in the experiments of Ref. [13], $H = 1$ kOe, $4\pi M = 1.76$ kG, exchange parameter $D = 5.4 \times 10^{-9}$ Oe cm$^2$ we obtain, numerically, from the dispersion relation $\lambda_z \approx 0.55$ s$^{-1}$ cm$^2$. Considering a pumping power $p = 2p_{c2}$ and using $\eta_m = 5 \times 10^7\,\mathrm{s}^{-1}$, we find for the healing length $\xi \approx 1\,\mu\mathrm{m}$. This value is very small compared to the typical length scale of the spatial variation of the driving field used in experiments to map the BEC wave function [32].

Equation (7.48) can be used to calculate the spatial distribution of the condensate for a nonuniform pumping power $p(z)$, which implies a spatially varying coefficient $b(z)$. Consider the *rf* field used for parallel pumping generated by a microwave current in a thin wire with axis in the $x$-direction at a distance $s$ from the YIG film. In this case, the power carried by the $z$-component of the field in the film plane, obtained with Ampere's law, is

$$\frac{p(z)}{p_{max}} = \left[\frac{h(z)}{h_{max}}\right]^2 = \frac{s^2}{(z^2 + s^2)^2}, \tag{7.51}$$

where $z = 0$ is the projection of the wire axis on the film plane. The profiles of the driving field and the power, given by Eq. (7.51), are shown in the inset of Fig. 7.8b for $s = 32\,\mu\mathrm{m}$, which is the distance inferred from the figures in Ref. [32]. Equation (7.48) can be integrated by numerical methods to find the wave function $\chi(z)$ for a spatially varying pumping power $p(z)$. This has been done for the power profile of a wire antenna, given by Eq. (7.51), and the corresponding $b(z)$ given by (7.49), assuming that $p_{max}$ is the applied microwave power $p$ [25]. Figure 7.8a shows the result obtained for the magnon condensate density, expressed by the wave function squared, calculated for $s = 32\,\mu\mathrm{m}$, critical power for BEC formation $p_{c2} = 0.8\,\mathrm{W}$, inferred from the data of Ref. [32], with two values of the pumping power, $p = 1\,\mathrm{W}$ and 6 W. The magnon density is peaked at $z = 0$ where the power is maximum and falls to zero outside the active region defined by $p(z) > p_{c2}$. The shape of the magnon density agrees qualitatively with the spatial distribution of the condensate measured

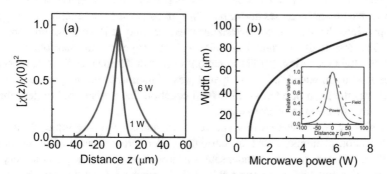

**Fig. 7.8** (a) Condensate magnon density expressed by the BEC wave function squared, calculated numerically with Eq. (7.48), for microwave driving with a wire at a distance $s = 32\,\mu\mathrm{m}$ from the YIG film, for two values of pumping power $p = 1\,\mathrm{W}$ and 6 W. (b) Width of the active BEC pumping region as a function of microwave power for driving with a wire at a distance $s = 32\,\mu\mathrm{m}$ from the YIG film. Inset shows the field and power profiles used in the calculation [25]. Reprinted with permission from S.M. Rezende, Phys. Rev. B **81**, 020414 (R) (2010). Copyright (2010) by the American Physical Society

experimentally with Brillouin light scattering [32]. The agreement is especially good in the central part of the distribution.

The main panel in Fig. 7.8b shows the width $w$ of the active region calculated with $p(w/2) = p_{c2}$ as a function of the microwave power for the same wire configuration considered before. The increase of the width with pumping power is in agreement with the experimental measurements of the spatial width of the condensate [32]. However, the measured widths are larger than the theoretical values by tens of $\mu m$. This is attributed to the fact that the parametric and secondary magnons propagate away from the pumping region before and during the process of BEC formation [25].

It is important to note that the magnon condensation in a microwave driven YIG film provides a unique situation in which the condensate wave function is governed by a Gross–Pitaevskii equation that was derived from a microscopic treatment of the boson dynamics in wave vector space. This is not common to other BEC systems. We also note that other treatments of the Bose–Einstein condensation of magnons rely on phenomenological models adapted from other BEC systems which require the introduction of ad hoc terms or parameters to connect them to the experimental situation [32–34].

To summarize this chapter, we have presented a theoretical model for the quasi-equilibrium Bose–Einstein condensation of magnons in a microwave driven YIG film that provides the basic requirements for the characterization of a BEC, namely: (a) The onset of the BEC is characterized by a phase transition that takes place as the microwave power $p$ is increased and exceeds a critical value $p_{c2}$; (b) The magnons in the condensate are in coherent states and as such they have non-zero small-signal transverse magnetization $m^+ \propto (p - p_{c2})^{1/4}$ that is the order parameter of the BEC; (c) For $p > p_{c2}$ the magnon system separates in two parts, one in thermal equilibrium with the reservoir and another one with $N_0 \propto (p - p_{c2})^{1/2}$ coherent magnons having frequencies and wave vectors in a very narrow region of phase space. As the microwave power increases above $p_{c2}$, $N_0$ approaches the total number of magnons pumped into the system, characterizing unequivocally a Bose–Einstein condensation. The results of the model fit quite well the experimental data obtained with Brillouin light scattering; (d) The equations of motion for the magnon operators in $k$-space lead to an equation for the spatial wave function of the magnon condensate which has the form of a Gross–Pitaevskii equation that is also used to describe the BEC wave function in other systems.

In closing, we note that this chapter does not cover some topics in magnon condensation phenomena that are currently under active research. One of them is the Bose–Einstein magnon condensation in low-dimensional magnets at very low temperatures and high magnetic fields [10]. In room temperature magnon BEC in YIG films under parallel pumping, two challenging phenomena are the nonlinear evaporative mechanism [20] and the formation of magnon supercurrents driven by a phase gradient, analogous to the supercurrents observed at low temperatures in superconductivity and superfluidity [35].

## Problems

7.1 Using the relations for the transformation of the spin operator into magnon operators for a ferromagnetic film studied in Sect. 4.2, show that the Hamiltonian of parallel pumping in a YIG film is given by the expressions (7.4) and (7.5).

7.2 Consider a YIG film magnetized in the plane by a field such that the minimum frequency for magnons with wave vector along the field is 4.0 GHz. Calculate the value of the chemical potential, in mK, for which the magnon system undergoes Bose–Einstein condensation.

7.3 From the maximum of the distribution function (7.36) in the condensed phase, calculate the number of BEC magnons in the state with minimum energy and compare to the result in Eq. (7.23).

7.4 Consider the magnon BEC in a YIG film driven with parallel pumping by a field generated by a wire antenna parallel and distant 20 μm from the film surface. Considering the critical power for BEC formation $p_{c2} = 1$ W, calculate the width of the BEC wave function for a pumping power of 8 W in the wire.

## References

1. Bose, S.N.: Plancks Gesetz und Lichtquantenhypothese. Z. Phys. **26**, 178 (1924)
2. Einstein, A.: Quantentheorie des einatomigen idealen Gases. Part I. Sber. Preuss. Akad. Wiss. **22**, 261 (1924)
3. Einstein, A.: Quantentheorie des einatomigen idealen gases. Part II. Sber. Preuss. Akad. Wiss. **1**, 3 (1925)
4. Leggett, A.J.: Bose-Einstein condensation in the alkali gases: Some fundamental concepts. Rev. Mod. Phys. **73**, 307 (2001)
5. Anderson, M.H., Ensher, J.R., Matthews, M.R., Wieman, C.E., Cornell, E.A.: Observation of Bose-Einstein condensation in a dilute atomic vapor. Science. **269**, 198 (1995)
6. Davis, K.B., Mewes, M.-O., Andrews, M.R., van Druten, N.J., Durfee, D.S., Kurn, D.M., Ketterle, W.: Bose-Einstein condensation in a gas of sodium atoms. Phys. Rev. Lett. **75**, 3969 (1995)
7. Vannucchi, F.S., Vasconcellos, A.R., Luzzi, R.: Dynamics of a Bose-Einstein condensate of excited magnons. Eur. Phys. J. B. **86**, 463 (2013)
8. Kasprzak, J., et al.: Bose-Einstein condensation of excitons polaritons. Nature. **443**, 409 (2008)
9. Snoke, D., Littlewood, P.: Polariton condensates. Phys. Today. **63**, 42 (2010)
10. Yin, L., Xia, J.S., Zapf, V.S., Sullivan, N.S., Paduan-Filho, A.: Direct measurement of the Bose-Einstein condensation universality class in $NiCl_2$-$4SC(NH_2)_2$ at ultralow temperatures. Phys. Rev. Lett. **101**, 187205 (2008)
11. Fröhlich, H.: Bose condensation of strongly excited longitudinal electric modes. Phys. Lett. A. **26**, 402 (1968)
12. Mesquita, M.V., Vasconcellos, A.R., Luzzi, R.: Positive-feedback-enhanced Fröhlich's Bose-Einstein-like condensation in biosystems. Int. J. Quant. Chem. **66**, 177 (1998)
13. Demokritov, S.O., Demidov, V.E., Dzyapko, O., Melkov, G.A., Serga, A.A., Hillebrands, B., Slavin, A.N.: Bose–Einstein condensation of quasi-equilibrium magnons at room temperature under pumping. Nature. **443**, 430 (2006)
14. Kalafati, Y.D., Safonov, V.L.: Thermodynamic approach in the theory of paramagnetic resonance of magons. Sov. Phys. JETP. **68**, 1162 (1989)

15. Kalafati, Y.D., Safonov, V.L.: Theory of quasiequilibrium effects in a system of magnons excited by incoherent pumping. Sov. Phys. JETP. **73**, 836 (1991)
16. Melkov, G.A., Safonov, V.L., Taranenko, A.Y., Sholom, S.V.: Kinetic instability and Bose condensation of nonequilibrium magnons. J. Magn. Magn. Mater. **132**, 180 (1994)
17. Demidov, V.E., Dzyapko, O., Demokritov, S.O., Melkov, G.A., Slavin, A.N.: Observation of spontaneous coherence in Bose–Einstein condensate of magnons. Phys. Rev. Lett. **100**, 047205 (2008)
18. Demidov, V.E., et al.: Magnon kinetics and Bose–Einstein condensation of nonequilibrium magnons studied in the phase space. Phys. Rev. Lett. **101**, 257201 (2008)
19. Chumak, A.V., Melkov, G.A., Demidov, V.E., Dzyapko, O., Safonov, V.L., Demokritov, S.O.: Bose–Einstein condensation of magnons under incoherent pumping. Phys. Rev. Lett. **102**, 187205 (2009)
20. Serga, A.A., et al.: Bose–Einstein condensation in an ultra-hot gas of pumped magnons. Nat. Commun. **5**, 3452 (2014)
21. Sandweg, C.W., Jungfleisch, M.B., Vasyuchka, V.I., Serga, A.A., Clausen, P., Schultheiss, H., Hillebrands, B., Kreisel, A., Kopietz, P.: Wide-range wavevector selectivity of magnon gases in Brillouin light scattering spectroscopy. Rev. Sci. Instrum. **81**, 073902 (2010)
22. Snoke, D.: Coherent questions. Nature. **443**, 403 (2006)
23. Rezende, S.M.: Theory of coherence in Bose-Einstein condensation phenomena in a microwave-driven interacting magnon gas. Phys. Rev. B. **79**, 174411 (2009)
24. Rezende, S.M.: Crossover behavior in the phase transition of the Bose-Einstein condensation in a microwave-driven magnon gas. Phys. Rev. B. **80**, 092409 (2009)
25. Rezende, S.M.: Wave function of a microwave-driven Bose-Einstein magnon condensate. Phys. Rev. B. **020414**(R), 81 (2010)
26. Rezende, S.M.: Magnon coherent states and condensates. Chapter 4. In: Demokritov, S.O., Slavin, A.N. (eds.) Magnonics-from Fundamentals to Applications. Springer, Berlin (2013)
27. Zakharov, V.E., L'vov, V.S.,Starobinets, S.S.: Spin-wave turbulence beyond the parametric excitation threshold. Usp. Fiz. Nauk., **114**, 609 (1974). [Sov. Phys. Usp. , **17**, 896 (1975)]
28. de Araújo, C.B.: Quantum-statistical theory of the nonlinear excitation of magnons in parallel pumping experiments. Phys. Rev. B. **10**, 3961 (1974)
29. Zagury, N., Rezende, S.M.: Theory of macroscopic excitations of magnons. Phys. Rev. B. **4**, 201 (1971)
30. Slavin, A.N., Kalinikos, B.A., Kovshikov, N.G.: Spin-wave envelope solitons in magnetic films. In: Wigen, P.E. (ed.) Nonlinear Phenomena and Chaos in Magnetic Materials. Singapore, World Scientific (1994). Chap. 9
31. Zvezdin, A.K., Popkov, A.F.: Contribution to the nonlinear theory of magnetostatic spin waves. Zh. Eksp. Teor. Fiz. **84**, 606 (1983). [Sov. Phys. JETP 57, 350 (1983)]
32. Malomed, B.A., Dzyapko, O., Demidov, V.E., Demokritov, S.O.: Ginzburg-Landau model of Bose-Einstein condensation of magnons. Phys. Rev. B. **81**, 024418 (2010)
33. Troncoso, R.E., Nunez, A.S.: Dynamics and spontaneous coherence of magnons in ferromagnetic thin films. J. Phys. Condens. Matter. **24**, 036006 (2012)
34. Li, F., Saslow, W.M., Pokrovsky, V.L.: Phase diagram for magnon condensate in yttrium iron garnet film. Sci. Rep. **3**, 1372 (2013)
35. Bozhko, D.A., Serga, A.A., Clausen, P., Vasyuchka, V.I., Heussner, F., Melkov, G.A., Pomyalov, A., L'vov, V.S., Hillebrands, B.: Supercurrent in a room-temperature Bose–Einstein magnon condensate. Nat. Phys. **12**, 1057 (2016)

## Further Reading

Gurevich, A.G., Melkov, G.A.: Magnetization Oscillations and Waves. CRC, Boca Raton (1994)
Feynman, R.P.: Statistical Mechanics: A Set of Lectures. W.A. Benjamin, Reading, Massachusetts (1972)

Inguscio, M., Stringari, S., Wieman, C.E.: Bose Einstein condensation in atomic gases. In: Proceedings of the International School of Physics "Enrico Fermi". IOS, Amsterdam (1999)

Landau, L.D., Lifshitz, E.M.: Statistical Physics. Pergamon Press, Oxford (1980)

Leggett, A.J.: Quantum Liquids: Bose Condensation and Cooper Pairing in Condensed-Matter Systems. Oxford University Press, Oxford (2006)

L'vov, V.S.: Turbulence Under Parametric Excitation, Applications to Magnets. Springer, Berlin, Heidelberg (1994)

Moskalenko, S.A., Snoke, D.W.: Bose Einstein Condensation of Excitons and Biexcitons. Cambridge University Press, Cambridge (2000)

Pitaevskii, L., Stringari, S.: Bose Einstein Condensation. Claredon Press, Oxford (2003)

# Magnon Spintronics

**8**

Spintronics is an emerging and very active field of science and technology that is devoted to the investigation of basic phenomena and device application based on the electron spin in addition to its charge. Spintronic devices are already used in high-density nonvolatile magnetic storage, and intense research efforts are under way worldwide to develop new functionalities for transport and processing of information not available in conventional electronics. In this chapter we study topics of spintronics in which magnons play a direct role. Initially we introduce the concept of spin current in nonmagnetic metals, which is essential to understand the spin Hall effect, the Rashba–Edelstein effect, and other spintronic phenomena. Then we present the phenomenon of spin transfer torque in magnetic multilayers and study the excitation of magnons by electric currents. Next we study the spin pumping effect, one of the most important in spintronics. In the following section we study the concepts of magnon accumulation and magnonic spin currents, and derive their important governing equations. Finally, we study the spin Seebeck effect in magnetic insulators that is based on the magnonic spin current.

## 8.1 Spin Current in Nonmagnetic Metals

A key concept in spintronics is that of the spin current. In metals, spin currents are transported by the conduction electrons, while in magnetic insulators they are carried by magnons. In this section we shall consider only metallic films. The concept of spin current was introduced several decades ago by Silsbee and co-workers [1, 2] under another name, namely, magnetization current. However, only in late 1990s the importance of the concept was understood in the context of newly discovered spintronic phenomena.

When electrons flow in some average direction in a conductor, they carry electric charge and also spin. If the electrons are unpolarized, with their spins in arbitrary directions, or pointing up and down relative to an applied magnetic field, the electron flow transports only charge current. This is pictorially illustrated in Fig. 8.1a for the

© Springer Nature Switzerland AG 2020
S. M. Rezende, *Fundamentals of Magnonics*, Lecture Notes in Physics 969,
https://doi.org/10.1007/978-3-030-41317-0_8

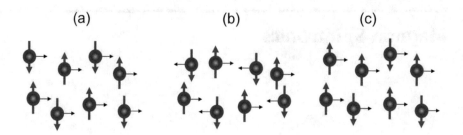

**Fig. 8.1** Pictorial illustration of various types of electronic currents. Black arrows indicate the direction of the electron motion and red arrows represent the electron spin. (**a**) Charge current. (**b**) Pure spin current. (**c**) Spin-polarized current

case of electrons with spins in a magnetic field. However, if by some means, fully polarized electrons with spins in one direction are made to flow in the opposite direction of electrons in the same number and with opposite spins, they carry a pure spin current but no net charge current. This is illustrated in Fig. 8.1b. A third possibility, shown in Fig. 8.1c, corresponds to the flow in one direction of a larger concentration of up-spin electrons than down-spin electrons, in which case the electron flow transports both charge and spin currents. This case is referred to as a spin-polarized electronic current.

One method to produce a spin-polarized electronic current in a magnetic multi-layer is illustrated in Fig. 8.2, where a metallic ferromagnetic (FM) layer between two nonmagnetic metals (NM) acts as a spin filter. Two mechanisms are responsible for the spin polarization, interfacial spin-dependent scattering and bulk polarization. In the electronic current traversing the FM layer, electrons with spin in the same direction as the magnetization have a smaller interfacial scattering cross section than the ones with spin in the opposite direction (we assume that the electron magnetic moment has the same direction as the spin) [3]. As a result, the electric resistance for electrons with up-spin is smaller than for electrons with down-spin, so that the electron flow traversing the FM layer has more spins in one direction than in the opposite, as in Fig. 8.1c.

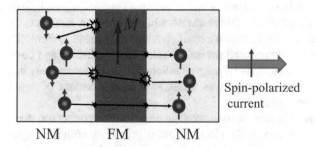

**Fig. 8.2** Schematic illustration of the method to generate a spin-polarized electronic current by means of a spin filter made of a ferromagnetic (FM) layer between two nonmagnetic (NM) metals

In order to understand the contribution to the current polarization from the bulk of the ferromagnetic metallic layer we have to recall some properties of a free electron gas in metals. Electrons are fermions and as such they obey the *Pauli exclusion principle* that says that in a system of identical particles there can never be more than one particle in the same quantum state. In a metal with free electrons having energy $\varepsilon(k)$, at a temperature $T$, the occupation of the energy states is governed by the Fermi–Dirac distribution function, given by

$$f(\varepsilon, T) = \frac{1}{e^{(\varepsilon - \mu_c)/k_B T} + 1},$$ (8.1)

where $\mu_c$ is the *chemical potential*, also called electrochemical potential, defined as the energy necessary to add one particle to the system. At zero-temperature Eq. (8.1) has a step-like behavior, i.e., $f(\varepsilon < \mu_c, T = 0) = 1$ and $f(\varepsilon > \mu_c, T = 0) = 0$, meaning that the probability is unity for electrons to occupy the energy levels below $\mu_c$, and it is zero for $\varepsilon > \mu_c$. Thus, at $T = 0$, the chemical potential is equal to the Fermi energy $\mu_c(T = 0) = \varepsilon_F$, defined as the energy of the highest occupied quantum state at $T = 0$. At finite temperatures some electrons are thermally excited to states with energy $\varepsilon > \varepsilon_F$, so that Eq. (8.1) deviates from a step-like function only in an energy range on the order of $k_B T$ around $\varepsilon_F$. The particle number density of the free electron system can be calculated with

$$n = \int_{-\infty}^{\infty} N(\varepsilon) f(\varepsilon) d\varepsilon,$$ (8.2)

where $N(\varepsilon)$ is the density of states and $N(\varepsilon)d\varepsilon$ is the number of electrons per unit volume with energies between $\varepsilon$ and $\varepsilon + d\varepsilon$. We assume that the temperature is relatively low so that the Fermi–Dirac distribution is very close to a step function. We also assume that the conduction electrons have a parabolic energy band, so that the density of states is $N(\varepsilon) \propto \varepsilon^{1/2}$ [see Kittel]. Considering that electrons have spin, and that there two spin states, we actually have to consider two energy bands, one for up-spin electrons and one for down-spin electrons. In a nonmagnetic metal (NM) and in the absence of an external magnetic field, the energy does not depend on the spin, so that the number of up- and down-spin electrons are the same. This is shown schematically by the shaded areas in Fig. 8.3a with the plots of the energy versus density of states. In the ferromagnetic (FM) layer there is a splitting in the bands, with the energy band of up-spin electrons shifted to lower energies relative to the down-spin band because of the exchange interaction, as illustrated in Fig. 8.3b. Thus, when the electrons from a NM layer enter in the FM layer, electrons from the down-spin band go to the up-spin band, so that the number of up-spin electrons becomes larger than the number of down-spin electrons. This results in a spin-polarized current as in Fig. 8.1c. The asymmetry in the density of states at the Fermi energy is used to define the spin polarization of the current through the FM metal as

**Fig. 8.3** Density of states and occupation of the electronic states in the free electron parabolic band approximation. The shaded areas represent the occupied energy states up to the Fermi energy. (**a**) Nonmagnetic metal with up-spin and down-spin electrons having the same energies. (**b**) Ferromagnetic metal with band splitting due to the exchange interaction

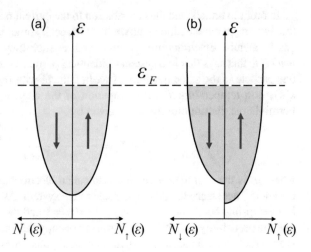

$$P = \frac{N_\uparrow(\varepsilon_F) - N_\downarrow(\varepsilon_F)}{N_\uparrow(\varepsilon_F) + N_\downarrow(\varepsilon_F)}. \qquad (8.3)$$

Note that this is a simple view of the polarization phenomenon based on an approximate band model for free electrons. Actually, the density of states in transition metals such as Ni, Fe, and Co, is more complicated because the Fermi energy overlaps with both the 4$s$ and 3$d$ energy bands.

For the study of pure spin currents in metals, initially we do not consider the electron spin. From Eqs. (8.1) and (8.2) one can see that if the electron density $n$ is increased by $\delta n$, the chemical potential changes from $\mu_c$ to $\mu_c + \delta\mu_c$, such that

$$\delta\mu_c = \frac{\delta n}{N(\varepsilon_F)}. \qquad (8.4)$$

When an electric voltage is applied at the ends of the metal, an electric field $\vec{E}$ is created that exerts on the conduction electrons a force $e\vec{E}$, where $e$ is the electron charge. This generates an electron drift motion and an associated charge current with density given by

$$\vec{J}_d = \sigma_e \vec{E}, \qquad (8.5)$$

where $\sigma_e = n\,e\,\tau/m^*$ is the electronic conductivity, $m^*$ is the effective mass and $\tau$ is the collision time. Besides the drift motion due to an electric field, electrons can also be driven by a spatial variation in their number density. The motion produced by a gradient in the concentration is called diffusive, and is described by the diffusion equation

$$\frac{\partial n(\vec{r},t)}{\partial t} = D\nabla^2 n(\vec{r},t), \tag{8.6}$$

where $D$ is the diffusion constant and $n(\vec{r},t)$ is the electron concentration in a position $\vec{r}$ at time $t$. The electron diffusion motion produces a current with density $\vec{J}_{diff}$, and since charge is conserved, it must satisfy the continuity equation

$$\frac{\partial \rho(\vec{r},t)}{\partial t} = -\nabla \cdot \vec{J}_{diff}(\vec{r},t). \tag{8.7}$$

where $\rho(\vec{r},t) = en(\vec{r},t)$ is the charge density. Equations (8.6) and (8.7) give an expression for the diffusion current density produced by a nonuniform electron concentration

$$\vec{J}_{diff}(\vec{r},t) = -eD\nabla n(\vec{r},t). \tag{8.8}$$

This relation is known as Fick's law, which states that particles flow from regions of higher concentration toward regions of lower concentration. Thus, in an electron gas with nonuniform concentration under an electric field, the current is the sum of the drift and diffusion components

$$\vec{J}(\vec{r},t) = \sigma_e \vec{E} - eD\nabla n(\vec{r},t). \tag{8.9}$$

The electric potential $\phi$ in the metal changes the electrochemical potential from $\mu_c$ to $\mu = \mu_c + e\phi$, so that with Eq. (8.4) and $\vec{E} = -\nabla\phi$ we have a relation between the field and the electrochemical potential gradient

$$\nabla\mu = -e\vec{E} + \frac{\nabla n}{N(\varepsilon_F)}. \tag{8.10}$$

If we make $\nabla\mu = 0$, this expression in Eq. (8.9) gives for the current density

$$\vec{J} = \left[\sigma_e - e^2 DN(\varepsilon_F)\right]\vec{E}, \tag{8.11}$$

which must be zero in thermal equilibrium. This condition yields the Einstein relation

$$\sigma_e = e^2 DN(\varepsilon_F). \tag{8.12}$$

Substitution of (8.10) and (8.12) in Eq. (8.9) gives an expression for the total current density in terms of the gradient of the chemical potential

$$\vec{J} = \frac{\sigma_e}{e} \nabla \mu. \tag{8.13}$$

The relations obtained thus far must now be generalized to account for the electron spin degree of freedom. Considering the spin state $\sigma = \uparrow, \downarrow$, and denoting by $\mu_\sigma$ and $\sigma_\sigma$ the chemical potential and the conductivity for electrons in each of the spin states, with Eq. (8.13) we can express the current density in the spin channel $\sigma$ as

$$\vec{J}_\sigma = \frac{\sigma_\sigma}{e} \nabla \mu_\sigma. \tag{8.14}$$

Now we can formally define the charge current density by $\vec{J}_C = \vec{J}_\uparrow + \vec{J}_\downarrow$ and the pure spin current density by $\vec{J}_S = \vec{J}_\uparrow - \vec{J}_\downarrow$, which with Eq. (8.14) become

$$\vec{J}_C = \frac{1}{e} \nabla (\sigma_\uparrow \mu_\uparrow + \sigma_\downarrow \mu_\downarrow), \tag{8.15}$$

$$\vec{J}_S = \frac{1}{e} \nabla (\sigma_\uparrow \mu_\uparrow - \sigma_\downarrow \mu_\downarrow), \tag{8.16}$$

where $\sigma_\uparrow$ and $\sigma_\downarrow$ are the conductivities for up- and down-spin electrons. In nonmagnetic metals we can write for the conductivities $\sigma_\uparrow = \sigma_\downarrow = \sigma_N/2$ and for the chemical potential $\mu = (\mu_\uparrow + \mu_\downarrow)/2$. With these relations Eq. (8.15) reproduces the expression for the charge current in (8.13), while the spin current density becomes

$$\vec{J}_S = \frac{\sigma_N}{2e} \nabla (\mu_\uparrow - \mu_\downarrow), \tag{8.17}$$

The difference in the chemical potentials for up- and down-spin electrons has a special significance because, as we will show shortly, it is proportional to the magnetization that appears in the nonmagnetic metal due to the injection of a spin current [1–3]. Thus, we define the *spin accumulation* by

$$\mu_S = (\mu_\uparrow - \mu_\downarrow)/2, \tag{8.18}$$

with which the spin current density in Eq. (8.17) can be written as

$$\vec{J}_S = \frac{\sigma_N}{e} \nabla \mu_S. \tag{8.19}$$

Two observations about this result have to be made. The gradient implies that the vector $\vec{J}_S$ has the direction of the variation of $\mu_S$. However, there is another important direction, namely, the one of the spin polarization. Thus, to be more rigorous in the notation, the spin current should be represented by a second rank tensor. To make it simple, we shall keep the vector notation and, when necessary, we shall use a superscript to represent the direction of the spin polarization. The second point is that in Eq. (8.19) the spin current density has the same dimension as the charge current, that is, charge per unit time per unit area. Since the angular momentum of the electron is $\hbar/2$, when convenient we shall use the spin current in units of angular momentum per unit time per unit area, that is

$$\vec{J}_S = \frac{\hbar \sigma_N}{2e^2} \nabla \mu_S. \tag{8.20}$$

As will be seen in the study of the spin pumping and spin Seebeck effects in the next sections, in a bilayer made of a ferromagnetic or antiferromagnetic film in contact with a nonmagnetic (NM) metallic layer, the spin current produced in the magnetic film is injected into the NM layer through the interface. In order to obtain an equation of motion for the spin accumulation in the NM layer, we use the fact that without an electric field, Eq. (8.4) gives a relation between the density of electrons with spin $\sigma$ and the corresponding chemical potential, $\delta n_\sigma = N_o(\varepsilon_F)\delta\mu_\sigma$. With this result in Eq. (8.6) we obtain a diffusion equation for the spin accumulation

$$\frac{\partial \mu_S(\vec{r},t)}{\partial t} = D \nabla^2 \mu_S(\vec{r},t). \tag{8.21}$$

Actually, in the time dependence of the spin accumulation, one must consider that as the electrons diffuse in the metal and undergo collisions, there is a finite probability to have a spin flip. This effect can be taken into account by introducing in Eq. (8.21) a relaxation term

$$\frac{\partial \mu_S(\vec{r},t)}{\partial t} = D \nabla^2 \mu_S(\vec{r},t) - \frac{\mu_S(\vec{r},t)}{\tau_{sf}}, \tag{8.22}$$

where $\tau_{sf}$ is the spin-flip relaxation time. Assuming that the spin accumulation varies in space only along a direction $y$ perpendicular to the interface, the solution of Eq. (8.22) in steady state ($\partial/\partial t = 0$) is

$$\mu_S(y) = A e^{-y/\lambda_N} + B e^{y/\lambda_N}, \tag{8.23}$$

where $\lambda_N = (D\tau_{sf})^{1/2}$ is the spin diffusion length of the metal, and $A$ and $B$ are coefficients determined by the boundary conditions. The spin diffusion length varies from a new nanometers in heavy metals like Pt, Pd, and W, to hundreds of nanometers in Cu and Au.

Two important quantities can be related to the spin accumulation in Eq. (8.18). The first is the spin density (spin per unit volume) given by $\vec{s} = (\hbar/2)(n_\uparrow - n_\downarrow)\vec{\sigma}$, where $\vec{\sigma}$ is the unit vector in the direction of spin quatization. Using the relation obtained from Eq. (8.4), $\delta n_{\uparrow,\downarrow} = (1/2)N(\varepsilon_F)\delta\mu_{\uparrow,\downarrow}$, we have for the spin density

$$\vec{s} = (\hbar/2)N(\varepsilon_F)\vec{\mu}_S. \tag{8.24}$$

The other quantity of interest is the magnetization created in the metal by the injection of the spin current. Since the magnetization is the magnetic moment per unit volume, it is obtained simply by multiplying Eq. (8.24) by $g\mu_B/\hbar$. With $g = 2$ we have

$$\vec{\delta m_N} = \mu_B N(\varepsilon_F)\vec{\mu}_S. \tag{8.25}$$

Clearly, since both $\vec{s}$ and $\vec{\delta m_N}$ are proportional to $\vec{\mu}_S$, they also obey a diffusion equation like Eq. (8.22).

## 8.2    Conversions Between Charge and Spin Currents

The conversions between charge and spin currents are essential operations in any spintronic device [4]. As we saw in the previous section, one method to create a spin current from a charge current is to use spin filtering by a magnetic layer. More efficient conversions between charge and spin currents are obtained with phenomena based on the electron spin–orbit coupling. Charge currents can be converted into pure spin currents by the spin Hall effect (SHE), while the spin-charge current conversion takes place by the inverse spin Hall effect (ISHE) [5, 6]. In two-dimensional (2D) electron systems, similar conversions can be obtained by the Rashba–Edelstein effect (REE) and its inverse (IREE), associated with the locking between the spin and charge degrees of freedom [4]. In this section we shall present the basic mechanisms of these effects.

### 8.2.1    Spin Hall Effects

The spin Hall effects are phenomena that originate in the spin–orbit coupling (SOC) of the conduction electrons in metals and semiconductors, which make possible the interconversion of charge and spin currents [5, 6]. These conversions are more efficient in heavy metals, that have strong SOC, such as Pt, Pd, Ta, and W. In the direct spin Hall effect (SHE), illustrated in Fig. 8.4a, electrons with opposite spins in a charge current are scattered by spin-orbit mechanisms in opposite directions. This gives rise to a spin current in the transverse direction and to spin accumulation at the two edges of the film. In the reciprocal effect, called inverse spin Hall effect (ISHE), illustrated in Fig. 8.4b, the same scattering mechanisms result in partial conversion of a spin current into a charge current. This generates a voltage between the two edges of the film, allowing the electric detection of spin currents.

The possibility of creating spin polarization in a charge current in a material with spin–orbit coupling was initially proposed by Dyakonov and Perel in 1971 [7, 8]. This proposal received very little attention until 1999, when Hirsch formulated a simple model for this effect and introduced the term spin Hall effect

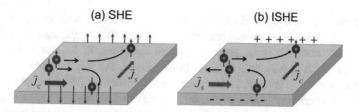

**Fig. 8.4** Illustration of the spin Hall effects in a metallic film. (**a**) In the SHE, electrons in a charge current are scattered to the left (spin-up electrons) and to the right (spin-down), creating a spin current transverse to the direction of spin quantization. (**b**) In the inverse spin Hall effect (ISHE), the same scattering mechanism partially converts a spin current into a charge current

[9], by which the phenomenon is known today. The paper by Hirsch quickly stimulated theoretical work on the spin Hall effects and several detailed models were proposed for the spin-orbit based scattering mechanisms [5, 6, 10–12]. Only a few years later the spin Hall effect was observed experimentally by the detection of the spin polarization with optical techniques in semiconductors [13], and by direct electronic measurements in metallic structures [14].

We shall present here a simplified description of the mechanisms that give origin to the spin Hall effects. Consider unpolarized electrons incident on a scattering center without spin, with a potential given by

$$V(\vec{r}) = V_C(\vec{r}) + V_{so}(\vec{r})\,\vec{\sigma} \cdot \vec{L}, \tag{8.26}$$

where $\vec{\sigma}$ and $\vec{L}$ are, respectively, the spin and orbital angular momentum of the electrons, $V_C(\vec{r})$ is the Coulomb potential and $V_{so}(\vec{r})$ is the spin-orbit scattering potential. Due to the last term in Eq. (8.26), up- and down-spin electrons are scattered toward opposite directions, so as to describe trajectories with opposite angular momenta, as illustrated in Fig. 8.4. There are two effects of this spin-orbit scattering process. In one of them, the amplitude of the scattered electron wave as it propagates away from the scattering center is different for up-spin and down-spin electrons. This is called *skew scattering*. In the other, the apparent center from which the scattered wave propagates is different for up-spin and down-spin electrons, characterizing a *side-jump scattering*. Both mechanisms result in a motion of the scattered electrons transverse to the direction of the incoming flow, that produces a spin accumulation and a diffusive spin current. Using the Boltzmann transport and the diffusion equations, it can be shown that the spin current density generated from the charge current density $\vec{J}_C$ by means of the spin Hall effect is [see White]

$$\vec{J}_S = \theta_{SH}\,\vec{\sigma} \times \vec{J}_C, \tag{8.27}$$

where $\vec{\sigma}$ is the unit vector along the spin polarization direction and $\theta_{SH}$ is a proportionality factor called *spin Hall angle*, which has a magnitude and sign that depend on $V_{so}(\vec{r})$. If $\vec{J}_S$ in Eq. (8.27) is expressed with the same unit as the charge current, $\theta_{SH}$ is a dimensionless parameter. Its value for Pt, Pd, and W, which are heavy-metal materials with large spin–orbit coupling, is on the order of 0.05 [5, 6]. The value in β-Ta has comparable magnitude, but it is negative.

As previously mentioned, in the inverse spin Hall effect (ISHE) a spin current is partially converted into a charge current. Since the scattering mechanisms involved in ISHE are the same as in the SHE, the fraction of the spin current converted into charge current is the same as in the reciprocal effect in SHE. Thus, the charge current density generated by the ISHE is given by

$$\vec{J}_C = \theta_{SH} \, \vec{J}_S \times \vec{\sigma}. \tag{8.28}$$

The parameter $\theta_{SH}$ here is the same as in Eq. (8.27), but the symbols of the current densities have different meanings. Here $\vec{J}_S$ represents the incoming spin current, while in (8.27) it represents the output spin current. Similarly, in (8.27) $\vec{J}_C$ represents the incoming charge current, while here it represents the generated charge spin current. As mentioned earlier, the ISHE makes possible the electric detection of spin currents and it is a very important effect for the integration of spintronics with electronics. As will be described in this chapter, the ISHE is also a key phenomenon for the study of the spin pumping and the spin Seebeck effects.

## 8.2.2   Rashba–Edelstein Effects

Recent developments in the techniques for thin-film growth and characterization have revealed that topology plays an important role in spintronic phenomena. In thin-film materials and heterostructures with heavy-metal compounds, the properties of 2D electron systems arising from the spin–orbit coupling gives rise to new processes for the conversion of charge and spin currents, namely, the Rashba–Edelstein (RE) effects. As the composite name implies, these effects are based on phenomena involving two mechanisms. The first is the Rashba effect that arises from SOC and broken inversion symmetry at material surfaces and interfaces [15, 16]. As shown in White's book, a relativistic treatment of the Hamiltonian for an electron with charge $e$, subject to static electric $(\vec{\varepsilon})$ and magnetic $(\vec{H})$ fields, gives two terms that depend on the electron spin. One of them is

$$H_Z = -\frac{e\hbar}{2mc} \vec{\sigma} \cdot \vec{H}, \tag{8.29}$$

that corresponds to the Zeeman interaction of Eq. (1.4), and the other is

$$-\frac{e\hbar}{4m^2c^2} \vec{\sigma} \cdot \vec{\varepsilon} \times \vec{p}, \tag{8.30}$$

where $\vec{p}$ is the linear momentum of the electron. Considering a spherically symmetric scalar potential $V(r)$, using $\vec{\varepsilon} = -\hat{r}\,\partial V/\partial r$ and $\hbar\vec{L} = \vec{r} \times \vec{p}$ in Eq. (8.30), we obtain

$$H_{SOC} = \frac{e\hbar^2}{4m^2c^2}\frac{1}{r}\frac{\partial V}{\partial r}\vec{\sigma}\cdot\vec{L}, \tag{8.31}$$

which represents the spin–orbit coupling energy. In the case of films and multilayers, the broken symmetry of the crystalline potential $V$ at surfaces and interfaces generates an electric field $\vec{\varepsilon} = -\hat{z}\,\partial V/\partial z$, where $z$ is the coordinate normal to the surface. Using this expression in Eq. (8.30), with $\vec{p} = \hbar\vec{k}$, we obtain the Rashba energy

$$H_R = \alpha_R(\hat{z} \times \vec{k})\cdot\vec{\sigma}, \tag{8.32a}$$

where

$$\alpha_R = \frac{e\hbar^2(\partial V/\partial z)}{4m^2c^2} \tag{8.32b}$$

is the Rashba parameter [15, 16]. Thus, the Hamiltonian for an electron in two dimensions, with kinetic and Rashba energies, and without particle interactions is

$$H = \frac{\hbar^2k^2}{2m} + \alpha_R(\hat{z} \times \vec{k})\cdot\vec{\sigma}, \tag{8.33}$$

where $\vec{k}$ is the electron wave vector in the plane. The first term in this Hamiltonian gives an energy dispersion quadratic in wave number, while the second term gives an energy that varies linearly with wave number and also depends on the spin. The energy dispersion is then

$$E^{(\pm)} = \frac{\hbar^2k^2}{2m} \pm \alpha_R k, \tag{8.34}$$

where the $\pm$ signs correspond to the two spin orientations in the plane in a direction perpendicular to $\vec{k}$. The linear term in (8.34) displaces the parabolic subbands sideways in the $\vec{k}$ plane and also downwards in energy. The displacements in wave number are given by the positions of minimum energy. With $dE^{(\pm)}/dk = 0$ we find that the positions of the minima are

$$k_0^{(\pm)} = \mp\alpha_R\frac{m}{\hbar^2}. \tag{8.35}$$

Substitution of these values in Eq. (8.34) gives for the minimum energy of both branches the value $E_{min}^{(\pm)} = -\alpha_R^2 m/2\hbar^2$ . The dispersion relations $E^{\pm}(k_x, k_y)$ in

**Fig. 8.5** (a) Schematic illustration of the spin-polarized energy dispersion of a 2D electron gas with Rashba interaction in equilibrium. (b) Fermi circles exhibiting the spin–momentum locking

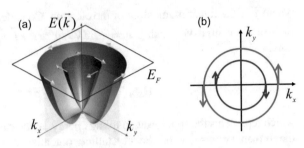

Eq. (8.34) are represented by the surfaces shown in Fig. 8.5a, formed by the rotation about the energy axis of the spin-split parabolas for the two spin orientations. A most important consequence of the Rashba effect in a 2D electron gas is the locking of spin and momentum degrees of freedom to each other [17, 18]. The spin–momentum locking is clearly observed in the Fermi circles shown in Fig. 8.5b, obtained by the cross section of the energy surfaces with the plane for constant energy at the Fermi level, as in Fig. 8.5a. There is another manner to see the mechanism of the spin–momentum locking. The second term in Eq. (8.33) can be seen as the interaction of the spin with an effective Rashba magnetic field $\vec{H}_R \propto \vec{k} \times \hat{z}$. This field aligns the spin perpendicularly to $\vec{k}$, as in Fig. 8.5. The spin–momentum locking is a key feature of the Rashba interaction that enables the spin–charge conversion in 2D electron systems.

The second mechanism involved in the Rashba–Edelstein effects is a consequence of the behavior of the spin-polarized Fermi circles in nonequilibrium. In a 2D electrons gas in equilibrium, as in Fig. 8.5b, all states in $k$-space are occupied, so that the total momentum is null and there is no net spin density. If an electric field is applied, say in the -$x$ direction, a force acts on the electrons in the $x$ direction (for the electron charge negative), producing a change $\hbar \delta k_x$ in the momenta of all electrons. Thus, in a time $\tau$ equal to the collision time, the two Fermi circles are displaced along the $k_x$ axis by $\delta k_x = e \varepsilon_x \tau / \hbar$, as illustrated in Fig. 8.6a. In this situation, the number of occupied states with positive $k_x$ is larger than with negative $k_x$, so that a charge current flows in the -$x$ direction. In the presence of the Rashba interaction, the displacement of the two Fermi circles also results in a spin accumulation, since the number of occupied states with positive spin in the $y$-direction is larger than with negative spin. The association of the spin polarization with the charge current in a 2D electron gas with Rashba interaction was first pointed out by Edelstein [19], and for this reason the RE effects are also called simply Edelstein effects [4, 18].

The mechanism of the inverse Rashba–Edelstein (IREE) is the reciprocal of the one in the direct RE effect. As illustrated in Fig. 8.6b, a pure spin current with spin polarization in the $y$-direction, injected in the plane of the 2D electron gas with Rashba interaction, creates a spin accumulation in the plane. This produces a displacement of the two Fermi circles in opposite directions along $k_x$ to account for the spin imbalance, generating a charge current in the -$x$ direction. Note that since the charge current flows in the plane, it is quantified by a surface current density $j_{Cx}$, with dimension of charge/(time.length).

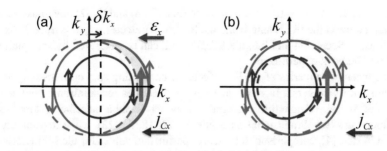

**Fig. 8.6** Schematics of the mechanisms for the Rashba–Edelstein effects in a 2D electron gas with Rashba interaction. The dashed contours are the Fermi circles in equilibrium, while the solid ones correspond to nonequilibrium. (**a**) Direct REE: An electric field in the -x direction exerts a force on the electrons resulting in a displacement of the Fermi circles, producing a charge current in the -x direction and the associated spin accumulation. (**b**) Inverse REE: A spin current with y polarization injected perpendicularly to the plane creates a spin accumulation that induces a displacement of the Fermi circles, resulting in a charge current in the -x direction

For the calculation of the charge current generated by the IREE, we consider initially that, in the absence of the Rashba interaction, a displacement of the Fermi circle (radius $k_F$) by $+\delta k_x$ increases the density of occupied states with $k_x > 0$ by $\delta n_x/2$ and decreases the density with $k_x < 0$ by the same amount. This results in a charge current with surface density $j_{Cx} = -e\, v_F \delta n_x$, $v_F$ being the Fermi velocity. With $v_F = \hbar k_F/m$ the current density becomes $j_{Cx} = -(e\hbar/m)\,\delta n_x k_F$. Thus, in the presence of the Rashba interaction, the current densities for the two spin states are given by $j_{Cx}^{\pm} = -(e\hbar/m)\delta n_x^{\pm} k_F^{\pm}$, where $k_F^{\pm}$ is the radius of the Fermi circle for $\pm$ spin. Examination of the surfaces in Fig. 8.6a shows that $k_F^{\pm} = k_F \pm k_0$, so that with Eq. (8.35) the total current density $j_C = j_C^+ + j_C^-$ becomes

$$j_C = \alpha_R \frac{e}{\hbar}\delta n, \tag{8.36}$$

where, in first order in $\alpha_R$, we have considered $\delta n_x^+ = \delta n_x^- = \delta n/2$. In the inverse RE effect, the incremental density of states $\delta n$ results from the spin imbalance created by the spin current injected in the plane of the 2D electron gas. Hence, the magnitude of the 3D spin current density $J_S$, in units of charge current, is

$$J_S = e\frac{\delta n}{\tau_S}, \tag{8.37}$$

where $\tau_S$ is the effective relaxation time of the spin density in the 2D electron gas in the presence of spin–momentum locking [20]. Thus, we have a direct relation between the densities of the 2D charge current generated in the plane and the injected 3D spin current

$$j_C = \lambda_{IREE}\, J_S, \tag{8.38}$$

where $\lambda_{IREE}$ is called the IREE length. With Eqs. (8.36) and (8.37) one finds a simple relation between the IREE length and the Rashba parameter, $\lambda_{IREE} = \alpha_R \tau_S / \hbar$. As will be shown in Sects. 8.4 and 8.6, the IREE length can be measured by spin pumping and spin Seebeck experiments.

The recent discoveries of the RE effects for converting spin currents into charge currents, and vice-versa, have opened new possibilities for spintronic applications with the so-called quantum materials, such as Rashba interfaces, topological insulators and two-dimensional materials [4, 17, 18, 21–23]. Most importantly, as shown in Ref. [4], comparison of typical experimental values for the spin Hall angle in heavy metals and the IREE length in 2D electrons of topological insulators, shows that the conversion efficiency of the IREE can be one order of magnitude larger than of the ISHE.

## 8.3   Magnon Excitation by Spin Transfer Torque

### 8.3.1   Spin Transfer Torque

An important phenomenon that occurs in magnetic trilayer structures traversed by an electric current in the configuration perpendicular to the plane (CPP) is the spin transfer torque (STT). The STT arises from the conservation of angular momentum when a spin-polarized current is injected in a magnetic layer, making possible to alter the magnetization state in that layer. In nanostructures the effect of the STT created by an electric current can be much larger than the one due to the Oersted–Ampère magnetic field created by the current. The discovery of this effect made possible the development of STT magnetoresistive random access memories (MRAM), that have several advantages over semiconductor RAMs [24], tunable microwave oscillators [25, 26], as well as other nanomagnonic devices [27, 28].

The concept of the spin transfer torque was proposed theoretically by Slonczewski [29] and independently by Berger [30]. To explain the mechanism underlying the STT we consider the trilayer in Fig. 8.7a, made of two ferromagnetic (FM) layers with macrospin vectors $\vec{S}_1$ and $\vec{S}_2$, separated by a thin nonmagnetic metal (NM) layer. Electrons of the spin-polarized current produced in the thicker FM layer 1 (FM1) are injected through the NM layer into the second FM layer. By conservation of angular momentum, each spin reaching FM2 changes the spin angular momentum of that layer in a direction perpendicular to the spin and with magnitude $(\hbar/2) \sin \theta$, where $\theta$ is the angle between $\vec{S}_1$ and $\vec{S}_2$. The rate of change of the spin is proportional to the electron flow, which is determined by the electric current. Considering the current density $\vec{J}$ with polarization $P$, defined as in Eq. (8.3), the angular momentum transferred to the layer FM2 per unit volume, per unit time, is then $\hbar \sin \theta P J / (2e\, t_2)$, where $t_2$ is the thickness of FM2. According to Newton's law, the time derivative of the spin angular momentum corresponds to an effective torque, so that the spin transfer torque per unit volume acting on $\vec{S}_2$ can be written as

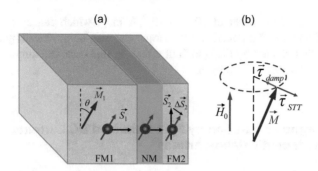

**Fig. 8.7** (a) Pictorial representation of the mechanism of spin transfer torque in a trilayer made of two ferromagnetic (FM) layers intercalated by a nonmagnetic metallic (NM) layer, traversed by an electric current. (b) Illustration of the action of the spin transfer and damping torques on the magnetization of the FM2 layer

$$\vec{\tau}_{STT} = \frac{\hbar P J}{2e\,t_2}\widehat{s}_2 \times (\widehat{s}_1 \times \widehat{s}_2), \qquad (8.39)$$

where $\widehat{s}_1$ and $\widehat{s}_2$ are the unit vectors of the macrospins in the two FM layers, related to the magnetizations by $\widehat{s}_\alpha = \vec{M}_\alpha/M_\alpha$. This result can also be obtained [31] with a simple quantum mechanical treatment of the spin and orbital wave functions of the electrons in the trilayer [see also Stancil and Prabhakar]. Adding the STT torque to the Landau–Lifshitz–Gilbert Equation (1.40) and using the fact that the magnetization is related to the angular momentum per unit volume through the gyromagnetic ratio $\gamma$, we obtain the equation of motion for $\vec{M}_2$

$$\frac{\mathrm{d}\vec{M}_2}{\mathrm{d}t} = \gamma \vec{M}_2 \times \vec{H}_{eff} - \frac{\alpha}{M_2}\vec{M}_2 \times \frac{\mathrm{d}\vec{M}_2}{\mathrm{d}t} + \frac{\hbar\gamma P J}{2e t_2}\widehat{s}_2 \times (\widehat{s}_1 \times \widehat{s}_2), \qquad (8.40)$$

where $\alpha$ is the Gilbert damping parameter. The STT torque has the same form as the Landau–Lifshitz damping term, which can be shown to be equivalent to the Gilbert damping [See Guimarães]. Figure 8.7b shows the direction of the damping torque and the STT acting on the magnetization of the FM2 layer for a current in the sense of Fig. 8.7b, corresponding to $J > 0$. While the damping torque pushes the magnetization toward the equilibrium direction, determined by the applied magnetic field, the STT acts as an anti-damping torque, pulling the magnetization away from the equilibrium direction. Thus, the STT can either reverse the magnetization direction, or set the magnetization to precess about the field at microwave frequencies. However, if the direction of the current is reversed, the STT produces additional magnetization damping. Note that the magnitude of the torque in Eq. (8.39) is proportional to the current per unit area, and inversely proportional to the magnetic layer thickness. For this reason, the STT phenomenon is important only in magnetic nanostructures. Typically, the current density required to switch the magnetization in a magnetic layer with thickness of a few nanometers and lateral dimensions of tens of nanometers is on the order of $10^{10}$ to $10^{12}$ A m$^{-2}$, which can be obtained with currents of few mA. Of course, a spin torque can also be produced by a pure spin current, as that created by the spin Hall effect. In this case, the torque is given by Eq. (8.39), with $PJ$ replaced by $J_S$.

### 8.3.2  Magnon Excitation by Spin-Polarized DC Currents in Magnetic Nanostrutures

The first experimental evidences of the magnon excitation by spin transfer torque were obtained with the injection of currents into magnetic multilayers by means of point contacts [32–34]. In one class of experiments the traces of spin wave excitation were observed by changes in the sample magnetoresistance with increasing injected DC current density [32, 34]. By varying the current intensity at a fixed value of the external magnetic field, sudden changes in the resistance were observed that were attributed to the onset of spin wave growth at certain critical values of the current. In another experiment, AC current was injected into a magnetic trilayer by means of two point contacts and the excited spin waves were observed directly with Brillouin light scattering [33]. By changing the position of the laser beam relative to the point contacts, it was possible to separate the contribution of the STT to the spin wave excitation from the one due to the Oersted–Ampère field. However, in both experiments, the evidence of spin wave excitation by the STT was very indirect and no information on the spatial nature of the excited modes was provided. Subsequently, direct observation of microwave oscillations due to spin waves excited by STT created by spin-polarized DC currents in magnetic multilayers were reported by various groups [35–38]. These developments stimulated intense research efforts to understand in detail the mechanisms of the oscillations [39–44] and to develop efficient microwave spin torque nano-oscillators [25–28].

Microwave spin torque nano-oscillators (STNO) can utilize either fully metallic GMR-type spin valves or magnetic tunnel junctions (MTJ), having a thin dielectric spacer between the two FM layers. Let us consider the simple structure for a STNO illustrated in Fig. 8.8, magnetized by a magnetic field applied either parallel or perpendicular to the film plane. Electrons of the electric current injected through the contacts become spin polarized at the thick FM pinned layer and reach the free FM layer, separated by a thin metallic or insulating layer. The spin-polarized electrons exert an anti-damping STT at the free-layer magnetization that tend to drive its dynamics. Since the lateral dimensions of the film are much larger than the thickness, we may consider the magnon normal modes as in the films studied in Chap. 4. We shall study the magnon excitation with the quantum approach used in previous chapters.

Consider initially the magnon system in a FM film described by the free Hamiltonian in Eq. (3.74). The effect of the STT on the spins can be calculated with Eq. (8.39) using Newton's law $\mathrm{d}(\hbar \vec{S}_i)/\mathrm{d}t = V \vec{\tau}_{STT}$, where $V$ is the film volume. We obtain for the spin raising operator at site $i$

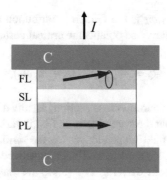

**Fig. 8.8** Schematic structure of a STNO. The current is injected through contacts C. Electrons flowing upward are spin polarized by the magnetization of the thick ferromagnetic pinned layer (PL), traverse the thin metallic or insulating nonmagnetic spacer layer (SL), and drive the magnetization of the ferromagnetic free layer (FL)

$$\frac{dS_i^+}{dt} = \frac{\beta J}{S} S_i^z S_i^+,$$

(8.41a)

where $J$ is the current density and the coefficient $\beta$ is given by

$$\beta = \frac{P\hbar\gamma}{2teM},$$

(8.41b)

where $t$ is the thickness of the FM film and $M$ is saturation magnetization. Using Eqs. (3.22) and (3.31) to transform the spin operators into the magnon operators, and neglecting the effect of the dipolar interaction in the transformations, one obtains from (8.41a) in the linear approximation

$$\frac{dc_k}{dt} = \beta J c_k.$$

(8.42)

Adding this contribution to the equation for $c_k$ obtained from the Heisenberg equation with the Hamiltonian (3.74), and introducing the relaxation rate $\eta_k$ in a phenomenological manner, we obtain

$$\frac{dc_k}{dt} = -i\omega_k c_k - (\eta_k - \beta J)c_k.$$

(8.43)

The solution of this equation of motion is straightforward

$$c_k(t) = c_k(0)\, e^{-i\omega_k t}\, e^{-(\eta_k - \beta J)t}.$$

(8.44)

This result means that when the current density exceeds the critical value $J_c = \eta_k/\beta$, the effect of the anti-damping STT overcomes the damping, and the magnon mode with lowest relaxation rate grows exponentially. Writing the relaxation rate as

$\eta_k = \eta_0 + \alpha_G \omega_k$, where $\alpha_G \omega_k$ is the Gilbert contribution and $\eta_0$ is a residual value independent of the frequency, one obtains the critical current for magnon excitation

$$I_c = \frac{2A_c\, etM}{P\hbar\gamma}\left(\eta_0 + \alpha_G\,\omega_k\right), \tag{8.45}$$

where $A_c$ is the area of the current cross section, assumed to be uniform in a certain region of the film. Note that the second term in (8.45) varies with magnetic field and is responsible for the field dependence of the critical current observed in experiments. Equation (8.45) applies to films magnetized either parallel or perpendicular to the plane [41]. In deriving Eq. (8.45) we have assumed that both the current and the spin waves are confined to the same region. This is valid for the nanopillar configuration of Fig. 8.8, employed in some experiments [35, 38]. In this case the excited magnons are standing wave modes that satisfy the boundary conditions at the edges of the film. In experiments using point contacts [36], the radiation of spin waves away from the excitation region represents an additional loss that must be overcome, causing an increase in the critical current [45]. Also, in the perpendicular configuration, the excited magnetic modes might be *standing self-localized spin-wave bullets* [46], not simply standing waves of the magnon modes studied in Chap. 4, or gyrating magnetic vortices [47].

The microwave power emitted by a single STNO is very small, on the order of tens of μW. One scheme used to increase the device output power consists in fabricating arrays of nano-oscillators in close proximity [25–27]. With the appropriate array design, one can have a coupling between neighboring oscillators such that they operate in phase-locking conditions, resulting in increased power emission [48–50].

The nonlinear interactions, not considered so far, have three important effects on the excitation of spin waves by spin-polarized currents. Two of them were studied in Chap. 6 in connection with parametrically excited magnons, saturation of the spin wave amplitude above threshold and quantum coherence. The third effect manifests only in samples with nanometric dimensions, where the magnon population can be comparable to the number of spins. As the magnon population increases with increasing current, the oscillation frequency shifts due to the magnon energy renormalization. We shall study the effects of the magnon interactions considering only the four-magnon interaction.

The starting point is the nonlinear equation of motion for the magnon operator including terms with three operators. This is obtained by adding to Eq. (8.43) the term with two boson operators from the transformation of $S_i^z$ in (8.31), given by Eq. (3.22c), and the contribution from the four-magnon interaction, given by (6.69). Neglecting the effect of the dipolar interaction in the transformations, and considering that the expectation values of the operators $c_k^\dagger$ and $c_k$ can be treated as classical variables, the equation of motion for $c_k$ becomes

$$\frac{dc_k}{dt} = -i\omega_k c_k - (\eta_k - \beta J)c_k - \frac{3\beta J}{2SN}c_k^* c_k c_k - iT_k c_k^* c_k c_k, \tag{8.46}$$

where $T_k$ is the coefficient of the four-magnon interaction arising from the surface dipolar and surface anisotropy energies. For standing wave modes in the nanopillar FM film this is different from the one used in Chap. 6, and is given by [41].

$$T_k \approx \pm \frac{\gamma 6\pi M_{eff}}{NS}, \tag{8.47}$$

where the plus sign applies for the magnetization perpendicular to the film, and the minus sign for the magnetization parallel to the film plane [41]. In order to interpret the roles of the nonlinear interactions we rewrite (8.46) using $n_k = c_k^* c_k$ for the number of magnons

$$\frac{dc_k}{dt} = -i(\omega_k + T_k n_k)c_k - [\eta_k - \beta J(1 - 3n_k/2SN)]c_k. \tag{8.48}$$

This equation shows that when the current traversing the FM film is above the critical value, the spin-wave amplitude initially grows exponentially while the number of magnons $n_k$ is negligible. However, as $n_k$ increases and approaches $SN$, the STT driving decreases due to the negative nonlinear term and its effect is balanced by the relaxation, so that the spin-wave amplitude saturates. With Eq. (8.48) and its complex conjugate one can obtain a simple equation for the magnon number in the steady state regime ($dn_k/dt = 0$), valid at times $t \gg 1/\eta_k$

$$n_{ss} = \frac{2SN}{3}\frac{(I - I_c)}{I}, \tag{8.49}$$

where $I_c = J_c A_c$ is the critical current, $A_c$ being area of the current cross section. Note that contrary to the role it plays in the parallel pumping instability process, here the four-magnon interaction does not limit the growth of the spin-wave amplitude. The reason for this is that the origin of the driving in the parallel pumping process is fundamentally different from the one with a DC current. There, the instability results from the generation of magnon pairs due to the conversion of photons into magnons, whereas here the spin wave system becomes unstable because the STT pulls the magnetization away from the equilibrium direction and balances the effect of the relaxation. It turns out that as the spin wave amplitude grows, there is a reduction in the $z$-component of the spin, causing a downward deviation of the STT from linearity. This is the nonlinear effect that limits the growth of the spin waves generated by the spin-polarized DC current. The effect of the 4-magnon interaction here is to change the magnon frequency, i.e. energy renormalization. Using Eqs. (8.47) and (8.49) in the first term on the right-hand side of (8.48), we obtain the frequency shift as a function of current for $I \geq I_c$

$$\Delta \omega_k = \pm \gamma \, 4\pi M_{eff} \frac{I - I_c}{I}. \tag{8.50}$$

This result shows that as the current is increased beyond threshold, the oscillation frequency shifts upwards (blue shift) for a field perpendicular to the film plane, and shifts downwards (red shift) for a field parallel to the film plane, in full agreement with experiments [36–38].

In order to compare theory with experiments, the equations for the real and imaginary parts of $c_k$ obtained from (8.46) have been solved numerically [41] using conditions and material parameters of Krivorotov et al. [38]. The experiments were carried out with a nanopillar-shaped sample made of two 4 nm thick permalloy (Py $=$ Ni$_{80}$Fe$_{20}$) layers separated by a 8 nm thick Cu spacer layer, on top of an antiferromagnetic underlayer for pinning the bottom Py layer by exchange bias. The current is applied to the multilayer through Cu electrodes on both sides, so that it is uniformly distributed over the lateral dimension. As argued in Ref. [41], the measurements of the oscillations induced by DC currents in multilayers do not allow a clear-cut identification of the excited spin-wave mode. However, the fact that in order to radiate efficiently the mode should have the smallest possible value of $k$, strongly suggests that the driven mode is the lowest order standing wave mode allowed by the structure, i.e., the mode with wave vector in the plane, with $k = \pi/L$, where $L$ is the relevant lateral dimension. Using this value and the other material parameters discussed in Ref. [41], the time evolution of $c_k(t)$ was calculated numerically for $I > I_c = 1.25$ mA to obtain the Fourier spectra. Figure 8.9a shows spectra of the $x$-component of the magnetization (that is proportional to $c_k$) calculated for the same values of the driving current of the experiments of Ref. [38]. Figure 8.9b shows the comparison of the calculated frequency vs current with the experimental data, demonstrating that the magnon excitation theory presented here is in very good agreement with experiments.

To close this section, we shall study the quantum nature of the spin waves excited by the DC current, using the same formalism described in Chaps. 6 and 7 to study the quantum coherence of parametrically excited spin waves and Bose–Einstein condensed magnons. The theory is based on a quantum spin wave formalism to treat the magnetic excitations described by the following system Hamiltonian

$$H = H_S + H_R + H_{RS}, \tag{8.51}$$

where $H_S$, $H_R$, and $H_{RS}$ are, respectively, the Hamiltonians for the spin-wave system, the heath-bath reservoir, and the reservoir-system interaction. As considered earlier, the system Hamiltonian can be written as $H_S = H_0 + H^{(4)}$, while the heath-bath reservoir and the reservoir-system Hamiltonians are given in Eq. (6.71). Using (8.51) in the Heisenberg equation, and introducing the effects of the relaxation and the STT, we obtain the following equation of motion for the magnon destruction operator [51].

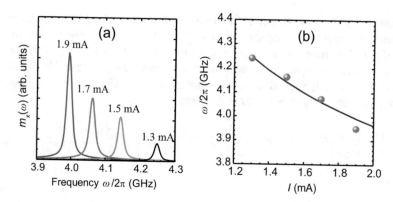

**Fig. 8.9** (a) Spectra of the $m_x$ component of the magnetization obtained by the numerical solution of Eq. (8.46) for several values of the driving current to compare with the data of Ref. [38]. (b) The solid line shows the calculated frequency vs current while the symbols represent the data of Ref. [38]

$$\frac{dc_k}{dt} = -i\omega_k c_k - (\eta_k - \beta J)c_k - \frac{3\beta J}{2SN}c_k^\dagger c_k c_k - iT_k c_k^\dagger c_k c_k + F_k(t), \qquad (8.52)$$

where $\eta_k = \pi f(\omega_k)|\beta_{k,\,p}|^2$ is the magnon relaxation rate, $f(\omega_k)$ is the density of heat bath modes, and

$$F_k(t) = -i\sum_p \beta_{k,p} R_p e^{-i\omega_p t} \qquad (8.53)$$

represents a Langevin random force with correlators of Markoffian systems type. As in Chaps. 6 and 7, the next step consists in transforming the magnon operators into a new variable in the representation of coherent magnon states $|\alpha_k\rangle$. Considering the coherent state eigenvalue in a frame rotating with frequency $\omega_k$, defined by $c'_k|\alpha_k\rangle = \alpha_k(t)e^{-i\omega_k t}|\alpha_k\rangle$, where $c'_k = c_k/(SN)^{1/2}$ is the normalized operator, from Eq. (8.52) one can show that equation of motion for the new variable is [51, 52].

$$\frac{d\alpha_k(t)}{dt} + iT'_k|\alpha_k|^2\alpha_k(t) - \frac{3\eta_k R}{2}\left(\frac{2}{3}\frac{R-1}{R} - |\alpha_k|^2\right)\alpha_k(t) = F_k(t)e^{i\omega_k t}, \qquad (8.54)$$

where $R = I/I_c$ is a driving parameter and $T'_k = SN T_k$ is a renormalized coefficient. Equation (8.54) contains all the information carried by the equations of motion for the magnon operators. It is similar to the nonlinear Langevin equation which appeared in previous chapters, but contains the nonlinear frequency shift term in $T'_k$. Equation (8.54) reveals that the magnon modes with amplitude $\alpha_k$ are driven thermally by the heat bath modes and also by the STT produced by the spin-polarized DC current. The solutions of Eq. (8.54) confirm the previous analysis. For $I < I_c$ the driving parameter $R$ is less than 1 and the magnon amplitudes are essentially the ones of the thermal reservoir. For $I > I_c$, $R > 1$, and the magnon amplitudes grow exponentially and are limited by the effect of the nonlinear interaction. Well above threshold the steady-state solution of Eq. (8.54) gives for the normalized number of magnons $n'_k = n_k/SN = |\alpha_k|^2$, an expression identical to Eq. (8.49). The final step to obtain information about the coherence of the excited mode is to find an equation for the probability density $P(\alpha_k)$, defined in Eq. (3.98), that is stochastically equivalent to the Langevin equation. Using $\alpha_k = a_k \exp(i\varphi_k)$ we obtain a Fokker–Planck equation in the form

$$\frac{\partial P}{\partial t'} + \frac{1}{x}\frac{\partial}{\partial x}\left[(A - x^2)x^2 P\right] = \frac{1}{x}\frac{\partial}{\partial x}\left(x\frac{\partial P}{\partial x}\right) + \frac{2T'_k}{3R\eta_k}x^2\frac{\partial P}{\partial \varphi_k} + \frac{1}{x^2}\frac{\partial^2 P}{\partial \varphi_k^2}, \quad (8.55)$$

where

$$t' = \left[(3R/2)n'_k\right]^{1/2}\eta_k t, \quad x = (3R/2n'_k)^{1/4}a_k, \quad (8.56)$$

represent normalized time and magnon amplitude, and $A$ is related to the driving parameter $R$ by

$$A = \left(\frac{2}{3R\overline{n}'_k}\right)^{1/2}(R - 1). \quad (8.57)$$

Equation (8.55) describes the full dynamics of the magnon excitation in the film by the DC spin-polarized current. Here we are interested in the stationary solution of Eq. (8.55) independent of $\varphi_k$ which has the form

$$P(x) = C\exp\left(\frac{1}{2}Ax^2 - \frac{1}{4}x^4\right), \quad (8.58)$$

where $C$ is a normalization constant such that the integral of $P(x)$ in the range of $x$ from zero to infinity is equal to unity. Note that Eq. (8.58) is different from (6.81) and (7.36) obtained in Chaps. 6 and 7, but its behavior is similar to the previous ones.

Figure 8.10 shows plots of $P(x)$ for four values of $R = I/I_c$ for material parameters as in the experiments of Ref. [38], $\omega_k/2\pi = 4\,\mathrm{GHz}$ and $T = 300\,\mathrm{K}$. For $R = 0.8$ the coefficient $A$ is negative $(-2.9)$ and $P(x)$ behaves as a Gaussian distribution, characteristic of systems in thermal equilibrium and described by incoherent magnon states. For $R = 1$, $A = 0$ and $P(x)$ also looks a broad Gaussian function indicating a noisy oscillation. For $R = 1.2$, $A$ is positive but small (2.4), and the stationary state consists of two components, a coherent one convoluted with a smaller fluctuation with Gaussian distribution. Finally, for $R = 2.0$, $A$ is positive and large (9.2) and $P(x)$ becomes similar to a Poisson distribution, characterizing the dominance of a coherent state. Since the variance of $P(x)$ is proportional to $A^{-1}$, for $A \gg 1$ the function $P(x)$ becomes a delta-like distribution, characteristic of a pure coherent magnon state. For $A > 0$ it is easy to show that $P(x)$ given by Eq. (8.58) has a peak at $x_0 = A^{1/2}$. Thus, for large values of $A$, $P(x)$ represents a coherent state with an average number of magnons given by $x_0^2 = A$. This gives for the magnon number an expression identical to Eq. (8.49).

**Fig. 8.10** Plots of the probability density $P(x)$ as a function of the normalized magnon amplitude $x$ for four values of the driving parameter $R = I/I_c$ for material parameters as in the experiments of Ref. [38]

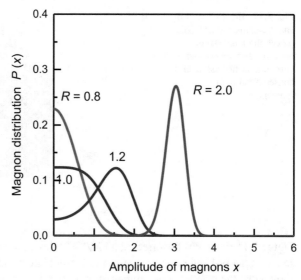

## 8.4    Spin Pumping by $k = 0$ Magnons in Magnetic Bilayers

### 8.4.1    Spin Pumping Damping

The conservation of angular momentum in the interaction between electric currents and the magnetization in metallic magnetic multilayers gives rise to several striking phenomena. In one of them, the spin transfer torque, studied in the previous section, spin currents can drive the magnetization dynamics without the intermediation of magnetic fields. The Onsager reciprocal of the STT phenomenon, corresponding to the generation of currents by the magnetization dynamics, was first proposed long time ago by Silsbee and co-workers [1, 2]. This effect, now called *spin pumping*, attracted more attention only after the rigorous theoretical work of Tserkovnyak, Brataas and Bauer [53, 54], who gave the effect the name by which it is known nowadays. The main concern of the work in Refs. [53, 54] was the additional magnetization damping of a ferromagnetic (FM) film produced by the atomic contact with a nonmagnetic metallic (NM) layer. Those authors showed that the precessing magnetization in the FM film generates a spin current that flows across the FM/NM interface and carries angular momentum out of the magnetization dynamics, producing additional damping. Interestingly, the spin pumping effect attracted much more attention after the experimental observation of the ferromagnetic resonance in magnetic multilayers by means of the electric signal generated by the spin-pumped spin currents [55, 56], not considered in the theoretical proposal of the spin pumping process.

The spin pumping process consists in the emission of a spin current by a time-varying magnetization, such as the precession in ferromagnetic resonance.

**Fig. 8.11** Illustration of the
spin pumping process. The
precessing magnetization in
the FM layer injects a spin
current into the NM layer by
conservation of angular
momentum

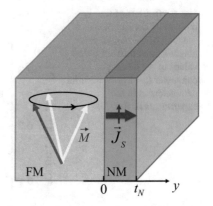

Figure 8.11 illustrates the effect in a FM/NM bilayer, where the precessing magnetization in the FM layer injects a pure spin current into an adjacent NM layer. It can be
shown that the magnetization dynamics pumps a spin current into the NM layer, that
at the FM/NM interface has density given by [53, 54, 57]

$$\vec{J}_S^p = \frac{\hbar}{4\pi M} \left( \frac{g_r^{\uparrow\downarrow}}{M} \vec{M} \times \frac{d\vec{M}}{dt} - g_i^{\uparrow\downarrow} \frac{d\vec{M}}{dt} \right), \qquad (8.59)$$

where $g^{\uparrow\downarrow} = g_r^{\uparrow\downarrow} + ig_i^{\uparrow\downarrow}$ is the spin-mixing conductance (dimension of 1/area) that
has real and imaginary parts given by

$$g_r^{\uparrow\downarrow} = \frac{1}{2} \sum_{mn} [|r_{mn}^{\uparrow} - r_{mn}^{\downarrow}|^2 + |t_{mn}^{\uparrow} - t_{mn}^{\downarrow}|^2], \qquad (8.60a)$$

$$g_i^{\uparrow\downarrow} = \text{Im} \sum_{mn} [r_{mn}^{\uparrow}(r_{mn}^{\downarrow})^* + t_{mn}^{\uparrow}(t_{mn}^{\downarrow})^*], \qquad (8.60b)$$

where $r_{mn}^{\uparrow}$ $(r_{mn}^{\downarrow})$ is the reflection coefficient (per unit area) for spin-up (spin-down)
electrons at the FM/NM interface, $t_{mn}^{\uparrow}$ $(t_{mn}^{\downarrow})$ is the transmission coefficient (per unit
area) for spin-up (spin-down) electrons, and $m$ and $n$ denote the electron modes
involved at the Fermi energy in the normal metal. The spin-mixing conductance
expresses the transparency of the interface to the flow of spin current from the FM to
the NM. Its real part is on the order of $k_F^2/4\pi$, where $k_F$ is the wave number at the
Fermi energy of the normal metal, that gives $g_r^{\uparrow\downarrow} \sim 10^{15} \text{ cm}^{-1}$. The imaginary part
$g_i^{\uparrow\downarrow}$ vanishes for ballistic and diffusive contacts, so that in real interfaces it is orders
of magnitude smaller than the real part [53].

Clearly, the first term in Eq. (8.59), with the real part of the spin-mixing
conductance, represents a damping torque, similar to the Gilbert damping torque
in Eq. (8.40). The second term in Eq. (8.59) represents an out-of-plane torque and is
known as field-like torque. Since $g_i^{\uparrow\downarrow} \ll g_r^{\uparrow\downarrow}$, we will no longer consider the second
term. To study in detail the effect of the damping torque we have to consider that the

spin current pumped at the interface leads to a spin accumulation in the NM layer. This produces a spin current that flows back into the FM layer, so that the net spin current flowing out of the FM layer is smaller than the first term in Eq. (8.59). As shown in [54, 57], the back-flow spin current density at the interface has amplitude

$$J_S^{back} = \frac{g_r^{\uparrow\downarrow}}{4\pi} \mu_S, \tag{8.61}$$

where $\mu_S$ is the spin accumulation in the NM at the interface.

To find the back-flow spin current we have to consider the finite width of the NM layer and to solve the boundary condition problem at the FM/NM interface and at the NM outer surface. The starting point is the diffusion equation for the spin accumulation in the NM layer. We consider the bilayer in Fig. 8.11 with infinite lateral dimensions, so that the spin accumulation diffuses only in the $y$-direction. Assuming a harmonic time dependence, $\mu_S(y, t) = \mu_S(y) \exp(i\omega t)$, where $\omega$ is the frequency of the magnetization precession in the FM layer, Eq. (8.22) leads to

$$i\omega \mu_S(y) = D \frac{\partial^2 \mu_S(y)}{\partial y^2} - \frac{\mu_S(y)}{\tau_{sf}}. \tag{8.62}$$

In anticipation of the boundary conditions, instead of writing the solution of this equation in terms of exponential functions as in (8.23), we use the related hyperbolic functions

$$\mu_S(y) = A \cosh \kappa(y - t_N) + B \sinh \kappa(y - t_N), \tag{8.63}$$

where $\kappa = (1 + i\omega\tau_{sf})^{1/2}/\lambda_N$ is a wave number, $\lambda_N = (D\tau_{sf})^{1/2}$ and $t_N$ are, respectively, the spin diffusion length and the thickness of the NM layer, and $A$ and $B$ are coefficients to be determined by the boundary conditions. Using Eq. (8.20) with Einstein's relation (8.12), $\vec{J}_S = (1/2)\hbar D N(\varepsilon_F)\nabla\mu_S$, the spin current density (in units of angular momentum/area.time) obtained from (8.63) is

$$\vec{J}_S(y) = \hat{y}\frac{1}{2} \hbar D N(\varepsilon_F)\kappa [A \sinh \kappa(y - t_N) + B \cosh \kappa(y - t_N)]. \tag{8.64}$$

The boundary conditions are determined by conservation of angular momentum, which implies continuity of the spin current at the surfaces and interfaces. Thus, at the FM/NM interface we have $\vec{J}_S(y = 0) = \vec{J}_S^{net}$, where $\vec{J}_S^{net}$ is the net spin current pumped at the interface (considering the back-flow). At the surface of the NM layer, since the spin current vanishes outside, we have $\vec{J}_S(y = t_N) = 0$. Inspection of Eq. (8.64) shows that the second boundary condition implies $B = 0$, so that with the first one we easily find the coefficient $A$. The spin accumulation then becomes

$$\mu_S(y) = -\frac{2J_S(0)}{\hbar N(\varepsilon_F)D\kappa}\frac{\cosh\kappa(y-t_N)}{\sinh\kappa t_N},\tag{8.65}$$

while the spin current density is

$$\vec{J}_S(y) = \vec{J}_S(0)\frac{\sinh\kappa(t_N-y)}{\sinh\kappa t_N}.\tag{8.66}$$

Subtracting from the spin current pumped into the NM by the FM layer, given by Eq. (8.59), the back-flow current in (8.61), it can be shown that the net spin current density flowing into the NM layer is, at the interface [54, 57]

$$\vec{J}_S(y=0) = \frac{\hbar g_{eff}^{\uparrow\downarrow}}{4\pi M^2}\left(\vec{M}\times\frac{d\vec{M}}{dt}\right),\tag{8.67}$$

where

$$g_{eff}^{\uparrow\downarrow} \approx \frac{g_r^{\uparrow\downarrow}}{1+(2\sqrt{\xi/3}\tanh\kappa t_N)^{-1}}\tag{8.68}$$

is the real part of the effective spin-mixing conductance that takes into account the effect of the back-flow current. Equation (8.68) is an approximate result obtained assuming that the electron elastic scattering time $\tau_{el}$ is much smaller than the spin-flip time $\tau_{sf}$, so that the parameter $\xi = \tau_{el}/\tau_{sf}$ is small. In practice this condition is easily satisfied in metals with large atomic number $Z$ that have large spin–orbit coupling.

One of the important consequences of the spin pumping process is the additional damping of the magnetization of a FM film when placed in contact with a NM layer. This is a result of the conservation of angular momentum, since the spin current flowing out of the FM layer due to spin pumping subtracts angular momentum from the magnetization dynamics. To quantify this effect we use the Landau–Lifshitz–Gilbert equation (1.40) for the magnetization of the FM layer and subtract the contribution of the angular momentum pumped out. This is given by Eq. (8.67) divided by the FM layer thickness, to have angular momentum per unit volume, and multiplied by the gyromagnetic ratio $\gamma$. Thus, we have

$$\frac{d\vec{M}}{dt} = \gamma\vec{M}\times\vec{H}_{eff} - \frac{\alpha}{M}\vec{M}\times\frac{d\vec{M}}{dt} - \frac{\gamma\hbar}{4\pi M^2 t_{FM}}g_{eff}^{\uparrow\downarrow}\vec{M}\times\frac{d\vec{M}}{dt}.\tag{8.69}$$

Thus, the total Gilbert damping parameter of the FM magnetization becomes

$$\alpha_T = \alpha + \frac{\gamma\hbar g_{eff}^{\uparrow\downarrow}}{4\pi M t_{FM}}.\tag{8.70}$$

It is useful to have an expression for the FMR linewidth including the contribution from the spin pumping. Using the relation between the linewidth and the relaxation rate in Eq. (1.32), expressed in terms of the Gilbert parameter, $\Delta H = \omega \alpha / \gamma$, we obtain for the total linewidth

$$\Delta H_T = \Delta H_{FM} + \frac{\hbar \omega \, g_{eff}^{\uparrow\downarrow}}{4 \pi M \, t_{FM}}, \tag{8.71}$$

where $\omega$ is the FMR frequency and $\Delta H_{FM}$ is the linewidth of the FM film without the NM layer. The additional linewidth due to spin pumping is proportional to the frequency and to the spin-mixing conductance of the interface because the flow of angular momentum out of the FM layer is proportional to these quantities. Also, it varies inversely with the FM layer thickness because the spin pumping is an interface phenomenon and the magnetization damping is determined by the out-flow angular momentum per volume. For this reason, the spin pumping damping in metallic FM/NM bilayers is important only for FM film thickness up to a few tens of nanometers. The linear dependence of the linewidth on the frequency and the variation with the inverse of the FM layer thickness are considered the signatures of the spin pumping damping [57, 58]. Equation (8.71) leads to an expression for the spin-mixing conductance in FM/NM bilayers in terms the incremental FMR linewidth with deposition of the NM layer

$$g_{eff}^{\uparrow\downarrow} = \frac{4 \pi M \, t_{FM}}{\hbar \omega} (\Delta H_T - \Delta H_{FM}). \tag{8.72}$$

Although this expression is widely used to determine the spin-mixing conductance from measurements of the FMR linewidths, it lacks a factor arising from the fact that in films magnetized in the plane, the relation between the linewidth and the relaxation rate is given by Eq. (4.53), and not simply $\eta = \gamma \, \Delta H$. At frequencies in the X-band microwave range, this factor is approximately 2 for permalloy and 1 for YIG. Another common problem is that, as studied in Sect. 4.3, in ultrathin FM films there is another damping mechanism that depends on the film thickness, namely two-magnon surface scattering. Thus, in order to clearly identify the contribution of the spin pumping damping to the FMR linewidth in measurements with varying FM layer thickness, it is necessary to take into account the two-magnon scattering damping. This has been clearly demonstrated in measurements of the FMR linewidth in bare permalloy (Py) films and in Py/Pt bilayers with varying Py layer thickness. Figure 8.12a shows the field derivative of the absorption spectra measured in a Py (7 nm) bare film and in a Py (7 nm)/Pt (10 nm) bilayer, using a microwave cavity at 8.6 GHz with the field in the plane [59]. The dotted red line is the spectrum for the bare Py film that can be fitted with a Lorentzian derivative function yielding a peak-to-peak linewidth $\Delta H_{pp} = 32.2$ Oe, which is substantially larger than the value of 22.4 Oe for a 200 nm Py film that has negligible two-magnon damping. The solid blue line is the spectrum for Py (7 nm)/Pt (10 nm) that has $\Delta H_{pp} = 51.8$ Oe, showing the substantial increase in the damping due to spin pumping.

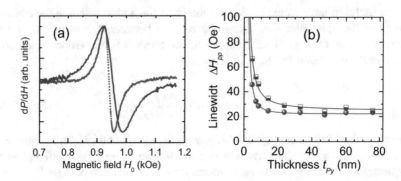

**Fig. 8.12** (a) Field derivatives of the FMR absorption measured at 8.6 GHz in a bare Py (7 nm) film (dotted red line) and in Py (7 nm)/Pt (10 nm) bilayer (solid blue line). (b) Data for the FMR linewidth measured at 8.6 GHz in Py films with varying thickness $t_{Py}$. The (red) circles are the data for the Py bare films and the (blue-white) squares are the data for the Py films covered with a 10 nm thick Pt layer. The lines are least-square-deviation fits of Eqs. (8.73) and (8.74) to data, from which the parameters are determined [59]

Figure 8.12b shows the variation with Py film thickness of the measured FMR linewidths for bare Py films and for films covered with a 10 nm thick Pt layer. In the bare Py films there are two sources for the FMR linewidth, the intrinsic Gilbert damping that does not depend on the film thickness, and the two-magnon scattering that varies with thickness as in Eq. (4.55). The measured linewidth versus thickness can be fitted with the expression

$$\Delta H = \Delta H_{int} + \frac{C_{2m}}{t_{Py}^2}. \tag{8.73}$$

The least-square deviation fit of Eq. (8.73) to the bare film data, represented by the lower solid line in Fig. 8.12b, yields the following parameters, $\Delta H_{int} = 22.15$ Oe and $C_{2m} = 520.5$ Oe nm$^2$. In order to interpret the data for the Py/Pt bilayers we consider three contributions to the linewidth, one term independent of the thickness ($\Delta H_0$) that includes the intrinsic damping, the two-magnon scattering linewidth, and the spin pumping damping given by Eq. (8.71). The two-magnon scattering contribution is assumed to be the same obtained with the fit of Eq. (8.73) to the bare Py film data, because the variation of the resonance field with Py thickness is nearly the same as in the Py/Pt bilayer, indicating that the Pt layer does not change the surface anisotropy. Thus, the total linewidth can be written as

$$\Delta H = \Delta H_0 + \frac{C_{2m}}{t_{Py}^2} + \frac{C_{SP}}{t_{Py}}. \tag{8.74}$$

The upper (blue) solid line in Fig. 8.12b represents the fit of Eq. (8.74) to the Py/Pt data considering $C_{2m} = 520.5$ Oe nm$^2$, from which one obtains $C_{SP} = 94.5$ Oe nm and $\Delta H_0 = 24.45$ Oe. Using the value of $C_{SP}$ obtained from the

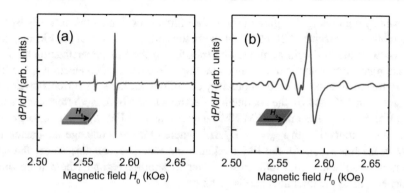

**Fig. 8.13** Field scan microwave absorption spectra at a frequency 9.4 GHz of 28 μm thick YIG film with lateral dimensions 2 × 3 mm with the magnetic field applied parallel to the film plane. In (**a**) the YIG film is bare while in (**b**) and it is covered with a 6 nm Pt layer [Adapted from Ref. 61]. Reproduced from S. M. Rezende et al., Appl. Phys. Lett. **102**, 012402 (2013), with the permission of AIP Publishing

fit and $4\pi M = 11.5\,\text{kG}$ for Py, one finds with Eq. (8.71) (multiplied by $2/\sqrt{3}$ to account for the peak-to-peak linewidth) a value of $g_{eff}^{\uparrow\downarrow} = 1.6 \times 10^{15}\,\text{cm}^{-2}$ for the spin-mixing conductance. This is smaller than the values obtained for Py/Pt without considering the effect of the two-magnon scattering damping [5, 6, 60].

Although in metallic FM/NM bilayers the spin pumping damping is important only for FM film thickness up to tens of nanometers, in bilayers where the FM film is an insulator, the spin pumping damping can be relevant in films with thickness in the micrometer range. This has been demonstrated in FMR experiments with bare single-crystal YIG films of thickness in the range 4–28 μm and with YIG films covered with a Pt layer [61]. Figure 8.13 shows the field derivative d$P$/d$H$ of the microwave absorption spectra for a 28 μm thick YIG film obtained with the field parallel to the film plane, measured with the rectangular waveguide at a fixed microwave frequency of 9.4 GHz. The spectrum in Fig. 8.13a measured before deposition of the Pt layer allows a clear identification of the main resonance (FMR) and the standing wave magnetostatic spin-wave modes that have quantized in-plane wave numbers due to the boundary conditions at the edges of the film. The lines to the left of the FMR correspond to hybridized surface modes, whereas those to the right are hybridized volume magnetostatic modes. As shown in Fig. 8.13b, the deposition of a 6 nm thick Pt layer on the YIG film produces a pronounced broadening of the lines for all modes. They have nearly the same peak-to-peak linewidth $\Delta H_{pp} \approx 5.4\,\text{Oe}$, corresponding to a half-width at half-maximum (HWHM) $\Delta H = \sqrt{3}\Delta H_{pp}/2 \approx 4.6\,\text{Oe}$, which is nearly ten times larger than in bare YIG. Additional evidences of the spin pumping nature of the increased linewidth with the Pt covering are provided by the linear variation with the FMR frequency and the out-of-plane angle dependence of the linewidths measured in three samples with YIG film thickness of 8, 15 and 28 μm [61].

The spin pumping damping in thick YIG films is explained theoretically by two alternative approaches [62]. In one of them, the coupled motion of the FM magnetization with the NM spin accumulation transfers to the FM magnetization an additional relaxation from the overdamped motion of the conduction electron spins. The other treatment is based on the boundary conditions for the spin currents carried by magnons in YIG and by the conduction electrons in the NM. Both treatments show that Eq. (8.71) applies only to FM/NM bilayers that have FM layer thickness smaller than a coherence length $t_{coh} = \sqrt{D/\Delta H}$, where $D$ is the exchange parameter and $\Delta H$ is the linewidth. As the FM thickness increases beyond this value, the spin pumping additional linewidth crosses over to another regime where it becomes independent of the thickness and is given by

$$\Delta H_{SP} \approx \frac{\hbar \omega g_{eff}^{\uparrow\downarrow}}{4\pi M t_{coh}}. \tag{8.75}$$

For YIG/Pt, with $D = 5 \times 10^{-9}$ Oe cm$^2$ and $\Delta H \approx 2$ Oe, one obtains $t_{coh} \approx 500$ nm, whereas for Py, with $D = 2 \times 10^{-9}$ Oe cm$^2$ and $\Delta H \approx 20$ Oe, $t_{coh} \approx 100$ nm.

### 8.4.2    Electric Spin Pumping Effect

The spin pumping was conceived in 2002 as a mechanism for the magnetization damping in FM/NM bilayers [53, 54]. However, a more interesting effect of the spin pumping is the generation of an electric voltage in the NM layer, first observed in 2004 by Azevedo et al. [55]. The initial experimental observation of the electric spin pumping effect was made in FM/NM/FM trilayers, because the authors of Ref. [55] were actually trying to detect a DC voltage predicted to be generated in a magnetic multilayer under FMR [63]. The observed electric signal associated with the FMR was interpreted as produced by the spin current generated in the spin pumping process, but the origin of the conversion was not identified. Subsequently, the mechanisms involved in the electric detection of the FMR were clearly identified by Saitoh and co-workers [56], and confirmed in studies of the spin pumping in FM/NM bilayers by several groups [60, 64–67].

The initial studies of the electric spin pumping effect (SPE) were concentrated in bilayers with metallic ferromagnets. In structures with metallic FMs, in addition to the voltage produced by the SPE-ISHE in the NM layer one has to consider the current generated in the FM layer by the classical induction effect due to the *rf* variation of the in-plane magnetization. This contributes to the total voltage with a component referred to as galvanomagnetic, or anisotropic magnetoresistance voltage. This represents a spurious effect in the electric detection of the SPE, that has to be disentangled from the measured voltage to obtain the SPE contribution [60, 66–68]. For this reason, the research on the spin pumping effect gained additional impetus with the observation in 2010 of the SPE in bilayers with yttrium iron garnet [69]. The advantage of using insulating ferromagnets in the studies of the SPE is that

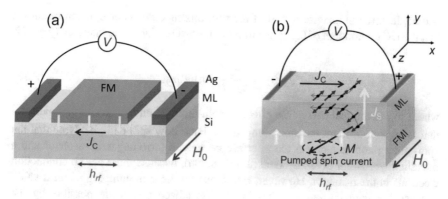

**Fig. 8.14** Schematic illustrations of sample arrangements used in the studies of the electric spin pumping effect. (**a**) The metallic FM film is deposited on top of the metallic layer (ML). (**b**) The ML is deposited on top of the ferro-ferrimagnetic insulator (FMI) film. The microwave magnetic field that drives the FMR is perpendicular to the static magnetic field $H_0$ that defines the direction of the spin polarization of the spin current. The contacts at the edges are used to measure the total DC voltage induced by the precessing magnetization

they are free from the spurious effects of metallic FMs. In recent years the spin pumping effect has become a most important tool for producing spin currents used in studies of spin-charge current conversion in 3D and 2D electron systems.

The electric spin pumping effect in a bilayer made of a ferromagnetic (FM) film and a metallic layer (ML), or a 2D electron system, is based on two distinct phenomena. In the first, the magnetization precession in the FM layer, or $k = 0$ magnon, driven by a microwave field under FMR conditions, produces a spin current by the spin pumping process. The spin current flows into the adjacent layer, where it is partially converted into a charge current by means of the inverse spin Hall or Rashba–Edelstein effects, generating a voltage proportional to the magnetization response. Figure 8.14 shows illustrations of typical sample arrangements used to investigate the electric spin pumping effect, either with metallic or insulating FM layers. In Fig. 8.14a, the ML (denoting a metallic layer or a 2D electron system) is deposited on the substrate, and the metallic FM film, in the form of an island, is deposited on top. Figure 8.14b shows a commonly used arrangement for the SPE in which the ML is deposited on top of the ferro-ferrimagnetic (FMI) insulator film, grown on some crystalline substrate. In both cases the FMR is excited by the *rf* field in a microwave cavity, or hollow waveguide, or microstrip line, and the voltage generated between the two contacts at the ends of ML is measured directly with a DC nanovoltmeter.

As discussed in the previous section, the precessing magnetization in the FM layer injects a pure spin current into the adjacent layer, that at the interface has current density given by Eq. (8.67). Writing the time dependence of the magnetization in the FM layer as $\vec{M}(t) = \hat{z}M_z + (\hat{x}m_x + \hat{y}m_y)\exp(i\omega t)$, where $m_x, m_y << M_z$, and neglecting the imaginary part of the spin-mixing conductance, it can be shown that the DC component of the spin current in the $y$-direction at the interface ($y = 0$) is

$$J_S(0) = \frac{\hbar \omega g_{eff}^{\uparrow\downarrow}}{4\pi M^2} \, \text{Im}\,[m_x^* m_y],$$  (8.76)

where $g_{eff}^{\uparrow\downarrow}$ is the real part of the effective spin-mixing conductance at the interface, that takes into account the spin pumped and back-flow spin currents. Note that the spin-mixing conductance at the interface between a ferromagnetic insulator and a metallic layer is not given by Eqs. (8.60a, 8.60b), because there are no conduction electrons in the insulator. However, Eq. (8.59) for the spin pumping remains valid, because the spin current transmission to the adjacent is made possible by the interfacial exchange coupling between the spins of $d$-electrons in the FMI side and of $s$-electrons in the NM side [62, 69, 70].

In order to calculate the spin-pumped spin current we need to find the *rf* components of the magnetization in the FM film driven by the *rf* magnetic field $\vec{h} = (\hat{x} h_y + \hat{y} h_y) \exp(i\omega t)$. Using the Landau–Lifshitz Eq. (1.10) with the coordinate system in Fig. 8.14b, we obtain the linearized equations for the magnetization components

$$i\omega m_x = -\gamma H_0 m_y + \gamma M h_y^{dip} + \gamma M h_y,$$  (8.77a)

$$i\omega m_y = \gamma H_0 m_x - \gamma M h_x^{dip} - \gamma M h_x,$$  (8.77b)

where $h_x^{dip}, h_y^{dip}$ are the components of the *rf* dipolar field. Considering the *rf* field in the $x$-direction, as in Fig. 8.14, using for the geometry of a film with the static field in the plane, $h_{dip}^y = -4\pi m_y$ and $h_{dip}^x = 0$, and introducing the relaxation rate $\eta_0$ in a phenomenological manner, we obtain

$$m_x = \frac{\gamma^2 (H_0 + 4\pi M) M}{(\omega_0^2 - \omega^2) + i2\omega_0\eta_0} h_{rf},$$  (8.78a)

$$m_y = -i \frac{\omega \gamma M}{(\omega_0^2 - \omega^2) + i2\omega_0\eta_0} h_{rf},$$  (8.78b)

where $h_{rf}$ denotes the amplitude of the *rf* field and $\omega_0$ is the FMR frequency. The surface anisotropy, which is important in ultrathin films, can be accounted for in Eqs. (8.78a, 8.78b) replacing $4\pi M$ by $4\pi M_{eff} = 4\pi M - 2K_S/M t$. For films magnetized in the plane, the FMR frequency given by Eq. (4.43) is $\omega_0 = \gamma [H_0(H_0 + 4\pi M_{eff})]^{1/2}$. For measurements done with fixed frequency $\omega$ and varying field $H_0$, it is necessary to express $\omega_0$ and $\eta$ in terms of $H_0$, and $\omega$ in terms of the field for resonance $H_r$ (field at which the resonance is at the frequency $\omega$). Equation (4.53) gives a relation between the relaxation rate and the linewidth

$$\eta_0 = \gamma \Delta H \frac{(2H_r + 4\pi M_{eff})}{2(H_r^2 + H_r 4\pi M_{eff})^{1/2}}. \tag{8.79}$$

Substitution of Eqs. (8.78a, 8.78b) and (8.79) in (8.76) leads to an expression for the spin current density pumped through the FMI/ML interface

$$J_S(0) = \frac{\hbar \omega g_{eff}^{\uparrow\downarrow} p}{16\pi} \left(\frac{h_{rf}}{\Delta H}\right)^2 L(H - H_r), \tag{8.80a}$$

where

$$L(H - H_r) = \frac{\Delta H^2}{(H - H_r)^2 + \Delta H^2}, \tag{8.80b}$$

is the Lorentzian function, and

$$p = \frac{4\omega (H_r + 4\pi M_{eff})}{\gamma (2H_r + 4\pi M_{eff})^2} \tag{8.80c}$$

is the ellipticity factor. Eqs. (8.80a–8.80c) for the spin-pumped spin current density at the interface is valid for any kind of adjacent layer.

### 8.4.3 Electric Spin Pumping Experiments with 3D Metallic Layers

The first experiments with the electric spin pumping effect were carried out with FM/ML bilayers, in which the metallic layer converts the spin-pumped spin current into a 3D charge current by the inverse spin Hall effect [55, 56, 60–76]. The sample arrangements are the ones shown in Fig. 8.14, in which the ML is usually made of a heavy-metal compound with thickness larger than the spin diffusion length $\lambda_N$. In this case, the pumped spin current flows through the FM/ML interface, producing a spin accumulation that diffuses into the ML with the associated pure spin current. The variation of the spin current along the $y$-direction is given by Eq. (8.66). Considering $\omega \tau_{sf} \ll 1$, so that $\kappa \approx 1/\lambda_N$, we can write for the spin current density

$$J_S(y) = J_S(0) \frac{\sinh \left[(t_N - y)/\lambda_N\right]}{\sinh (t_N/\lambda_N)}, \tag{8.81}$$

where $t_N$ is the thickness of the ML. Then, due to the ISHE, a fraction of the charge carriers undergoes spin-orbit scattering generating a transverse charge motion with current density $\vec{J}_C$ given by $\vec{J}_C = \theta_{SH}(2e/\hbar) \vec{J}_S \times \vec{\sigma}$, where $\theta_{SH}$ is the spin Hall angle and $\vec{\sigma}$ is the spin polarization. This produces a DC voltage along the length $L$ of the NM layer that is measured at the contacts placed at the ends, as illustrated in

Fig. 8.14. Integration of the charge current density along $x$ and $y$ gives for the spin pumping voltage [68].

$$V_{SPE} = R w \lambda_N \frac{2e}{\hbar} \theta_{SH} \tanh\left(\frac{t}{2\lambda_N}\right) J_S(0) \cos \phi, \qquad (8.82)$$

where $R$ and $w$ are, respectively, the resistance and width of the ML, and $\phi$ is the angle between the field (that defines the direction of the spin polarization) and the direction normal to the charge current, as in Fig. 8.14b. Finally, using Eqs. (8.80a–8.80c) in (8.82) we obtain for the spin pumping-ISHE voltage

$$V_{SPE} = \frac{1}{4} R w \lambda_N e \theta_{SH} f g_{eff}^{\uparrow\downarrow} p_{xz} \tanh\left(\frac{t_N}{2\lambda_N}\right) \left(\frac{h_{rf}}{\Delta H}\right)^2 \cos \phi L(H - H_r), \qquad (8.83)$$

where $f = \omega/2\pi$ and $p_{xz}$ is a factor that expresses the ellipticity and the spatial variation of the $rf$ magnetization of the FMR mode. This result shows that in field scan measurements, the SPE voltage $V_{SPE}(H_0)$ has the shape of a Lorentzian function with peak value at $H = H_r$. For $\theta_{SH} > 0$, the voltage is maximum and positive for $\phi = 0°$, it is null for $\phi = 90°$, and it is negative for $\phi = 180°$. The peak value is proportional to the spin Hall angle and to the spin-mixing conductance, but it also depends on the spin diffusion length. Thus, if $g_{eff}^{\uparrow\downarrow}$ is known, for instance from linewidth measurements, one can determine $\theta_{SH}$ and $\lambda_N$ by fitting Eq. (8.83) to measurements of the SPE voltage as a function of the ML thickness $t_N$.

Experimental data for the voltages generated in FM/NM bilayers under microwave-driven FMR are shown in Fig. 8.15, for the FM conductor permalloy (Py) and for the FM insulator yttrium iron garnet (YIG). The Py/Pt bilayers were prepared by DC magneton sputtering on Si substrate slabs with lateral dimensions

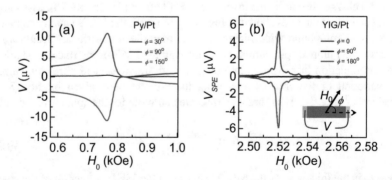

**Fig. 8.15** Field scans DC voltage measured in FM/NM bilayers, as indicated, for several values of the angle between the applied magnetic field and the sample axis. (a) Py (18.5 nm)/Pt (6.0 nm) driven at microwave frequency 8.6 GHz [Adapted from Ref. 68]. (b) YIG (6 μm)/Pt (4 nm), driven at 9.4 GHz [71]. Reprinted with permission from J. B. S. Mendes et al., Phys. Rev. B **89**, 140,406 (R) (2014). Copyright (2014) by the American Physical Society

$1.5 \times 3.0$ mm$^2$ with several thicknesses of both layers [68]. By using a shadow mask, the Py layer covers only the central part of the NM layer surface so that Ag electrodes can be attached to the edges of the NM layer. The samples, with two Cu electrodes attached with silver paint, are introduced through a small hole in the back wall of a rectangular microwave cavity tuned in the TE$_{102}$ mode at 8.6 GHz, in a nodal position of minimum $rf$ electric field and maximum $rf$ magnetic field. This precaution minimizes the generation of electric current by the $rf$ electric field. By scanning the applied magnetic field one can investigate the angular dependence of the DC voltage that develops along the NM layer. As shown in Fig. 8.15a, for an arbitrary angle of the field, the voltage $V(H_0)$ is not a Lorentzian function as in Eq. (8.80b). This is so because, as mentioned earlier, the classical induction effect due to the $rf$ variation of the in-plane magnetization in the Py layer produces a voltage that adds to the SPE voltage. The spectra in Fig. 8.15a for $\phi = 30, 150°$ can be fitted by a superposition of a Lorentzian function (symmetric about $H_r$) and a Lorentzian derivative (antisymmetric about $H_r$). The first is generated mainly by the spin pumping effect and the second is due to the classical induction. By measuring the angle dependencies of the symmetric and antisymmetric lines it is possible to separate the two contributions [66–68].

The spectra for the YIG/Pt bilayer shown in Fig. 8.15b were obtained in a sample made of a 4 nm thick Pt layer sputter deposited on a single-crystal YIG film with thickness 6 μm, grown by LPE on a GGG substrate, and cut with lateral dimensions $1.5 \times 3.0$ mm$^2$. The sample is mounted on the tip of a plastic rod that is introduced through a small hole in the back wall of a shorted rectangular microwave X-band waveguide in a position of maximum microwave magnetic field and zero $rf$ electric field. The shorted waveguide is used instead of a resonating microwave cavity to avoid detuning produced by the strong resonance of YIG and also nonlinear effects. The field scan spectra shown in Fig. 8.15b, obtained with frequency 9.4 GHz and input microwave power 80 mW, do not have the spurious antisymmetric contribution to the voltage because YIG is an insulator. For a field angle $\phi = 0°$ besides the strong peak corresponding to the $k = 0$ magnon (FMR), one can see smaller peaks produced by magnetostatic modes. The peak lines are nearly Lorentzian and are all positive, consistent with the positive spin Hall angle of Pt. If the sample is rotated in the plane by 90° relative to the field, the voltage vanishes because the charge current generated by the ISHE is perpendicular to the direction of the spin polarization. For a field angle $\phi = 180°$ all peaks are negative, confirming the SPE origin of the voltage. Using in Eq. (8.83) the values of the $rf$ magnetic field, frequency, and resistance of the experiments, one obtains good agreement with the measured voltage peak with $\theta_{SH} = 0.05$ and $\lambda_N = 3.7 \times 10^{-7}$ cm for Pt, $p_{11} = 0.31$ appropriate for the FMR (1,1) mode in YIG at 9.4 GHz [71], and $g_{eff}^{\uparrow\downarrow} = 1 \times 10^{14}$ cm$^{-2}$ for the YIG/Pt interface, all consistent with the values in Refs. [70–76]. Experiments with YIG covered by other metallic layers are also well explained by Eq. (8.83) [71, 72].

### 8.4.4  Electric Spin Pumping Experiments with 2D Electron Systems

The spin pumping experiments using 2D electron systems are performed with sample arrangements similar to those in Fig. 8.14. The microwave-driven FMR magnetization precession in the FM film injects a 3D spin current into the 2D layer, that is converted by the IREE into a charge current with density given by Eq. (8.38). Since the spin current density in Eqs. (8.80a–8.80c) is expressed in units of angular momentum/(time.area), the current produced by the IREE has a 2D density, in units of charge current, given by $j_C = (2e/\hbar)\lambda_{IREE}J_S$. Thus, the 2D charge current density is

$$j_C = \lambda_{IREE} \frac{e\,\omega\, g_{eff}^{\uparrow\downarrow} P}{8\pi} \left(\frac{h_{rf}}{\Delta H}\right)^2 L(H - H_r). \tag{8.84}$$

Since the current intensity generated in the 2D layer is $I_C = w j_C$, where $w$ is the width of the layer, the peak voltage at the field $H = H_r$, is

$$V_{SPE}^{peak} = \frac{1}{4}\lambda_{IREE} R\, w\, e f\, g_{eff}^{\uparrow\downarrow} P \left(\frac{h_{rf}}{\Delta H}\right)^2, \tag{8.85}$$

where $R$ is the layer resistance and $f$ is the microwave frequency $f = \omega/2\pi$. Since the spin-mixing conductance can be determined by linewidth measurements, this equation allows a direct determination of the IREE length from the voltage measured in spin pumping experiments. Of course, if the FM layer is metallic, care must be taken to disentangle the spin pumping voltage, as discussed in the previous section.

The first reported electric spin pumping experiments with spin-to-charge current conversion in a 2D electron gas were carried out using a heterostructure with a Bi/Ag interface that has a strong Rashba interaction due to the large SOC of Bi [20]. Subsequently, several experiments were reported with 2D electron systems in other Rashba interfaces [77, 78], topological insulators [79–83], graphene [84], and 2D semiconductors [85]. We shall briefly present here one of these experiments.

Topological insulators are electronic materials that have a bulk band gap, like an ordinary insulator, but have conducting states on their edges or surfaces with energy that varies linearly with wave vector [86]. Thus, the energy dispersion surface $E(k_x, k_y)$ has the shape of a cone, called Dirac cone. In topological insulators with strong spin–orbit coupling, the Fermi contours of the surface or interface Dirac cones are spin-polarized, enabling spin–charge conversion by the Rashba–Edelstein effects [17, 18, 79–83]. Figure 8.16a shows the sample arrangement used to observe spin-to-charge current conversion in the topological insulator (TI) structure α-Sn/Ag in spin pumping experiments [80]. As described in Ref. [80], bulk Sn is a metal, but a α-Sn (001) layer under strain or quantum-size effects can have a band gap as an insulator. Using angular-resolved photoelectron spectroscopy (ARPES) to study the energy band structure of α-Sn (001) films with few atomic layers, covered with various materials, it was found that α-Sn/Ag is a TI with spin–momentum locking.

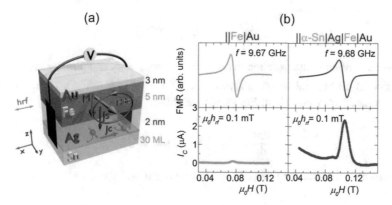

**Fig. 8.16** (a) Schematic illustration of the sample arrangement for spin pumping into α-Sn structures by ferromagnetic resonance (FMR) of a Fe layer. (b) Field scans FMR absorption derivative and DC current measured in the samples with the structures indicated [Adapted from Ref. 80]. Reprinted with permission from J.-C. Rojas-Sánchez et al., Phys. Rev. Lett. **116**, 096602 (2016). Copyright (2016) by the American Physical Society

The α-Sn layers are grown on substrates (S) of InSb(001) by molecular beam epitaxy, and the ferromagnetic material used to generate the FMR spin-pumped spin current is Fe, so that the heterostructure is S/α-Sn/Ag/Fe/Au, with the Au capping layer used for protection. Figure 8.16b shows the FMR and induced current ($I_C = V_{SPE}/R$) field spectra measured in two samples with microwave pumping frequency as indicated. Clearly, only the sample with α-Sn/Ag shows a significant spin-to-charge conversion by the IREE. The measured efficiency of the conversion process, expressed by $\lambda_{IREE} = 2.1$ nm, is the highest obtained so far in any type of 2D electron system. This value would correspond to a spin Hall angle of 0.62, if the conversion mechanism were the ISHE, a value one order of magnitude larger than in Pt [80]. This result demonstrates the potential of topological insulators for the conversion of spin currents into charge currents for the use in spintronic devices [4, 80].

## 8.5 Magnonic Spin Current, Magnon Accumulation, and Diffusion of Magnons

Although they have no conductions electrons, magnetic insulators can also carry spin currents. The mechanism of spin transport in these materials is based on the conservation of angular momentum and the properties of magnons. The starting point to derive the magnonic spin current is the Landau–Lifshitz equation for the magnetization (1.10) having only the effective exchange field (2.18). Using the CGS system and considering the magnetization in the same direction as the spin we have

$$\frac{d\vec{M}}{dt} = -\gamma \frac{D}{M} \vec{M} \times \nabla^2 \vec{M}. \tag{8.86}$$

We can rewrite this equation in the form of a continuity equation by means of the relation of differential vector analysis $\vec{A} \times \nabla^2 \vec{A} = \nabla \cdot (\vec{A} \times \nabla \vec{A})$. Using this relation in the right-hand side of Eq. (8.86) we obtain

$$\frac{d\vec{M}}{dt} = -\gamma \frac{D}{M} \nabla \cdot (\vec{M} \times \nabla \vec{M}).$$

This equation has the form of a continuity equation

$$\frac{d\vec{M}}{dt} = -\nabla \cdot \vec{J}_M, \tag{8.87}$$

in which

$$\vec{J}_M = \frac{\gamma D}{M} \vec{M} \times \nabla \vec{M} \tag{8.88}$$

is the *magnetization current* [1, 2, 69], that has dimension of magnetic moment/ (area.time). Equation (8.88) has an important consequence in the study of collective spin excitations in magnetic multilayers. Conservation of angular momentum requires that, at any interface with a ferromagnetic film we have

$$\frac{D}{M} \vec{M} \times \frac{\partial \vec{M}}{\partial n} + \vec{\tau}_s = 0, \tag{8.89}$$

where $n$ denotes the coordinate normal to the interface and $\vec{\tau}_s$ is the surface torque density exerted on the interface, arising from forces other than ferromagnetic exchange, such as the surface anisotropy. Equation (8.89) is known as the Rado–Weertman boundary condition, proposed much before the concept of spin currents was established [87], and extensively used in the study of magnetic multilayers [88].

The magnetization current with polarization in the equilibrium direction ($z$) is given by

$$\vec{J}_M^z = \frac{\gamma D}{M} (\vec{M} \times \nabla \vec{M})_z. \tag{8.90}$$

Writing the magnetization as $\vec{M} = \hat{x} m_x + \hat{y} m_y + \hat{z} M_z$, where the small-signal transverse components are $m_x, m_y << M_z \approx M$, Eq. (8.90) becomes

$$\vec{J}_M^z = -\frac{i\gamma D}{2M} (m^- \nabla m^+ - m^+ \nabla m^-), \tag{8.91}$$

where $m^\pm = m_x \pm im_y$. The circularly polarized magnetization components can be identified as a spin-wave wave function, $\psi(\vec{r},t) = m^+(\vec{r},t)$, so that Eq. (8.91) gives for the magnetization current

$$\vec{J}_M^z = \frac{i\gamma D}{2M}\left(\psi\nabla\psi^* - \psi^*\nabla\psi\right). \tag{8.92}$$

This equation has the form of the probability flux density of quantum mechanics, that expresses the probability current density of the particle with wave function $\psi$. Equations (8.89)–(8.92) show that in a state of uniform magnetization there is no exchange magnetization current. To have a current it is necessary that the magnetization varies is space, such as in a domain wall or in a spin wave.

In order to find an expression for the spin current from the equations above we use the relation between the magnetization and the magnon operator, obtained with Eqs. (3.22) and (3.30), and neglecting the dipolar interaction in the transformations, $m^+ = (2\gamma\hbar M/V)^{1/2}\sum_k e^{ikr}c_k$. Substitution in Eq. (8.91) gives for the magnetization current operator

$$\vec{J}_M^z = \frac{2\gamma D\gamma\hbar}{V}\sum_k c_k^\dagger c_k \vec{k}. \tag{8.93}$$

Using the spin wave dispersion relation $\omega_k = \gamma H + \gamma D\vec{k}\cdot\vec{k}$, for which the group velocity is $\vec{v}_k = 2\gamma D\vec{k}$, and considering that the magnetization is related to the angular momentum per unit volume through the gyromagnetic ratio $\gamma$, the expectation value of Eq. (8.93) gives for the spin current density

$$\vec{J}_S^z = \frac{\hbar}{V}\sum_k n_k \vec{v}_k, \tag{8.94}$$

where $n_k$ is the number of magnons [69]. This equation expresses the spin current carried by magnons, either in coherent or incoherent states. It is called spin-wave spin current, or magnonic spin current. In FM conductors the spin-wave spin current adds to the spin current carried by the conduction electrons. In FM insulators the magnonic spin current is the only mechanism for spin transport. Denoting by $n_k^0$ the number of magnons in thermal equilibrium, given by the Bose–Einstein distribution (3.86), using $\sum_k \vec{v}_k n_k^0 = 0$, and replacing the sum over wave vectors by an integral in the Brillouin zone, we obtain from Eq. (8.94)

$$\vec{J}_S^z = \frac{\hbar}{(2\pi)^3}\int d^3k\, \vec{v}_k\left(n_k - n_k^0\right). \tag{8.95}$$

In this equation, $\delta n_k = n_k - n_k^0$ represents the number of magnons in excess of equilibrium. It is useful to define the magnon accumulation [89–91].

$$\delta n_m = \frac{1}{(2\pi)^3} \int d^3k\, \delta n_k = \frac{1}{(2\pi)^3} \int d^3k \left(n_k - n_k^0\right). \tag{8.96}$$

Magnons are quasiparticles whose motion can be driven by concentration gradients, temperature gradients and external forces. Consider that a magnon system described by a distribution function that depends of the wave vector, on the position and on time $f(\vec{k}, \vec{r}, t)$. The transport equation for magnons is obtained by taking the total derivative with respect to time

$$\frac{df}{dt} = \frac{\partial f}{\partial t} + \frac{\partial f}{\partial \vec{r}}\frac{d\vec{r}}{dt} + \frac{\partial f}{\partial \vec{k}}\frac{d\vec{k}}{dt} = \left(\frac{\partial f}{\partial t}\right)_{scatt}, \tag{8.97}$$

where $(\partial f/\partial t)_{scatt}$ denotes the change in the magnon number due to scattering processes. We assume that the scattering processes are represented by a mechanism in which the distribution for $k$-magnons in excess of equilibrium decays exponentially with a relaxation time $\tau_k$

$$\delta f(\vec{k}, \vec{r}, t) = \delta f(\vec{k}, \vec{r}, 0)\, e^{-t/\tau_k}, \tag{8.98}$$

which implies that

$$\left(\frac{\partial f}{\partial t}\right)_{scatt} = -\frac{f - f_0}{\tau_k}, \tag{8.99}$$

where $f_0$ is the distribution in thermal equilibrium. Using in Eq. (8.97) the expression (8.99) and $\vec{v}_k = d\vec{r}/dt$, we obtain

$$\frac{\partial f}{\partial t} = -\vec{v}_k \cdot \nabla_r f - \frac{1}{\hbar}\vec{F}_k \cdot \nabla_k f - \frac{f - f_0}{\tau_k}, \tag{8.100}$$

where $\vec{F}_k = \hbar(d\vec{k}/dt)$ represents the external forces acting on the $k$-magnons. This is the Boltzmann transport equation for magnons or other quasiparticles governed by the $f(\vec{k}, \vec{r}, t)$ distribution function. In the steady state ($\partial f/\partial t = 0$) and in the absence of external forces Eq. (8.100) gives

$$f(\vec{k}, \vec{r}) = f_0(\vec{k}, \vec{r}) - \tau_k \vec{v}_k \cdot \nabla_r f(\vec{k}, \vec{r}). \tag{8.101}$$

Denoting the distribution function $f(\vec{k}, \vec{r})$ for magnons by $n_k(\vec{r})$, we rewrite Eq. (8.101) as

$$n_k(\vec{r}) - n_k^0 = -\tau_k \vec{v}_k \cdot \nabla \delta n_k(\vec{r}), \tag{8.102}$$

where we have considered that the temperature is uniform so that $\nabla_r n_k = \nabla \delta n_k$. Using this result in Eq. (8.95) we obtain for the spin current

$$\vec{J}_{Sdiff}^z = -\frac{\hbar}{(2\pi)^3} \int d^3k\, \tau_k\, \vec{v}_k [\vec{v}_k \cdot \nabla \delta n_k(\vec{r})] \tag{8.103}$$

which is due to a gradient in the magnon concentration, hence it is a diffusion current. This current can be expressed in terms of the magnon accumulation by using an approximate solution of Boltzmann equation. In the spirit of linear response theory, we write the excess magnon number as the sum of the equilibrium distribution plus a small deviation [see Reif]

$$n_k(\vec{r}) = n_k^0 + n_k^0 \lambda_k g(\vec{r}), \tag{8.104}$$

where the parameter $\lambda_k$ in lowest order of energy is chosen as to eliminate the singularity at $\varepsilon_k = 0$. Thus, we have [92]

$$n_k(\vec{r}) = n_k^0 + n_k^0 \varepsilon_k g(\vec{r}), \tag{8.105}$$

where $g(\vec{r})$ is a spatial distribution to be determined by the solution of the boundary value problem. Substitution in Eq. (8.96) gives for the magnon accumulation

$$\delta n_m(\vec{r}) = I_0 g(\vec{r}), \tag{8.106}$$

where

$$I_0 = \frac{1}{(2\pi)^3} \int d^3k\, n_k^0 \varepsilon_k \tag{8.107}$$

is an integral over the first Brillouin zone that is well behaved for a gapless magnon dispersion. Using Eqs. (8.105) and (8.106) in Eq. (8.103) leads to a magnon diffusion current

$$\vec{J}_{Sdiff}^z(\vec{r}) = -\hbar D_m \nabla \delta n_m(\vec{r}), \tag{8.108}$$

where

$$D_m = \frac{1}{(2\pi)^3 I_0} \int d^3k\, \tau_k\, v_{ky}^2\, n_k^0 \varepsilon_k \tag{8.109}$$

is the magnon diffusion coefficient and $v_{ky}$ denotes the component of the magnon velocity in the direction of the gradient. Considering that the magnon accumulation relaxes into the lattice with a magnon-phonon relaxation time $\tau_{mp}$, conservation of angular momentum implies that $\nabla \cdot \vec{J}_{Sdiff}^{z} = -\hbar\,\delta n_m/\tau_{mp}$. Using this relation in Eq. (8.108) leads to a diffusion equation for the magnon accumulation

$$\nabla^2 \delta n_m(\vec{r}) = \frac{\delta n_m(\vec{r})}{l_m^2}, \qquad (8.110)$$

where $l_m = (D_m\tau_{mp})^{1/2}$ is the magnon diffusion length. The magnitude of the diffusion length depends strongly on the quality of the material. In yttrium iron garnet films grown by sputtering, it varies from tens to hundreds of nanometers at room temperature, and it increases at lower temperatures due to the longer relaxation times [92]. In good single-crystal LPE grown YIG it reaches several micrometers [93]. As will be studied in the next section, the magnonic spin current produced by the diffusion of magnons plays an important role in the spin Seebeck effect in magnetic insulators.

Another formulation for the transport of spin currents in FMI employs the concept of the magnon chemical potential $\mu_m(\vec{r})$, used to characterize the nonequilibrium magnon BE distribution

$$n_k(\vec{r}) = \frac{1}{e^{(\varepsilon_k - \mu_m)/k_B T} - 1}, \qquad (8.111)$$

and find the evolution of $\mu_m(\vec{r})$ [94]. As we will show here, this approach is completely equivalent to the one based on the magnon accumulation. Consider for the magnon number a small deviation from the equilibrium distribution in the form

$$n_k(\varepsilon_k, \vec{r}) = n_k^0[1 + \varphi_k(\varepsilon_k, \mu_m, \vec{r})], \qquad (8.112)$$

where the spatial variation of $\varphi_k$ is determined by the external perturbations, such as thermal gradients, and the boundary conditions. In order to relate the function $\varphi_k$ with the chemical potential we write the density of magnons as

$$n_m(\vec{r}) = \frac{1}{(2\pi)^3} \int d^3k\, n_k(\varepsilon_k, \vec{r}) = \int d\varepsilon\, D(\varepsilon)\, n_k(\varepsilon), \qquad (8.113)$$

where $n_k(\varepsilon) = n_k(\varepsilon_k, \vec{r})$ is related to the chemical potential as in Eq. (8.111) and $D(\varepsilon)$ is the magnon density of states [95]. Then, the density of magnons in excess of equilibrium is, from Eq. (8.113)

$$\delta n_m(\vec{r}) = \int d\varepsilon D(\varepsilon)[n_k(\varepsilon) - n_k^0]. \tag{8.114}$$

If we consider as above, a situation not far removed from equilibrium, we can write $\delta n_m(\vec{r}) = \int d\varepsilon D(\varepsilon)(-\partial n_k^0/\partial\varepsilon)\delta\mu$. As the derivative $(-\partial n_k^0/\partial\varepsilon)$ changes appreciably only within a few $k_B T$ around $\mu_m$, we can use the approximate expression [see Ashcroft and Mermin]

$$\int d\varepsilon D(\varepsilon)(-\partial n_k^0/\partial\varepsilon)\delta\mu \approx -D(\mu_m)n_k^0\big|_{-\infty}^{\infty}\delta\mu_m = D(\mu_m)\delta\mu_m. \tag{8.115}$$

Then, from Eq. (8.114) we write

$$\delta n_m(\vec{r}) = D(\mu_m)\delta\mu_m(\vec{r}). \tag{8.116}$$

Replacing Eqs. (8.112) and (8.113) in Eq. (8.96) we obtain,

$$\delta n_m(\vec{r}) = \frac{1}{(2\pi)^3} \int_{-\infty}^{\infty} d\varepsilon D(\varepsilon)n_k^0(\varepsilon)\varphi_k(\varepsilon, \mu_m, \vec{r}), \tag{8.117}$$

so that from Eq.(8.116), and separating variables, one can write the functional form of $\varphi(\varepsilon, \mu_m, \vec{r})$ as

$$\varphi(\varepsilon, \mu_m, \vec{r}) = \varphi(\varepsilon)\delta\mu_m(\vec{r}), \tag{8.118}$$

where $\varphi(\varepsilon)$ can be expanded in powers of $\varepsilon$, which, in lowest order is $\varphi(\varepsilon) = \lambda\varepsilon$, where $\lambda$ is a parameter with dimension of energy$^{-2}$. Using Eq. (8.118) in (8.117) one obtains a simple relation between the magnon accumulation and the chemical potential

$$\delta n_m(\vec{r}) = \lambda I_0 \delta\mu_m(\vec{r}), \tag{8.119}$$

where the integral $I_0$ is the same as given by Eq. (8.107). Thus, the magnon accumulation is proportional to the chemical potential, so that the formulation of the magnonic spin current in terms of the magnon accumulation is entirely equivalent to the one based on the magnon chemical potential.

The magnon diffusion can also be used as the basic mechanism for transport of spin information in devices that operate with fully electrical excitation and detection of magnons [93]. Figure 8.17a shows the schematics of such device consisting of two nanometer wide platinum strips deposited on the surface of a single-crystal YIG film grown on a GGG substrate. One Pt strip functions as injector and the other as detector. When a charge current $I$ is sent through the injector, the spin Hall effect generates a transverse spin current. A spin accumulation then builds up at the Pt/YIG interface, pointing in the film plane, resulting in a magnon accumulation in the YIG

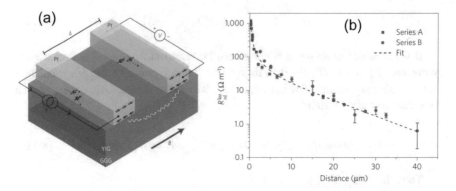

**Fig. 8.17** (**a**) Schematic illustration of the device used to transport spin information by magnon diffusion. A charge current *I* through the left Pt strip (injector) generates a spin accumulation at the Pt/YIG interface by means of the SHE. Through the exchange interaction at the interface, angular momentum is transferred to the YIG, creating a magnon accumulation that diffuses and reaches the right Pt strip (detector), where they are absorbed and a spin accumulation is generated. Through the ISHE the spin current is converted to a charge voltage *V*. (**b**) Measured nonlocal $R_{nl} = V/I$ as a function of the distance between the Pt strips and fit by the diffusion–relaxation model [93]. Reproduced from L.J. Cornelissen et al., Nat. Phys. **11**, 1022 (2015), with permission of Springer-Nature

film. The nonequilibrium magnons diffuse laterally in the YIG film producing a magnonic spin current from injector to detector. At the detector, the reciprocal process occurs: magnons interact at the interface, flipping the spins of electrons, and creating a spin accumulation in the Pt strip. Owing to the ISHE, the induced spin current is converted into a charge current, which under open-circuit conditions generates a voltage *V*. The device response is usually expressed by the nonlocal resistance, defined by $R_{nl} = V/I$. Figure 8.17b shows the nonlocal resistance measured in two series of devices, one with distances between the Pt strips of few micrometers and the other with distances of tens of micrometers.

The authors of Ref. [93] identify two spin transport regimes. At large distances, the signal decay is dominated by magnon relaxation and is characterized by an exponential decay. For shorter distances the signal is characteristic of diffusive transport, described by the diffusion of magnon accumulation. Assuming a one-dimensional motion, the solution of Eq. (8.110) is

$$\delta n_m(x) = A e^{-x/l_m} + B e^{x/l_m}, \tag{8.120}$$

where $x$ is the coordinate axis normal to the Pt strips and $A$, $B$ are coefficients to be determined by the boundary conditions. Assuming strong spin-magnon coupling between the Pt strips and the YIG film, the boundary conditions are $\delta n_m(0) = \delta n_0$ and $\delta n_m(d) = 0$, where $d$ is distance between the strips, and $\delta n_0$ is the injected magnon accumulation, which is proportional to the applied current and is determined by various material and interface parameters. These conditions imply that the injector acts as a low-impedance magnon source, and all magnon current is absorbed when it arrives at the detector. Using these conditions in Eq. (8.120), with (8.108) we obtain the magnonic diffusion spin current at the detector

$$J^z_{Sdiff}(x = d) = -2\hbar D_m \frac{\delta n_0}{l_m} \frac{e^{d/l_m}}{1 - e^{2d/l_m}}. \qquad (8.121)$$

The nonlocal resistance is proportional to the current (8.121), so that it can be written as

$$R_{nl} = \frac{C}{l_m} \frac{e^{d/l_m}}{1 - e^{2d/l_m}}, \qquad (8.122)$$

where $C$ is a distance-independent prefactor, which, together with the diffusion length are the two adjustable parameters of the model. The dashed line in Fig. 8.17b represents the best fit of Eq. (8.122) to the nonlocal resistance data, that yields a magnon diffusion length $l_m = 9.4 \pm 0.6\,\mu m$ [93].

## 8.6 Magnonic Spin Seebeck Effect

### 8.6.1 Model for the Magnonic Spin Seebeck Effect in Ferromagnetic Insulators

In this section we study a phenomenon that has attracted considerable attention in recent years, the spin Seebeck effect. The spin Seebeck effect (SSE) refers to the generation of a spin current in ferromagnetic or ferrimagnetic (FM) materials by a temperature gradient, and is the analog of the ancient thermoelectric Seebeck effect whereby a charge current is created by a temperature gradient in a metal or semiconductor [96–98]. Discovered in 2008 by Uchida et al. [99], the SSE was soon observed in a variety of materials and structures [96–104], arousing the field of spintronics and giving origin to the area of *spin caloritronics*. The SSE is usually detected by the voltage created in a metallic layer (ML) attached to the FM layer as a result of the conversion of the spin current into a charge current by means of the inverse spin Hall effect (ISHE). The FM material can be a metal, semiconductor, or an insulator, while the ML is made of a paramagnetic metallic material with strong spin–orbit coupling, such as Pt or Ta, or a FM material such as permalloy [104], or an antiferromagnetic metal such as IrMn [71, 72], or also a 2D electron system, such as a Rashba interface, a topological insulator, or a 2D semiconductor.

Depending on the experimental arrangement, the spin current generated by the SSE can be perpendicular or parallel to the temperature gradient, characterizing the so-called transverse or longitudinal configurations, respectively [96–101]. While the transverse SSE can be observed in both metallic and insulating magnetic materials, the longitudinal spin Seebeck effect (LSSE) is observed unambiguously only in

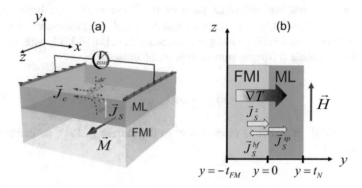

**Fig. 8.18** Ferromagnetic insulator (FMI)/metallic layer (ML) bilayer used to investigate the longitudinal spin Seebeck effect. (**a**) Illustration of the conversion of spin into charge current by the inverse spin Hall effect in the ML. (**b**) Side view of the FMI/ML bilayer with the coordinate axes used to calculate the spin currents generated by a temperature gradient perpendicular to the plane of the bilayer

insulators because they are free from the anomalous Nernst effect [96–98]. We shall restrict the attention here to the longitudinal configuration, that has proved to be more interesting for scientific research and for applications. This is illustrated in Fig. 8.18, and consists of a bilayer made of a ferro-ferrimagnetic insulator (FMI), such as yttrium iron garnet, and a metallic layer, subject to a thermal gradient perpendicular to the plane. The SSE has its origin in two separate mechanisms: (1) a magnonic spin current created by the action of a thermal gradient in the FMI/ML structure is pumped into the ML; (2) the spin current in the ML is converted into a charge current by means of the inverse spin Hall effect (ISHE), or the inverse Rashba–Edelstein effect (IREE), producing the electric voltage.

The first step to calculate the SSE voltage consists in finding the magnonic spin current created by the temperature gradient. In the absence of external forces and in the relaxation approximation, in steady state the Boltzmann transport equation for the magnon number has the solution given by Eq. (8.102). Since the temperature varies in space, in the calculation of the spin current we have to consider the term $\nabla n_k^0 = (\partial n_k^0 / \partial T) \nabla T$. Using Eq. (8.102) in (8.95) we see that the spin current is the sum of two parts, $\vec{J}_S^z = \vec{J}_{S\nabla T}^z + \vec{J}_{S\nabla n}^z$, where

$$\vec{J}_{S\nabla T}^z = -\frac{\hbar}{(2\pi)^3} \int d^3 k\, \tau_k \frac{\partial n_k^0}{\partial T} \vec{v}_k (\vec{v}_k \cdot \nabla T) \tag{8.123}$$

is the contribution of the drift of magnons due to the temperature gradient, while

$$\vec{J}_{S\nabla n}^z = -\frac{\hbar}{(2\pi)^3} \int d^3 k\, \tau_k\, \vec{v}_k [\vec{v}_k \cdot \nabla \delta n_k(\vec{r})] \tag{8.124}$$

is due to the spatial variation of the magnon accumulation, which is the diffusion current given by Eq. (8.108). With the temperature gradient normal to the plane, Eq. (8.123) gives the spin current in the $y$-direction

$$J_S^z = -S_S \nabla T, \tag{8.125}$$

$$S_S = \frac{\hbar}{(2\pi)^3 T} \int d^3 k \, \tau_k v_{ky}^2 \frac{e^x x}{(e^x - 1)^2}, \tag{8.126}$$

where $T$ is the average temperature and $x = \varepsilon_k / k_B T$ is the normalized magnon energy. We consider the magnon and phonon systems to have the same temperature $T$, as demonstrated experimentally [105]. Solution of the diffusion equation (8.110) gives for the spatial variation of the magnon accumulation

$$\delta n_m(y) = A \, e^{y/l_m} + B e^{-y/l_m}, \tag{8.127}$$

where $A$ and $B$ are coefficients to be determined by the boundary conditions. Using Eq. (8.127) in (8.103) and (8.125), we obtain the total $y$ component of the $z$-polarized magnonic spin current density in the FMI

$$J_S^z(y) = -S_S \nabla_y T - \hbar \frac{D_m}{l_m} A \, e^{y/l_m} + \hbar \frac{D_m}{l_m} B e^{-y/l_m}. \tag{8.128}$$

Next we consider the FMI layer in atomic contact with a ML having a strong spin-orbit scattering. The precessing spins associated with the magnon accumulation at the FMI/ML interface pump a spin current into the ML given by Eq. (8.67). One can show [91] that the sum of the $y$ components of the $z$-polarized spin pump and back-flow spin currents at the FMI/ML interface is

$$J_S^z(0^+) = -\frac{\hbar g_{eff}^{\uparrow\downarrow}}{4\pi M^2} \sum_k \omega_k (m_k^+ m_k^-), \tag{8.129}$$

where $m_k^+$ and $m_k^-$ are the circular polarized transverse components of the magnetization associated with the $k$-magnon and $g_{eff}^{\uparrow\downarrow}$ is the real part of the effective spin-mixing conductance that takes into account the spin-pumped and back-flow spin currents. Using the linear approximation $m_k^+ m_k^- \approx 2 M \gamma \hbar \, \delta n_k / V$, replacing the sum in (8.129) by an integral over the Brillouin zone, and using the relation (8.106), one can show that the spin current thermally pumped through the ML interface becomes

$$J_S^z(0^+) = -\frac{\gamma \hbar^2 g_{eff}^{\uparrow\downarrow}}{2\pi M} \frac{\delta n_m(0)}{I_0} \frac{1}{(2\pi)^3} \int d^3 k \, \omega_k n_k^0. \tag{8.130}$$

In order to calculate the coefficients $A$ and $B$ in Eq. (8.127), and the magnon accumulation $\delta n_m(0)$ at the FMI/ML interface, in terms of the temperature gradient $\nabla T$, we use the boundary conditions at $y = -t_{FM}$ and $y = 0$. They are determined by conservation of the angular momentum flow that requires continuity of the spin currents at the interfaces. Thus, considering at the substrate/FMI interface $J_S^z(-t_{FM}) = 0$ and at the FMI/ML interface $J_S^z(0^-) = J_S^z(0^+)$, we obtain with Eqs. (8.128) and (8.130) the magnon accumulation at the FMI/NM interface created by the temperature gradient $\nabla T$ in the $y$-direction

$$\delta n_m(0) = \frac{1 - \cosh^{-1}(t_{FM}/l_m)}{a\tanh(t_{FM}/l_m) + bg_{eff}^{\uparrow\downarrow}} S_S \nabla T, \qquad (8.131)$$

where

$$a = \hbar D_m/l_m, \quad b = \frac{\gamma\hbar^2}{2\pi M I_0} \frac{1}{(2\pi)^3} \int d^3k\, n_k^0 \omega_k. \qquad (8.132)$$

It can be shown that in YIG/Pt bilayers $a \gg bg_{eff}^{\uparrow\downarrow}$, so that the spin current density at the interface calculated with Eqs. (8.130)–(8.132) can be written approximately as

$$J_S^z(0) = -\frac{bg_{eff}^{\uparrow\downarrow}\rho}{a} S_S \nabla T, \qquad (8.133)$$

where $\rho$ is a factor that represents the effect of the finite thickness of the FMI layer given by

$$\rho = \frac{\cosh(t_{FM}/l_m) - 1}{\sinh(t_{FM}/l_m)}, \qquad (8.134)$$

such that $\rho \approx 1$ for $t_{FM} \gg l_m$, and $\rho \approx 0$ for $t_{FM} \ll l_m$. Equation (8.133) shows that the magnon spin current at the FMI/ML interface generated by a thermal gradient perpendicular to the bilayer plane is proportional to the temperature gradient and to the spin-mixing conductance of the interface. This means that the ML layer in contact with the FMI film that is used for the spin-to-charge current conversion is essential for the existence of the spin current because it provides continuity. In order to calculate the integrals in Eqs. (8.107), (8.126) and (8.130), one needs the dispersion relation expressing the spin wave angular frequency $\omega_k$ in terms of the wave number $k$. As discussed in Sect. 3.7, since the lowest magnon optical branch in YIG lies above the zone-boundary value, the calculation of the thermal properties in the presence of an applied field $H$ at temperatures up to 300 K can be done considering only the acoustic branch. We use for YIG the magnon dispersion relation given by Eq. (3.176)

$$\omega_k = \gamma H + \omega_{ZB}\left(1 - \cos\frac{\pi k}{2k_m}\right), \qquad (8.135)$$

where $\omega_{ZB}$ the zone boundary frequency and $k_m$ is the value of the maximum wave number assuming spherical Brillouin zone. From Eq. (8.135) one has for the magnon group velocity $v_k = \omega_{ZB}(\pi/2k_m) \sin(\pi k/2k_m)$. The total spin current density at the FMI/ML interface in Eq. (8.133) then becomes

$$J_S^z(0) = -S_S \nabla T, \tag{8.136}$$

where the spin Seebeck coefficient $S_S$, defined in analogy to the Seebeck thermopower coefficient as in $J_C = -S \nabla T$, is given by

$$S_S = C_S \rho g_{eff}^{\uparrow\downarrow}, \tag{8.137}$$

where the factor $C_S$ depends on material parameters and universal constants, given by

$$C_S = F \frac{B_1 B_s}{(B_0 B_2)^{1/2}}, \tag{8.138a}$$

$$F = \frac{\gamma \hbar k_B \tau_{mp}^{1/2} \tau_0^{1/2} k_m^2 \omega_{ZB}}{4\pi M \pi 2\sqrt{3}}, \tag{8.138b}$$

and the parameters $B$ in Eq. (8.138a) are given by the integrals

$$B_s = \int_0^1 dq\, q^2 \sin^2\left(\frac{\pi q}{2}\right) \frac{e^x x}{\eta_q (e^x - 1)^2}, \quad B_1 = \int_0^1 dq\, q^2 \frac{x^2}{e^x - 1}, \tag{8.139a}$$

$$B_0 = \int_0^1 dq\, q^2 \frac{x}{e^x - 1}, \quad B_2 = \int_0^1 dq\, q^2 \sin^2\left(\frac{\pi q}{2}\right) \frac{x}{\eta_q (e^x - 1)}. \tag{8.139b}$$

In Eqs. (8.139a, 8.139b) $q = k/k_m$ is a normalized wave number and $\eta_q = \eta_k/\eta_0$ is a dimensionless relaxation rate, related to the magnon lifetime by $\eta_q = \tau_0/\tau_k$, where $\tau_0$ is the lifetime of magnons near the zone center ($k \approx 0$). One can also express the diffusion parameter in terms of the integrals in (8.139a, 8.139b). From Eqs. (8.107) and (8.108) it follows that

$$D_m = \frac{\tau_0 \pi^2 \omega_{ZB}^2}{12 k_m^2} \frac{B_2}{B_0}, \tag{8.140}$$

that has the dimension consistent with the relation $D_m = l_m^2/\tau_{mp}$, as it should.

In the experiments with the LSSE one applies a temperature difference between the two sides of a FMI/ML bilayer to create a spin current across the structure using the arrangement illustrated in Fig. 8.18. Similarly to the electric spin pumping process, the conversion of the spin current into charge current depends on the topology of the electron gas in the ML. In the case of a 3D system, the conversion is based on the inverse spin Hall effect. The 3D spin current density $\vec{J}_s^z$ flowing into the ML generates a 3D charge current density given by $\vec{J}_C = \theta_{SH}(2e/\hbar)\,\vec{J}_s^z \times \vec{\sigma}$, where $\theta_{SH}$ is the spin Hall angle and $\vec{\sigma}$ is the spin polarization. If the magnetic field is applied in the plane and transverse to the long dimension of the ML, the resulting charge current flows along the long direction and produces a direct current (DC) by means of the ISHE resulting in a SSE voltage at the ends of the ML. Since the spin current at the FMI/ML interface diffuses into the ML with diffusion length $\lambda_N$, in order to calculate the voltage at the ends of the ML layer one has to integrate the charge current density along $x$ and $y$ so that the SSE voltage becomes

$$V_{SSE} = R_{ML}\, w\lambda_N \frac{2e}{\hbar}\theta_{SH}\tanh\left(\frac{t_N}{2\lambda_N}\right) J_S^z(0), \tag{8.141}$$

where $R_{ML}$, $t_N$ and $w$ are, respectively, the resistance, thickness and width of the ML strip. Thus, with Eqs. (8.136), (8.137) and (8.141) we have for the SSE voltage

$$V_{SSE} = R_{ML}\, w\lambda_N \frac{2e}{\hbar}\theta_{SH}\tanh\left(\frac{t_N}{2\lambda_N}\right) C_S \rho\, g_{eff}^{\uparrow\downarrow}\nabla T, \tag{8.142}$$

where the parameters $\rho$ and $F$ and the integrals $B$ are given by Eqs. (8.134), (8.138a, 8.138b) and (8.139a, 8.139b). The spin Seebeck coefficient is often defined with reference to the voltage measured in the ML. One disadvantage of this definition is that the voltage varies with the resistance, so that two samples made with the same material but with different NM layer thicknesses have different spin Seebeck voltage coefficients. Here we quantify the SSE by the current spin Seebeck coefficient, $S_{SSE} = I_{SSE}/\nabla T$, where $I_{SSE} = V_{SSE}/R_{ML}$ is the charge current in the ML layer produced by the temperature gradient $\nabla T$. Thus, with Eqs. (8.136) and (8.142) we obtain

$$S_{SSE} = w\lambda_N \frac{2e}{\hbar}\theta_{SH}\tanh\left(\frac{t_N}{2\lambda_N}\right) C_S \rho\, g_{eff}^{\uparrow\downarrow}. \tag{8.143}$$

Equation (8.142) contains the essential features of the SSE in FMI/3DML bilayers. The SSE voltage is proportional to the temperature gradient across the sample and its strength depends on the spin Hall angle, the spin-mixing conductance, the FM layer thickness (through $\rho$), the NM layer thickness and diffusion length, and on the temperature and applied magnetic field (through the integrals $B$). As will be shown in the next section, despite of the dependence on so many parameters, the model explains quite well the experimental data in YIG/Pt bilayers.

In the case the ML is made of a 2D electron system, the spin-to-charge current conversion takes place by the inverse Rashba–Edelstein effect. In this case, with Eqs. (8.38), (8.136) and (8.137), we obtain for the SSE voltage

$$V_{SSE} = R_{MLW} \frac{2e}{\hbar} \lambda_{IREE} C_S \rho \, g_{eff}^{\uparrow\downarrow} \nabla T. \tag{8.144}$$

This result shows that if the FMI material parameter $C_S$ is known from measurements in a bilayer with another ML, and the spin-mixing conductance $g_{eff}^{\uparrow\downarrow}$ is known from linewidth measurements, the IREE length can be determined from spin Seebeck experiments.

## 8.6.2 Magnonic Spin Seebeck Effect in Antiferromagnetic Insulators

The spin Seebeck effect has also been observed in bilayers made of an antiferromagnetic insulator (AFI) with a nonmagnetic metal [106–109]. In this case, the spin current created by the temperature gradient also has its origin in the magnonic spin transport. In a two-sublattice antiferromagnet the spin current is carried by the two magnon modes, studied in Chap. 5. In the AF phase, the total $z$-component of the spin angular momentum carried by magnons is given by $S^z = \sum_{i,j}(S_i^z + S_j^z)$. With Eqs. (5.29), (5.30) and (5.32) one can write the $z$-component of the spin angular momentum as

$$S^z = \sum_k - a_k^\dagger a_k + b_k^\dagger b_k. \tag{8.145}$$

Using the transformation to the magnon operators for easy-axis AFs given by Eqs. (5.35) and keeping only terms with magnon number operators we have

$$S^z = \sum_k (-\alpha_k^\dagger \alpha_k + \beta_k^\dagger \beta_k). \tag{8.146}$$

The opposite signs in the angular momenta of the two modes is consistent with the semiclassical picture of the spins precessing in opposite directions. Considering for each mode $\mu$ the group velocity $\vec{v}_{\mu k} = \hat{k} \partial \omega_\mu / \partial k$, the spin current density operator is

$$\vec{J}_S^z = \frac{\hbar}{V} \sum_k [-\vec{v}_{\alpha k} \alpha_k^\dagger \alpha_k + \vec{v}_{\beta k} \beta_k^\dagger \beta_k]. \tag{8.147}$$

In this equation, which is valid for both easy-axis and easy-plane antiferromagnets, $\alpha_k^\dagger \alpha_k$ and $\beta_k^\dagger \beta_k$ represent the number operators of magnons of each mode. In easy-axis antiferromagnets, such as $MnF_2$ and $FeF_2$, in the absence of an applied magnetic field, the two magnon modes are degenerate and thus have equal occupation numbers in thermal equilibrium. Since they also have equal group velocities, in easy-axis AFs the spin current created by thermal gradients vanishes in zero-field, as observed experimentally [106–108]. As pointed out in Sect. 5.4, in easy-plane antiferromagnets, such as NiO, the two magnon modes are not degenerate, even in zero-field, and have different group velocities. Thus, they exhibit spin Seebeck effect even in the absence of an external field [95, 109].

The calculation of the magnonic spin current created by the temperature gradient in the AFI layer is similar to that in FMI. Since the contributions of the two modes to the spin current have opposite signs, we define the magnon accumulation $\delta n_m(\vec{r})$ as [93, 110]

$$\delta n_m(\vec{r}) = \frac{1}{(2\pi)^3} \int d^3k [(n_{\alpha k} - n_{\alpha k}^0) - (n_{\beta k} - n_{\beta k}^0)], \tag{8.148}$$

so that the magnon spin current density in Eq. (8.147) becomes

$$\vec{J}_S^z = -\frac{\hbar}{(2\pi)^3} \int d^3k \, \vec{v}_{mk} \{ [n_{\alpha k}(\vec{r}) - n_{\alpha k}^0) - (n_{\beta k}(\vec{r}) - n_{\beta k}^0] \}, \tag{8.149}$$

where $\vec{v}_{mk}$ is the $k$-magnon velocity. The distribution of the magnon number under the influence of a thermal gradient can be calculated with Boltzmann transport equation. In the absence of external forces and in the relaxation approximation, in steady state we obtain for each magnon mode

$$n_{\mu k}(\vec{r}) - n_{\mu k}^0 = -\tau_{\mu k} \vec{v}_{mk} \cdot \nabla n_{\mu k}^0 - \tau_{\mu k} \vec{v}_{mk} \cdot \nabla [n_{\mu k}(\vec{r}) - n_{\mu k}^0], \tag{8.150}$$

where $\tau_{\mu k}$ is the $\mu k$-magnon relaxation time. Using Eq. (8.150) in Eq. (8.149) one can show that the spin current is the sum of two parts, $\vec{J}_S^z = \vec{J}_{S\nabla T}^z + \vec{J}_{S\nabla n}^z$, where

$$\vec{J}_{S\nabla T}^z = -\frac{\hbar}{(2\pi)^3} \int d^3k \left( \tau_{\alpha k} \frac{\partial n_{\alpha k}^0}{\partial T} - \tau_{\beta k} \frac{\partial n_{\beta k}^0}{\partial T} \right) \vec{v}_{mk} (\vec{v}_{mk} \cdot \nabla T) \tag{8.151}$$

is the contribution of the magnon drift due to the temperature gradient, and

$$\vec{J}_{S\nabla n}^z = -\frac{\hbar}{(2\pi)^3} \int d^3k (\tau_{\alpha k} \vec{v}_{mk} \vec{v}_{mk} \cdot \nabla \delta n_{\alpha k} - \tau_{\beta k} \vec{v}_{mk} \vec{v}_{mk} \cdot \nabla \delta n_{\beta k}) \tag{8.152}$$

is due to the spatial variation of the magnon accumulation, which is governed by the diffusion equation. With the temperature gradient normal to the film plane, Eq. (8.151) gives for the spin current in the $y$-direction [95, 110]

$$J_S^z = S_S^z \nabla T, \tag{8.153a}$$

$$S_S^z = \frac{\hbar^2}{6\pi^2 k_B T^2} \int dk\, k^2 v_{mk}^2 \left[\frac{e^{\hbar\omega_{\beta k}/k_B T}\omega_{\beta k}}{\eta_{\beta k}\left(e^{\hbar\omega_{\beta k}/k_B T} - 1\right)^2} - \frac{e^{\hbar\omega_{\alpha k}/k_B T}\omega_{\alpha k}}{\eta_{\alpha k}\left(e^{\hbar\omega_{\alpha k}/k_B T} - 1\right)^2}\right], \quad (8.153b)$$

where $T$ is the average temperature and $\eta_{\mu k} = 1/\tau_{\mu k}$ is the $\mu k$-magnon relaxation rate. Here also we consider the magnon and phonon systems to have the same temperature $T$. The full solution of the problem requires the use of the diffusion equation to find the magnonic spin current subject to the boundary conditions, as previously done for the FMI. However, Eqs. (8.153a, 8.153b) contain the most relevant dependencies of the SSE on the sample temperature and applied field. It shows, for instance, that in an easy-axis AF, the spin current, and hence the SSE, vanishes for $H = 0$, at any temperature, because the two modes have the same occupancy. In SSE experiments one employs a metallic layer in contact with the AFI layer, that is used to convert the spin current into a charge current by the ISHE or the IREE. Similarly to the case of the FMI/ML bilayer, one can obtain the full expression for the spin current in the AFI layer and calculate the fraction that is injected into the metallic layer to find the voltage that is measured in the spin Seebeck effect [95]. As shown in [95, 110] the magnonic spin current model explains well the measured dependencies of the SSE voltage with magnetic field and temperature in MnF$_2$ [107] and in FeF$_2$ [108].

### 8.6.3   Comparison of Theory with Experimental Results for the SSE in FMI

In this section we show that the magnonic spin current model for the SSE expressed by Eq. (8.142) explains quite well the dependencies of the SSE voltage on sample temperature, thickness and magnetic field, measured in YIG/Pt bilayers by different groups [91, 101, 111–115]. A typical experimental setup used to measure the SSE at room temperature is shown schematically in Fig. 8.19a. A commercial Peltier module is used to heat or cool the side of the Pt layer, while the other side of the sample is in thermal contact with a copper block maintained at room temperature. In the experiments of Ref. [91] the sample consists of a strip of single-crystal YIG (111) film grown by liquid phase epitaxy onto a 0.5-mm-thick [111]-oriented Gd$_3$Ga$_5$O$_{12}$ (GGG) substrate. The strip is 10 mm long and 2.3 mm wide, and the YIG film is 8 μm thick. The YIG strip is covered with a 6-nm-thick Pt layer deposited by magnetron sputtering. Two Cu wires attached with silver paint to the ends of the Pt layer are used to measure the DC SSE voltage directly with a nanovoltmeter. The external magnetic field $H$ is applied in the film plane in a direction perpendicular to the long dimension of the strip.

Figure 8.19b shows the dependence of the SSE voltage on the magnetic field intensity for three values of the temperature difference $\Delta T$ across the GGG/YIG/Pt structure: 4, 8, and 12 K. The change in the sign of the voltage with the reversal in the direction of the field is due to the change in the sign of the spin polarization. The data show that for a fixed field direction, in the low field range of the measurements, $V$ does not depend on the value of the magnetic field, in agreement with the model. Figure 8.19c shows the measured variation of the SSE voltage $V$ with temperature

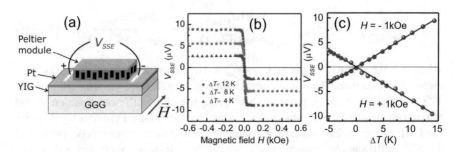

**Fig. 8.19** Measurement of the spin Seebeck effect in a YIG/Pt bilayer at room temperature. (**a**) Schematic illustration of the sample setup with the Peltier module used to heat or cool the Pt layer relative to the GGG substrate attached to a copper block. (**b**) Variation of the voltage $V_{SSE}$ created by the spin Seebeck effect with the magnetic field intensity measured in GGG (0.5 mm)/YIG (8 μm)/Pt (6 nm), for three values of the temperature difference $\Delta T$ between the two sides of the sample, $\Delta T = 4, 8$ and 12 K. $\Delta T$ positive corresponds to the temperature in the Pt layer higher than in the YIG film. (**c**) Variation of the SSE voltage with $\Delta T$ for the fields indicated [91]. Reprinted with permission from S. M. Rezende et al., Phys. Rev. B **89**, 014416 (2014). Copyright (2014) by the American Physical Society

difference $\Delta T$ for applied fields $H = \pm 1$ kOe. The solid lines represent fits with Eq. (8.142), where the temperature gradient is $\nabla T = \Delta T/d$, $d$ being the thickness of the GGG/YIG/Pt structures, which is approximately 0.5 mm.

To fit the theory to data it is necessary to evaluate numerically the integrals in Eqs. (8.139a, 8.139b). Using for YIG the dispersion relation in Eq. (8.135), with $H = 1$ kOe, $\omega_{ZB}/2\pi = 7.0$ THz and $k_m = 1.7 \times 10^7$ cm$^{-1}$, which are values used to calculate the magnon thermal properties of YIG as in Sect. 3.7.3, with the expression (3.188) for the magnon relaxation rate, one finds [92] for the integrals at $T = 300$ K: $B_S = 2.0 \times 10^{-4}$, $B_0 = 0.23$, $B_1 = 0.16$, and $B_2 = 6.9 \times 10^{-5}$. The magnon-phonon relaxation time $\tau_{mp}$ that enters in the factor (8.138a, 8.138b) has been estimated in Refs. [91, 92] to be $\tau_{mp} \approx 10^{-12}$ s, based on the value of the magnon diffusion length $l_m \approx 70$ nm obtained from the fit of Eq. (8.134) to the measured [115] thickness dependence of the SSE and the diffusion parameter calculated with Eq. (8.140). Note that the value of the magnon diffusion length varies with the quality of the sample because it depends strongly on the magnon relaxation rate. Considering $\rho = 1$ for $t_{FM} >> l_m$ and the spin-mixing conductance for the YIG/Pt interface [90, 93–96], $g_{eff}^{\uparrow\downarrow} = 10^{14}$ cm$^{-2}$, one obtains for the coefficient $S_S$ in Eq. (8.137) $S_s = 2.9 \times 10^{-10}$ erg/(K cm). With this value and the parameters for the Pt layer $\lambda_N = 3.7$ nm, $\theta_{SH} = 0.05$ [4, 5], $w = 0.2$ cm, and $t_N = 4$ nm, one finds [92] for the SSE voltage in YIG/Pt under a temperature gradient of $\nabla T = 200$ K/cm and a Pt layer with typical resistance $R_N = 200$ Ω, $V_{SSE} = 6.8$ μV, which is in good quantitative agreement with values measured in YIG/Pt [91, 112–115].

The magnonic spin current model also explains the measured dependence of the SSE voltage on the YIG thickness in YIG/Pt bilayers, observed with very thin YIG layers. Figure 8.20a shows the data of Ref. [115] obtained at room temperature with three series of samples with thickness varying in the range 20–300 nm. With fixed

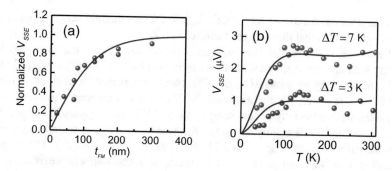

**Fig. 8.20** (a) Symbols represent data of Ref. [115] for the normalized SSE voltage for three series of YIG ($t_{FM}$)/Pt bilayers. The solid line is a least-quare deviation fit of the thickness factor given by Eq. (8.134). (b) Comparison of the experimental data of Ref. [91] with the calculated temperature dependence of the SSE voltage in YIG/Pt. The symbols represent the voltage measured with $\Delta T = 3$ K (blue) and 7 K (red) and the solid lines are calculated with the magnonic spin current model Eq. (8.142). Reprinted with permission from S.M. Rezende et al., Phys. Rev. B **89**, 014416 (2014). Copyright (2014) by the American Physical Society

temperature difference, the voltage tends to zero for very small YIG thickness and saturates at large thickness. This is precisely the behavior of the thickness factor in Eq. (8.134), where for $t_{FM} \ll l_m$ we have $\rho \ll 1$, while for $t_{FM} \gg l_m$, $\rho \approx 1$. The solid line is a fit of Eq. (8.134) to data, that yields a diffusion length $l_m \approx 70$ nm [91].

An important signature of a theoretical model for transport phenomena is the temperature dependence of physical quantities of interest. The symbols in Fig. 8.20b represent the temperature ($T$) dependence of the SSE voltage measured in the same sample described earlier, with $\Delta T = 3$ K and 7 K, under a transverse magnetic field of $H = 1$ kOe. Details of the experiments are presented in Ref. [91]. In order to compare the results of the model with data it is necessary to take into account the temperature dependences of the YIG magnetization and of the resistance of the Pt layer in the range $T = 0$–300 K. For the magnetization we use the factor $M(T)/M(0) = 1.0 - 0.3 \, (T/300)^2$, which represents well the experimental data for YIG [116]. The measured resistance of the Pt layer is fit with the expression obtained from the data, $R_N(T) \approx R_N(300) \, [0.7 + 0.3 \, (T/300)]$, where $R_N(300) = 166 \, \Omega$. The variation with temperature of the spin Hall angle and of the diffusion length $\lambda_N$ is neglected. In the thickness range of the experiments, $\lambda_N$ in the numerator of Eq. (8.142) approximately cancels out the one in the denominator of the tanh function. The curves in Fig. 8.20b, calculated with Eq. (8.142) adjusted to the measured SSE voltages at $T = 300$ K, show that the magnonic spin current model for the SSE describes quite well the temperature dependence of the experimental data below room temperature. The behavior of the SSE voltage below room temperature is explained by the competition between two factors. As $T$ is lowered the thermal magnon population decreases, but the magnon relaxation rate decreases faster, so that the voltage does not change much down to about 100 K. At low $T$ the exponential decrease in the magnon population dominates so that the overall behavior exhibits the bump at $\sim 100$ K, which is a characteristic feature of the model. At lower $T$ the thermal magnon population decreases faster so that the voltages vanishes at 0 K.

We note that the model also explains well the decrease in the measured SSE voltage with increasing magnetic field intensity $H$ at high fields [92], reported in Ref. [113]. At fields up to a few kOe, the SSE barely changes with field, as shown in Fig. 8.20a. However, for higher fields the SSE voltage decreases markedly with increasing field especially at low temperatures [113]. The suppression of the SSE with increasing field is mainly due to the decrease in the magnon thermal population with increasing energy gap created by the field, as in Eq. (8.135). The magnon relaxation rate does not have much influence at room temperature because the integrals in Eqs. (8.139a, 8.139b) are dominated by magnon states with high $q$, that have relaxation rate dominated by 4-magnon processes involving high $q$ modes. These have energies mostly due to the exchange interaction with small Zeeman contribution. On the other hand, at low temperatures the strong suppression in the SSE with field, also due to the decrease in the magnon thermal population with increasing energy gap, is not affected much by the relaxation because it is dominated by field independent processes characteristic of magnons with small wave numbers.

Finally, we note that the Onsager reciprocal of the spin Seebeck effect, the spin Peltier effect, has also been observed experimentally in structures with YIG and Pt junctions [117, 118]. By passing a spin current through the junction, the temperature of a magnetic junction increases or decreases, depending on the direction of a spin current. This effect has been explained theoretically with the same concepts presented here for the spin Seebeck effect [119].

## 8.7    Concluding Remarks and Perspectives for Magnon Spintronics and Related Fields

We conclude this chapter, and the book, with brief comments about other recent developments in magnon spintronics. As mentioned in the introductory remarks of this chapter, spintronics is an emerging field of science and technology that is very active and rapidly developing. New material structures and new spin phenomena are under intense investigation aiming the advance of knowledge and developments of future technologies of increased performance. Since electrons carry both spin and charge, they offer the possibility of combining the two and thereby adding a new degree of freedom to modern devices. For this reason, some authors predict that spintronic devices will make possible to extend the validity of Moore's law for semiconductor electronics, since these rely on the manipulation of charge only [28, 120, 121].

Due to space limitations, we have presented in this chapter only some basic and well understood concepts and phenomena in the field, such as spin-to-charge current conversion in 3D and 2D systems, spin torque, spin pumping and spin Seebeck effects. Two features are essential in all of them: (1) They manifest only in material structures with nanometric dimensions; (2) They require the existence of the spin–orbit coupling between the spin and momentum degrees of freedom of electrons.

In this book we have treated systems with nanometric dimensions in only one direction. In this case, the spin wave modes are made of plane waves, which are mathematically simple to treat. It turns out that spin waves in laterally confined structures are important for basic investigations and for applications, since integrated nanometric devices have reduced lateral dimensions. However, the concepts and ideas presented here constitute the basic background to understand magnonics in modern nanoscale systems. Details of specific systems can be found in research review books, such as the two volumes of *Spin Dynamics in Laterally Confined Structures*, edited by B. Hillebrands and K. Ounadjela. Among other topics they present: Spin waves in laterally confined structures; Laser-induced ultrafast demagnetization: Femtomagnetism, a new frontier; The dynamic response of magnetization to hot spins; Ultrafast magnetization and switching dynamics; and Laser-induced magnetization dynamics. In another series, *Spin Wave Confinement*, edited by S. O. Demokritov, one can find among other topics: Quantized spin-wave models due to lateral confinement; Nonuniform magnetization dynamics in ultrasmall ferromagnetic planar elements; Coupled spin waves in magnonic waveguides; Tuning of the spin wave band structure in nanostructured iron/permalloy nanowire arrays; Magnetization dynamics of reconfigurable 2D magnonic crystals; Spin wave optics in patterned garnet; and Spin waves on spin structures: Topology, localization, and nonreciprocity. Other good references for recent developments in nanomagnetism and nanomagnonics are the books *Principles of Nanomagnetism*, by A. P. Guimarães, and *Three-Dimensional Magnonics: Layered, Micro- and Nanostructures*, by G. Gubbiotti.

Another topic that would deserve a more extended treatment is about the consequences of the spin–orbit coupling, that plays an increasingly important role in magnetism. As briefly discussed in the initial chapters, the SOC is essential for the magneto-crystalline anisotropy and for the magnetoelastic interaction, that have important consequences in many magnetic properties and phenomena. In this chapter we showed that the spin Hall and Rashba–Edelstein effects have their origin in the SOC. However, there are other important effects of the SOC. One of them is the Dzyaloshinskii–Moriya interaction, proposed long ago to explain the weak ferromagnetism of some antiferromagnets [122] and the anisotropic superexchange interaction [123]. The exchange interaction between two spins $\vec{S}_1$ and $\vec{S}_2$, in Eq. (1.3), is proportional to $\vec{S}_1 \cdot \vec{S}_2$, so it is spatially isotropic and has time-reversal symmetry. In systems that lack inversion symmetry, the spin–orbit coupling induces an interaction in the form

$$H_{DMI} = \vec{D}_{12} \cdot (\vec{S}_1 \times \vec{S}_2), \tag{8.154}$$

which is called Dzyaloshinskii–Moriya interaction (DMI), where $\vec{D}_{12}$ is the Dzyaloshinskii–Moriya vector. The DMI is a chiral interaction that decreases or increases the energy of the spins depending on whether the rotation from $\vec{S}_1$ to $\vec{S}_2$ around $\vec{D}_{12}$ is clockwise or anticlockwise. If $\vec{S}_1$ and $\vec{S}_2$ are initially parallel, then the effect of a sufficiently strong DMI (with respect to exchange and anisotropy) is to introduce a tilt around $\vec{D}_{12}$. From Eq. (8.154) it is clear that in spin systems with inversion symmetry, the interchange of $\vec{S}_1$ and $\vec{S}_2$ does not change the interaction energy, and thus $D_{12} = 0$. The DMI was initially proposed as a superexchange interaction in bulk magnetic insulators [122, 123] and later extended to non-centrosymmetric magnetic metals [124]. In bulk magnetic materials, the presence of a DMI in additional to the exchange energy may not change the equilibrium spin configuration, but it modifies the magnon dispersion relation in a manner that can be calculated with the methods presented in Chap. 3.

The most important consequences of the DMI are observed in ferromagnetic films and multilayers, because the broken inversion symmetry at the interfaces of FM materials with strong SOC can result in a large DMI [125]. The existence of interfacial DMI was first demonstrated by the observation of spiral-like spatial modulations of the spin orientation with a winding periodicity related to the magnitude of the DMI [126]. More recently it was found that the DMI also enables the formation of other chiral spin structures, such as *skyrmions* and chiral domain wall. Skyrmions are swirling spin textures that resemble particles first described in the context of nuclear physics [127]. They are stable, topologically protected, particle-like objects that can have a motion driven by spin-polarized currents [127–129]. The recent upsurge of activity in materials properties and phenomena involving the spin–orbit coupling has led to the development of a subfield of spintronics, called *spin orbitronics*.

In the previous paragraphs we have mentioned only a partial list of topics of current interest in magnetic phenomena in nanoscale systems that we have not been able to treat in the book. The interest in nanomagnetics is not just a fashionable trend in science and engineering, but rather a natural consequence of technological progress. Novel physical phenomena are uncovered when magnetic systems are reduced to nanoscale, and new ideas come up for the development of future technologies of increased performance. However, it is not possible to predict which of the research directions will be the most successful for applications. All of them aim for low-energy means of magnetic excitation, manipulation and detection on the nanometer scale that might constitute next-generation physical processes for terahertz information storage and processing for computing applications. The most intriguing discoveries in the various subfields of spintronics and nanomagnetism might yet to be announced. Depending on future developments, new directions will be uncovered, adding to the complexity of the area. These will shape our preferences for certain topics, perhaps rendering others obsolete. It remains to be seen if and which emerging ideas will ultimately be successful in the very competitive technology market.

## Problems

8.1 Show that in a free electron gas in which the particle density $n$ is increased by $\delta n$, the chemical potential changes from $\mu$ to $\mu + \delta\mu$, where $\delta\mu = \delta n/N(\varepsilon_F)$ and $N(\varepsilon_F)$ is the density of states at the Fermi energy.

8.2 In a 2D electron gas with Rashba interaction, calculate the wave number position and the minimum energy for the two spin polarizations.

8.3 In a magnetic random access memory, the storage cell unit is made of a nanopillar in which the free layer is a 5 nm thick Py layer with lateral dimensions $50 \times 100$ nm$^2$. Using for the Py layer $4\pi M = 11$ kG and coercitive field $H_c = 10$ Oe, calculate the electric current intensity necessary to switch the magnetization, assuming 100% spin polarization.

8.4 The peak-to-peak FMR linewidth of a 10 nm thick Py film at a frequency of 10 GHz is 26 Oe. When a Pt layer is deposited on the Py film the linewidth increases due to the spin pumping damping. Calculate the total linewidth of the Py/Pt bilayer for Py film with thickness 100 nm, considering $4\pi M = 11$ kG for the magnetization and $g_{eff}^{\uparrow\downarrow} = 2 \times 10^{15}$ cm$^{-2}$ for the spin-mixing conductance of the Py/Pt interface.

8.5 The FMR linewidth (HWHM) of a 3 μm thick YIG film at a frequency of 10 GHz is 0.4 Oe and it increases to 1.0 Oe with the deposition of a Pt layer. Considering for YIG $4\pi M = 1.76$ kG and for the YIG/Pt interface $g_{eff}^{\uparrow\downarrow} = 1.5 \times 10^{14}$ cm$^{-2}$, find the coherence length $t_{coh}$ that determines the additional spin pumping linewidth and compare with the value calculated with $t_{coh} = \sqrt{D/\Delta H}$.

8.6 Show that the magnetization current defined in Eq. (8.79) has dimension of magnetic moment/(area.time).

8.7 In a microwave-driven spin pumping experiment with a YIG/Pt bilayer, the measured FMR linewidth (HWHM) at a frequency of 10 GHz is 1.0 Oe. Calculate the peak SPE voltage produced by a microwave field $h_{rf} = 5.0 \times 10^{-2}$ Oe, considering an ellipticity factor $p_{xz} = 0.3$ for the YIG film, a spin-mixing conductance $g_{eff}^{\uparrow\downarrow} = 10^{14}$ cm$^{-2}$ for the YIG/Pt interface, and $\lambda_N = 3.7$ nm, $\theta_{SH} = 0.05$, $w = 0.2$ cm, $t_N = 4$ nm, and $R_N = 150$ $\Omega$ for the Pt layer.

8.8 Using the expansion (8.95) for the magnon population in Eq. (8.93), show that the magnonic diffusion current is given by (8.97).

8.9 In a spin Seebeck experiment with the YIG/Pt bilayer sample of Problem 8.7, find the SSE voltage produced by a temperature difference of 15 K between the two sides of the 0.5 mm thick sample, considering for the coefficient $C_S$ in Eq. (8.137) $C_s = 2.9 \times 10^{-10}$ erg/(K cm).

# References

1. Silsbee, R.H., Janossy, A., Monod, P.: Coupling between ferromagnetic and conduction-spin-resonance modes at a ferromagnetic-normal metal interface. Phys. Rev. **B19**, 4382 (1979)
2. Johnson, M., Silsbee, R.H.: Coupling of electronic charge and spin at a ferromagnetic-paramagnetic metal interface. Phys. Rev. B. **37**, 5312 (1988)
3. Valet, T., Fert, A.: Theory of perpendicular magnetoresistance in magnetic multilayers. Phys. Rev. B. **48**, 7099 (1993)

4. Rojas-Sánchez, J.-C., Fert, A.: Compared efficiencies of conversions between charge and spin current by spin-orbit interactions in two- and three-dimensional systems. Phys. Rev. Applied. **11**, 054049 (2019)
5. Hoffmann, A.: Spin hall effects in metals. IEEE Trans. Magn. **49**, 5172 (2013)
6. Sinova, J., Valenzuela, S.O., Wunderlich, J., Back, C.H., Jungwirth, T.: Spin hall effects. Rev. Mod. Phys. **87**, 1231 (2015)
7. Dyakonov, M.I., Perel, V.I.: Possibility of orienting electron spins with current. Sov. Phys. JETP Lett. **13**, 467 (1971)
8. Dyakonov, M.I., Perel, V.I.: Current induced spin orientation of electrons in semiconductors. Phys. Lett. **35A**, 459 (1971)
9. Hirsch, J.E.: Spin hall effect. Phys. Rev. Lett. **83**, 1834 (1999)
10. Zhang, S.: Spin hall effect in the presence of spin diffusion. Phys. Rev. Lett. **85**, 393 (2000)
11. Sinova, J., Culcer, D., Niu, Q., Sinitsyn, N.A., Jungwirth, T., MacDonald, A.H.: Universal intrinsic spin hall effect. Phys. Rev. Lett. **85**, 393 (2000)
12. Engel, H.-A., Halperin, B.I., Rashba, E.I.: Theory of spin hall conductivity in n-doped GaAs. Phys. Rev. Lett. **95**, 166605 (2005)
13. Kato, Y.K., Myers, R.C., Gossard, A.C., Awschalom, D.D.: Observation of the spin hall effect in semiconductors. Science. **306**, 1910 (2004)
14. Valenzuela, S.O., Tinkham, M.: Direct electronic measurement of the spin hall effect. Nature. **442**, 176 (2016)
15. Rashba, E.: Properties of semiconductors with an extremum loop. 1. Cyclotron and combinational resonance in a magnetic field perpendicular to the plane of the loop. Sov. Phys. Solid State. **2**, 1109 (1960)
16. Bychkov, Y.A., Rasbha, E.I.: Oscillatory effects and the magnetic susceptibility of carriers in inversion layers. J. Phys. C Solid State Phys. **17**, 6039 (1984)
17. Manchon, A., Koo, H.C., Nitta, J., Frolov, S.M., Duine, R.A.: New perspectives for Rashba spin-orbit coupling. Nat. Mater. **14**, 871 (2015)
18. Soumyanarayanan, A., Reyren, N., Fert, A., Panagopoulos, C.: Emergent phenomena induced by spin-orbit coupling at surfaces and interfaces. Nature. **539**, 509 (2016)
19. Edelstein, V.M.: Spin polarization of conduction electrons induced by electric current in two-dimensional asymmetric electron systems. Solid State Commun. **3**, 233 (1990)
20. Rojas-Sánchez, J.C., Vila, L., Desfonds, G., Gambarelli, S., Attané, J.P., De Teresa, J.M., Magén, C., Fert, A.: Spin-to-charge conversion using Rashba coupling at the interface between non-magnetic materials. Nat. Commun. **4**, 2944 (2013)
21. Zhang, S., Fert, A.: Conversion between spin and charge currents with topological insulators. Phys. Rev. B. **94**, 184423 (2016)
22. Sklenar, J., Zhang, W., Jungfleisch, M.B., Jiang, W., Saglam, H., Pearson, J.E., Ketterson, J. B., Hoffmann, A.: Interface generation of spin-orbit torques. J. Appl. Phys. **120**, 180901 (2016)
23. Han, W., Otani, Y.C., Maekawa, S.: Quantum materials for spin and charge conversion. npj Quant. Mater. **3**, 27 (2018)
24. Apalkov, D., Dieny, B., Slaughter, J.M.: Magnetoresistive random access memory. Proc. IEEE. **104**, 1796 (2016)
25. Locatelli, N., Cros, V., Grollier, J.: Spin-torque building blocks. Nat. Mater. **13**, 11 (2014)
26. Chen, T., Dumas, R.K., Eklund, A., Muduli, P.K., Houshang, A., Awad, A.A., Dürrenfeld, P., Malm, B.G., Rusu, A., Akerman, J.: Spin-torque and spin-hall nano-oscillators. Proc. IEEE. **104**, 1919 (2016)
27. Urazhdin, S., Demidov, V.E., Ulrichs, H., Kendziorczyk, T., Kuhn, T., Leuthold, J., Wilde, G., Demokritov, S.O.: Nanomagnonic devices based on the spin-transfer torque. Nat Nanotechnol. **9**, 509 (2014)
28. Chumak, A.V., Vasyuchka, V.I., Serga, A.A., Hillebrands, B.: Magnon spintronics. Nat. Phys. **11**, 453 (2015)

29. Slonczewski, J.C.: Current-driven excitation of magnetic multilayers. J. Magn. Magn. Mater. **159**, L1 (1996)

30. Berger, L.: Emission of spin waves by a magnetic multilayer traversed by a current. Phys. Rev. B. **54**, 9353 (1996)

31. Stiles, M.D., Zangwill, A.: Anatomy of spin-transfer torque. Phys. Rev. B. **66**, 014407 (2002)

32. Tsoi, M., Jansen, A.G.M., Bass, J., Chiang, W.-C., Seck, M., Tsoi, V., Wyder, P.: Excitation of a magnetic multilayer by an electric current. Phys. Rev. Lett. **80**, 4281 (1998)

33. Rezende, S.M., de Aguiar, F.M., Lucena, M.A., Azevedo, A.: Magnon excitation by spin injection in thin Fe/Cr/Fe films. Phys. Rev. Lett. **84**, 4212 (2000)

34. Rippard, W.H., Pufall, M.R., Silva, T.J.: Quantitative studies of spin-momentum-transfer-induced excitations in Co/Cu multilayer films using point-contact spectroscopy. Appl. Phys. Lett. **82**, 1260 (2003)

35. Kiselev, S.I., Sankey, J.C., Krivorotov, I.N., Emley, N.C., Schoelkopf, R.J., Buhrman, R.A., Ralph, D.C.: Microwave oscillations of a nanomagnet driven by a spin-polarized current. Nature. **425**, 380 (2003)

36. Rippard, W.H., Pufall, M.R., Kaka, S., Russek, S.E., Silva, T.J.: Direct-current induced dynamics in $Co_{90}Fe_{10}/Ni_{80}Fe_{20}$ point contacts. Phys. Rev. Lett. **92**, 027201 (2004)

37. Kiselev, S.I., Sankey, J.C., Krivorotov, I.N., Emley, N.C., Rinkoski, M., Perez, C., Buhrman, R.A., Ralph, D.C.: Current-induced nanomagnet dynamics for magnetic fields perpendicular to the sample plane. Phys. Rev. Lett. **93**, 036601 (2004)

38. Krivorotov, I.N., Emley, N.C., Sankey, J.C., Kiselev, S.I., Ralph, D.C., Buhrman, R.A.: Time-domain measurements of nanomagnet dynamics driven by spin-transfer torques. Science. **307**, 228 (2005)

39. Rezende, S.M., de Aguiar, F.M., Azevedo, A.: Spin wave theory for the dynamics induced by direct currents in magnetic multilayers. Phys. Rev. Lett. **94**, 037202 (2005)

40. Russek, S.E., Kaka, S., Rippard, W.H., Pufall, M.R., Silva, T.J.: Finite-temperature modeling of nanoscale spin-transfer oscillators. Phys. Rev. B. **71**, 104425 (2005)

41. Rezende, S.M., de Aguiar, F.M., Azevedo, A.: Magnon excitation by spin-polarized direct currents in magnetic nanostructures. Phys. Rev. B. **73**, 094402 (2006)

42. Slavin, A., Tiberkevich, V.: Nonlinear auto-oscillator theory of microwave generation by spin-polarized current. IEEE Trans. Magn. **45**, 1875 (2009)

43. Demidov, V.E., Urazhdin, S., Ulrichs, H., Tiberkevich, V., Slavin, A., Baither, D., Schmitz, G., Demokritov, S.O.: Magnetic nano-oscillator driven by pure spin current. Nat. Mater. **11**, 1028 (2012)

44. Madami, M., Iacocca, E., Sani, S., Gubbiotti, G., Tacchi, S., Dumas, R.K., Akerman, J., Carlotti, G.: Propagating spin waves excited by spin-transfer torque: a combined electrical and optical study. Phys. Rev. B. **92**, 024403 (2015)

45. Slonczewski, J.C.: Excitation of spin waves by an electric current. J. Magn. Magn. Mater. **195**, L261 (1999)

46. Slavin, A., Tiberkevich, V.: Spin wave mode excited by spin-polarized current in a magnetic nanocontact is a standing self-localized wave bullet. Phys. Rev. Lett. **95**, 237201 (2005)

47. Ruotolo, A., Cros, V., Georges, B., Dussaux, A., Grollier, J., Deranlot, C., Guillemet, R., Bouzehouane, K., Fusil, S., Fert, A.: Phase-locking of magnetic vortices mediated by antivortices. Nat. Nanotechnol. **4**, 528 (2009)

48. Rezende, S.M., de Aguiar, F.M., Rodríguez-Suárez, R.L., Azevedo, A.: Mode locking of spin waves excited by direct currents in microwave nano-oscillators. Phys. Rev. Lett. **98**, 087202 (2007)

49. Georges, B., Grollier, J., Darques, M., Cros, V., Deranlot, C., Marcilhac, B., Faini, G., Fert, A.: Coupling efficiency for phase locking of a spin transfer nano-oscillator to a microwave current. Phys. Rev. Lett. **101**, 017201 (2008)

50. Li, Y., de Milly, X., Araujo, F.A., Klein, O., Cros, V., Grollier, J., de Loubens, G.: Probing phase coupling between two spin-torque nano-oscillators with an external source. Phys. Rev. Lett. **118**, 247202 (2017)

51. Rezende, S.M.: Quantum coherence in spin-torque nano-oscillators. Phys. Rev. B. **81**, 092401 (2010)
52. Haken, H.: Cooperative phenomena in systems far from thermal equilibrium and in nonphysical systems. Rev. Mod. Phys. **47**, 67 (1975)
53. Tserkovnyak, Y., Brataas, A., Bauer, G.E.W.: Enhanced Gilbert damping in thin ferromagnetic films. Phys. Rev. Lett. **88**, 117601 (2002)
54. Tserkovnyak, Y., Brataas, A., Bauer, G.E.W.: Spin pumping and magnetization dynamics in metallic multilayers. Phys. Rev. B. **66**, 22440 (2002)
55. Azevedo, A., Vilela Leão, L.H., Rodriguez-Suarez, R.L., Oliveira, A.B., Rezende, S.M.. Direct evidence of the spin-pumping effect, abstract HA-10. In: 49$^{th}$ Conference on Magnetism and Magnetic Materials, Jacksonville (2004); Dc Effect in Ferromagnetic Resonance: Evidence of the Spin-Pumping Effect?. J. Appl. Phys. **97**:10C715 (2005)
56. Saitoh, E., Ueda, M., Miyajima, H., Tatara, G.: Conversion of spin current into charge current at room temperature: Inverse spin-hall effect. Appl. Phys. Lett. **88**, 182509 (2006)
57. Tserkovnyak, Y., Brataas, A., Bauer, G.E.W., Halperin, B.: Nonlocal magnetization dynamics in ferromagnetic heterostructures. Rev. Mod. Phys. **77**, 1375 (2005)
58. Mizukami, S., Ando, Y., Miyazaki, T.: Effect of spin diffusion on Gilbert damping for a very thin permalloy layer in Cu/permalloy/Cu/Pt films. Phys. Rev. B. **66**, 104413 (2002)
59. Soares, M.M., Vilela-Leão, L.H., da Silva, G.L., Rodríguez-Suárez, R.L., Azevedo, A., Rezende, S.M., Ferromagnetic Resonance Linewidth and Spin Pumping in Permalloy/Platinum Bilayers. MMM Group Report, Universidade Federal de Pernambuco, Unpublished (2012)
60. Mosendz, O., Vlaminck, V., Pearson, J.E., Fradin, F.Y., Bauer, G.E.W., Bader, S.D., Hoffmann, A.: Detection and quantification of inverse spin Hall effect from spin pumping in permalloy/normal metal bilayers. Phys. Rev. B. **82**, 214403 (2010)
61. Rezende, S.M., Rodríguez-Suárez, R.L., Soares, M.M., Vilela-Leão, L.H., Domínguez, D.L., Azevedo, A.: Enhanced spin pumping damping in yttrium iron garnet/Pt bilayers. Appl. Phys. Lett. **102**, 012402 (2013)
62. Rezende, S.M., Rodríguez-Suárez, R.L., Azevedo, A.: Magnetic relaxation due to spin pumping in thick ferromagnetic films in contact with normal metals. Phys. Rev. B. **88**, 014404 (2013)
63. Berger, L.: Generation of dc voltages by a magnetic multilayer undergoing ferromagnetic resonance. Phys. Rev. B. **59**, 11465 (1999)
64. Costache, M.V., Sladkov, M., Watts, S.M., van der Wal, C.H., van Wees, B.J.: Electrical detection of spin pumping due to the precessing magnetization of a single ferromagnet. Phys. Rev. Lett. **97**, 216603 (2006)
65. Wang, X., Bauer, G.E.W., van Wees, B.J., Brataas, A., Tserkovnyak, Y.: Voltage generation by ferromagnetic resonance at a nonmagnet to ferromagnet contact. Phys. Rev. Lett. **97**, 216602 (2006)
66. Harder, M., Cao, Z.X., Gui, Y.S., Fan, X.L., Hu, C.-M.: Analysis of the line shape of electrically detected ferromagnetic resonance. Phys. Rev. B. **84**, 054423 (2011)
67. Ando, K., Takahashi, S., Ieda, J., Kajiwara, Y., Nakayama, H.Y.T., Harii, K., Fujikawa, Y., Matsuo, M., Maekawa, S., Saitoh, E.: Inverse spin-Hall effect induced by spin pumping in metallic system. J. Appl. Phys. **109**, 103913 (2011)
68. Azevedo, A., Vilela-Leão, L.H., Rodríguez-Suárez, R.L., Lacerda Santos, A.F., Rezende, S. M.: Spin pumping and anisotropic magnetoresistance voltages in magnetic bilayers: Theory and experiment. Phys. Rev. B. **83**, 144402 (2011)
69. Kajiwara, Y., Harii, K., Takahashi, S., Ohe, J., Uchida, K., Mizuguchi, M., Umezawa, H., Kawai, H., Ando, K., Takanashi, K., Maekawa, S., Saitoh, E.: Transmission of electrical signals by spin-wave interconversion in a magnetic insulator. Nature. **464**, 262 (2010)
70. Weiler, M., et al.: Experimental test of the spin mixing Interface conductivity concept. Phys. Rev. Lett. **111**, 176601 (2013)

71. Mendes, J.B.S., Cunha, R.O., Alves Santos, O., Ribeiro, P.R.T., Machado, F.L.A., Rodríguez-Suárez, R.L., Azevedo, A., Rezende, S.M.: Large inverse spin Hall effect in the antiferromagnetic metal $Ir_{20}Mn_{80}$. Phys. Rev. B. **89**, 140406(R) (2014)

72. Zhang, W., Jungfleisch, M.B., Jiang, W., Pearson, J.E., Hoffmann, A., Freimuth, F., Mokrousov, Y.: Spin Hall effects in metallic Antiferromagnets. Phys. Rev. Lett. **113**, 196602 (2014)

73. Qiu, Z., Ando, K., Uchida, K., Kajiwara, Y., Takahashi, R., Nakayama, H., An, T., Fujikawa, Y., Saitoh, E.: Spin mixing conductance at a well-controlled platinum/yttrium iron garnet interface. Appl. Phys. Lett. **103**, 092404 (2013)

74. Castel, V., Vlietstra, N., van Wees, B.J., Ben Youssef, J.: Yttrium iron garnet thickness and frequency dependence of the spin-charge current conversion in YIG/Pt systems. Phys. Rev. B. **90**, 214434 (2014)

75. Du, C., Wang, H., Chris Hammel, P., Yang, F.: $Y_3Fe_5O_{12}$ spin pumping for quantitative understanding of pure spin transport and spin Hall effect in a broad range of materials. J. Appl. Phys. **117**, 172603 (2015)

76. Jungfleisch, M.B., Chumak, A.V., Kehlberger, A., Lauer, V., Kim, D.H., Onbasli, M.C., Ross, C.A., Kläui, M., Hillebrands, B.: Thickness and power dependence of the spin-pumping effect in $Y_3Fe_5O_{12}$/Pt heterostructures measured by the inverse spin Hall effect. Phys. Rev. B. **91**, 134407 (2015)

77. Sangiao, S., De Teresa, J.M., Morellon, L., Lucas, I., Martinez-Velarte, M.C., Viret, M.: Control of the spin to charge conversion using the inverse Rashba-Edelstein effect. Appl. Phys. Lett. **106**, 172403 (2015)

78. Matsushima, M., Ando, Y., Dushenko, S., Ohshima, R., Kumamoto, R., Shinjo, T., Shiraishi, M.: Quantitative investigation of the inverse Rashba-Edelstein effect in Bi/Ag and Ag/Bi on YIG. Appl. Phys. Lett. **110**, 072404 (2017)

79. Baker, A.A., Figueroa, A.I., Collins-McIntyre, L.J., van der Laan, G., Hesjedal, T.: Spin pumping in Ferromagnet-topological insulator-Ferromagnet Heterostructures. Sc Rep. **5**, 7905 (2015)

80. Rojas-Sánchez, J.-C., et al.: Spin to charge conversion at room temperature by spin pumping into a new type of topological insulator: α-Sn films. Phys. Rev. Lett. **116**, 096602 (2016)

81. Kondou, K., Yoshimi, R., Tsukazaki, A., Fukuma, Y., Matsuno, J., Takahashi, K.S., Kawasaki, M., Tokura, Y., Otani, Y.: Fermi-level-dependent charge-to-spin current conversion by Dirac surface states of topological insulators. Nat. Phys. **12**, 1027 (2016)

82. Mendes, J.B.S., Santos, O.A., Holanda, J., Loreto, R.P., De Araujo, C.I.L., Chang, C.-Z., Moodera, J.S., Azevedo, A., Rezende, S.M.: *Dirac-surface-state-dominated spin to charge current conversion in the topological insulator* $(Bi_{0.22}Sb_{0.78})_2Te_3$ *films at room temperature.* Phys. Rev. B. **96**, 180415(R) (2017)

83. Mahendra, D.C., Liu, T., Chen, J.-Y., Peterson, T., Sahu, P., Li, H., Zhao, Z., Wu, M., Wang, J.-P.: Room-temperature spin-to-charge conversion in sputtered bismuth selenide thin films via spin pumping from yttrium iron garnet. Appl. Phys. Lett. **114**, 102401 (2019)

84. Mendes, J.B.S., Alves-Santos, O., Meireles, L.M., Lacerda, R.G., Vilela-Leão, L.H., Machado, F.L.A., Rodríguez-Suárez, R.L., Azevedo, A., Rezende, S.M.: Spin-to-charge-current conversion and magnetoresistance in yttrium iron garnet-graphene hybrid structure. Phys. Rev. Lett. **115**, 226601 (2015)

85. Mendes, J.B.S., Aparecido-Ferreira, A., Holanda, J., Azevedo, A., Rezende, S.M.: Efficient spin to charge current conversion in the 2D semiconductor $MoS_2$ by spin pumping from yttrium iron garnet. Appl. Phys. Lett. **112**, 242407 (2018)

86. Hasan, M.Z., Kane, C.L.: Topological insulators. Rev. Mod. Phys. **82**, 3061 (2010)

87. Rado, G.T., Weertman, J.R.: Spin-wave resonance in a ferromagnetic metal. J. Phys. Chem. Solids. **11**, 315 (1959)

88. Hillebrands, B.: Spin-wave calculations for multilayered structures. Phys. Rev. B. **41**, 530 (1990)

89. Zhang, S.S.-L., Zhang, S.: Magnon mediated electric current drag across a ferromagnetic insulator layer. Phys. Rev. Lett. **109**, 096603 (2012)
90. Zhang, S.S.-L., Zhang, S.: Spin convertance at magnetic interfaces. Phys. Rev. B. **86**, 214424 (2012)
91. Rezende, S.M., Rodríguez-Suárez, R.L., Cunha, R.O., Rodrigues, A.R., Machado, F.L.A., Fonseca Guerra, G.A., Lopez Ortiz, J.C., Azevedo, A.: Magnon spin-current theory for the longitudinal spin-Seebeck effect. Phys. Rev. B. **89**, 014416 (2014)
92. Rezende, S.M., Rodríguez-Suárez, R.L., Cunha, R.O., Lopez Ortiz, J.C., Azevedo, A.: Bulk magnon spin current theory for the longitudinal spin Seebeck effect. J. Magn. Magn. Mater. **400**, 171 (2016)
93. Cornelissen, L.J., Liu, J., Duine, R.A., Ben Youssef, J., van Wees, B.J.: Long-distance transport of magnon spin information in a magnetic insulator at room temperature. Nat. Phys. **11**, 1022 (2015)
94. Cornelissen, L.J., Peters, K.J.H., Bauer, G.E.W., Duine, R.A., van Wees, B.J.: Magnon spin transport driven by the magnon chemical potential in a magnetic insulator. Phys. Rev. **94**, 014412 (2016)
95. Rezende, S.M., Rodríguez-Suárez, R.L., Azevedo, A.: Magnon diffusion theory for the spin Seebeck effect in ferromagnetic and antiferromagnetic insulators. J. Phys. D. Appl. Phys. **51**, 174004 (2018)
96. Bauer, G.E.W., Saitoh, E., van Wees, B.J.: Spin caloritronics. Nat. Mat. **11**, 391 (2012)
97. Boona, S.R., Myers, R.C., Heremans, J.P.: Spin caloritronics. Energy Environ. Sci. **7**, 885 (2014)
98. Uchida, K., Ishida, M., Kikkawa, T., Kirihara, A., Murakami, T., Saitoh, E.: Longitudinal spin Seebeck effect: From fundamentals to applications. J. Phys. Condens. Matter. **26**, 343202 (2014)
99. Uchida, K., Takahashi, S., Harii, K., Ieda, J., Koshibae, W., Ando, K., Maekawa, S., Saitoh, E.: Observation of spin Seebeck effect. Nature. **455**, 778 (2008)
100. Uchida, K., Xiao, J., Adachi, H., Ohe, J., Takahashi, S., Ieda, J., Ota, T., Kajiwara, Y., Umezawa, H., Kawai, H., Bauer, G.E.W., Maekawa, S., Saitoh, E.: Spin Seebeck insulator. Nat. Mater. **9**, 894 (2010)
101. Uchida, K., Adachi, H., Ota, T., Nakayama, H., Maekawa, S., Saitoh, E.: Observation of longitudinal spin-Seebeck effect in magnetic insulators. Appl. Phys. Lett. **97**, 172505 (2010)
102. Jaworski, C.M., Yang, J., Mack, S., Awschalom, D.D., Heremans, J.P., Myers, R.C.: Observation of the spin-Seebeck effect in a ferromagnetic semiconductor. Nat. Mater. **9**, 898 (2010)
103. Slachter, A., Bakker, F.L., Adam, J.P., van Wees, B.J.: Thermally driven spin injection from a ferromagnet into a non-magnetic metal. Nat. Phys. **6**, 879 (2010)
104. Miao, B.F., Huang, S.Y., Qu, D., Chien, C.L.: Inverse spin Hall effect in a ferromagnetic metal. Phys. Rev. Lett. **111**, 066602 (2013)
105. Agrawal, M., Vasyuchka, V.I., Serga, A.A., Karenowska, A.D., Melkov, G.A., Hillebrands, B.: Direct measurement of magnon temperature: New insight into Magnon-phonon coupling in magnetic insulators. Phys. Rev. Lett. **111**, 107204 (2013)
106. Seki, S., Ideue, T., Kubota, M., Kozuka, Y., Takagi, R., Nakamura, M., Kaneko, Y., Kawasaki, M., Tokura, Y.: Thermal generation of spin current in an Antiferromagnet. Phys. Rev. Lett. **115**, 266601 (2015)
107. Wu, S.M., Zhang, W., Amit, K.C., Borisov, P., Pearson, J.E., Jiang, J.S., Lederman, D., Hoffmann, A., Bhattacharya, A.: Antiferromagnetic spin Seebeck effect. Phys. Rev. Lett. **116**, 097204 (2016)
108. Li, J., Shi, Z., Ortiz, V.H., Aldosary, M., Chen, C., Aji, V., Wei, P., Shi, J.: Spin Seebeck effect from antiferromagnetic magnons and critical spin fluctuations in epitaxial $FeF_2$ films. Phys. Rev. Lett. **122**, 217204 (2019)
109. Ribeiro, P.R.T., Machado, F.L.A., Gamino, M., Azevedo, A., Rezende, S.M.: Spin Seebeck effect in antiferromagnet nickel oxide in wide ranges of temperature and magnetic field. Phys. Rev. B. **99**, 094432 (2019)

110. Rezende, S.M., Rodríguez-Suárez, R.L., Azevedo, A.: Theory of the spin Seebeck effect in antiferromagnets. Phys. Rev. B. **93**, 014425 (2016)
111. Uchida, K., Ota, T., Adachi, H., Xiao, J., Nonaka, T., Kajiwara, Y., Bauer, G.E.W., Maekawa, S., Saitoh, E.: Thermal spin pumping and magnon-phonon-mediated spin-Seebeck effect. J. Appl. Phys. **111**, 103903 (2012)
112. Kikkawa, T., Uchida, K., Daimon, S., Shiomi, Y., Adachi, H., Qiu, Z., Hou, D., Jin, X.-F., Maekawa, S., Saitoh, E.: Separation of longitudinal spin Seebeck effect from anomalous Nernst effect: Determination of origin of transverse thermoelectric voltage in metal/insulator junctions. Phys. Rev. B. **88**, 214403 (2013)
113. Kikkawa, T., Uchida, K., Daimon, S., Qiu, Z., Shiomi, Y., Saitoh, E.: Critical suppression of spin Seebeck effect by magnetic fields. Phys. Rev. B. **92**, 064413 (2015)
114. Schreier, M., et al.: Sign of inverse spin Hall voltages generated by ferromagnetic resonance and temperature gradients in yttrium iron garnet platinum bilayers. J. Phys. D. Appl. Phys. **48**, 025001 (2015)
115. Kehlberger, A., et al.: Length scale of the spin Seebeck effect. Phys. Rev. Lett. **115**, 096602 (2015)
116. Gilleo, M.A., Geller, S.: Magnetic and crystallographic properties of substituted yttrium-iron garnet, $3Y_2O_3.xM_2O_3.(5-x)Fe_2O_3$. Phys. Rev. **110**, 73 (1958)
117. Flipse, J., Dejene, F.K., Wagenaar, D., Bauer, G.E.W., Youssef, J.B., van Wees, B.J.: Observation of the spin Peltier effect for magnetic insulators. Phys. Rev. Lett. **113**, 027601 (2014)
118. Daimon, S., Iguchi, R., Hioki, T., Saitoh, E., Uchida, K.I.: Thermal imaging of the spin Peltier effect. Nat. Commun. **7**, 13754 (2016)
119. Costa, S.S., Sampaio, L.C.: Magnon theory for the spin Peltier effect. J. Phys. D. Appl. Phys. In press (2020)
120. Moore, G.E.: Cramming more components onto integrated circuits. Electronics. **38**, 8 (1965)
121. Serga, A.A., Chumak, A.V., Hillebrands, B.: YIG Magnonics. J. Phys. D. Appl. Phys. **43**, 264002 (2010)
122. Dzyaloshinsky, I.: A thermodynamic theory of 'weak' ferromagnetism of antiferromagnetics. J. Phys. Chem. Solids. **4**, 241 (1958)
123. Moriya, T.: Anisotropic superexchange interaction and weak ferromagnetism. Phys. Rev. **120**, 91 (1960)
124. Fert, A., Levy, P.M.: Role of anisotropic exchange interactions in determining the properties of spin-glasses. Phys. Rev. Lett. **44**, 1538 (1980)
125. Fert, A.: Magnetic and transport properties of metallic multilayers. Mater. Sci. Forum. **59-60**, 439 (1990)
126. Bode, M., et al.: Chiral magnetic order at surfaces driven by inversion asymmetry. Nature. **447**, 190 (2007)
127. Nagaosa, N., Tokura, Y.: Topological properties and dynamics of magnetic skyrmions. Nat. Nanotechnol. **8**, 899 (2013)
128. Fert, A., Cros, V., Sampaio, J.: Skyrmions on the track. Nat. Nanotechnol. **8**, 152 (2013)
129. Jiang, W., Chen, G., Liu, K., Zang, J., te Velthuis, S.G.E., Hoffmann, A.: Skyrmions in magnetic multilayers. Phys. Rep. **704**, 1 (2017)

## Further Reading

Ashcroft, N.W., Mermin, N.D.: Solid State Physics. Holt, Rinehart and Winston, New York (1976)
Bandyopadhyay, S., Cahay, M.: Introduction to Spintronics, 2nd edn. CRC, Boca Raton (2015)
Datta, S.: Lessons from Nanoelectronics: A New Perspective on Transport- Basic Concepts, 2nd edn. World Scientific Pub. Co. Inc., Singapore (2017)
Demokritov, S.O. (ed.): Spin Wave Confinement. Pan Stanford, Singapore (2009)

Demokritov, S.O. (ed.): Spin Wave Confinement: Propagating Waves. Pan Stanford, Singapore (2017)

Demokritov, S.O., Slavin, A.N. (eds.): Magnonics- from Fundamentals to Applications. Springer, Heidelberg (2013)

Guimarães, A.P.: Principles of Nanomagnetism, 2nd edn. Springer, Berlin (2017)

Gubbiotti, G.: Three-Dimensional Magnonics: Layered, Micro- and Nanostructures. CRC Press, Taylor and Francis (2019)

Heinrich, B., Bland, J.A.C. (eds.): Ultrathin Magnetic Structures II. Springer, Heidelberg (1994)

Hillebrands, B., Ounadjela, K. (eds.): Spin Dynamics in Confined Magnetic Structures I. Springer, Heidelberg (2002)

Hillebrands, B., Ounadjela, K. (eds.): Spin Dynamics in Confined Magnetic Structures II. Springer, Heidelberg (2003)

Kittel, C.: Introduction to Solid State Physics, 8th edn. Wiley, New York (2004)

Reif, F.: Fundamentals of Statistical and Thermal Physics. Mc Graw-Hill Book Co, New York (2008)

Stancil, D.D., Prabhakar, A.: Spin Waves: Theory and Applications. Springer, New York (2009)

White, R.M.: Quantum Theory of Magnetism, 3rd edn. Springer, Berlin (2007)

Wu, M., Hoffmann, A. (eds.): Recent Advances in Magnetic Insulators – From Spintronics to Microwave Applications. Elsevier, San Diego (2013)

# Index

© Springer Nature Switzerland AG 2020
S. M. Rezende, *Fundamentals of Magnonics*, Lecture Notes in Physics 969,
https://doi.org/10.1007/978-3-030-41317-0

Printed in the United States
By Bookmasters